AEROMEDICAL PSYCHOLOGY

We dedicate this volume to our mentors in aeromedical psychology, Dr. Barton Pakull, Dr. Tony McDonald and Dr. Jeffrey Moore, who inspired us with their enthusiasm, passion for aviation and knowledge, and to our families and friends, for their support and sacrifices.

Aeromedical Psychology

Edited by

CARRIE H. KENNEDY
University of Virginia, USA

&

GARY G. KAY
Cognitive Research Corporation, USA

ASHGATE

Disclaimer
The views presented in this book are those of the authors and do not reflect the official policy or position of the U.S. Air Force, U.S. Army, U.S. Marine Corps, U.S. Navy, the Department of Defense, the U.S. Government, or any other institution with which the authors are affiliated.

Published by
Ashgate Publishing Limited
Wey Court East
Union Road
Farnham
Surrey, GU9 7PT
England

Ashgate Publishing Company
110 Cherry Street
Suite 3-1
Burlington, VT 05401-3818
USA

www.ashgate.com

British Library Cataloguing in Publication Data
Aeromedical psychology.
1. Airlines--Employees--Psychological testing. 2. Air pilots, Military--Psychological testing
3. Astronauts--Psychological testing. 4. Aviation psychology.
5. Aircraft accidents--Human factors.
I. Kennedy, Carrie H. II. Kay, Gary G.
629.1'325'019-dc23

ISBN: 978-0-7546-7590-7 (hbk)
978-0-7546-7594-5 (ebk)
978-1-4724-0125-0 (ePub)

The Library of Congress has cataloged the printed edition as follows:
Kennedy, Carrie H.
Aeromedical psychology / by Carrie H. Kennedy and Gary G. Kay.
p. cm.
Includes bibliographical references and index.
ISBN 978-0-7546-7590-7 (hardback) -- ISBN 978-0-7546-7594-5 (ebook) -- ISBN 978-1-4724-0125-0 (epub) 1. Aviation psychology. 2. Flight crews--Psychology. I. Kay, Gary G. II. Title.
TL555.K46 2013
629.132'52019--dc23

2012047567

Printed and bound in Great Britain by
TJ International Ltd, Padstow, Cornwall.

Contents

List of Figures

List of Tables

Preface

From a 12-second flight by the Wright Brothers in 1903 to the jet aircraft, space travel, and unmanned aerial systems of today, the field of aviation continues to develop. With these technological advances, so too has aerospace medicine advanced. Flight surgery dates back to World War I when the mental and physical health of fliers became paramount. For the last century aerospace medicine has grown from the practice solely of physicians to include psychologists, physiologists, and optometrists.

All of these specialists work together, to select the best individuals for flight status, to maintain their health in that status and to optimize their performance. Aeromedical psychology represents the practical application of psychology to the safety and effectiveness of recreational, commercial, and military aviation. For those psychologists working with individuals with flight status, this application may include mental health screenings or evaluations in the context of assessment and selection. When appropriateness for continued flight status is questioned, psychologists are engaged in the construction of treatment plans designed to return the aviator to flight status and which comply with a variety of aeromedical policies. In addition, aviation psychologists provide crisis response following flight mishaps in a culturally competent manner, and participate in mishap investigations.

With continuing developments in both technology and advancing medical/mental health options and consequent evolving policies for those with flight status, an edited work was required which tapped the expertise of a wide range of individuals and with an international perspective. Each contributor was selected in recognition of their individual expertise and their documented work in their specialty area and we are indebted to their commitment to the construction of this volume. They have enabled us to provide a comprehensive manual which includes guides on conducting mental health, substance abuse, and neuropsychological evaluations, policies regarding psychotropic medications, and issues related to fatigue, airsickness, fear of flying, and aging aviators. The manual concludes with a chapter on developing competency in this specialized area. We present this volume as a guide to the field and to the growing role of aeromedical psychologists.

Chapter 1
A History of Aeromedical Psychology

Tatana M. Olson, Mathew McCauley and Carrie H. Kennedy

Since the Wright Brothers conducted the first sustained flight with a powered, controlled aircraft on 17 December 1903, scientists have been fascinated with the role of the human in aviation. Scientific advancements in our knowledge of human cognition and physiology, as well as in aircraft capabilities, stimulated the need to better understand the human–aviation interface. History has shown that psychologists around the world significantly contributed to this understanding, providing analytical expertise and developing effective tools and methods.

The concept of aviation psychology has evolved over time. For example, in 1941 Razran and Brown (322) stated "the psychology of aviation comprises the selection, training, and maintenance of personnel (pilot, crew, and ground staff) and the care and education of passengers." Aeromedical psychology now represents an integration of aviation medicine and clinical psychology, and involves the application of clinical psychology principles, methods, and techniques to address individual and group issues within the aviation community (King 1999). The role of the aeromedical psychologist has also evolved over the years, encompassing traditional clinical concerns about mental disorders, personality dynamics, the assessment and selection of aviators, maintenance of mental health, and action concerning medical qualification or disqualification throughout an aviator's career (Bor and Hubbard 2006; Jones and Marsh 2001). Please note that for the purposes of this volume, aeromedical psychology, aerospace psychology, and aviation psychology are used interchangeably.

As an applied science, aerospace psychology focuses on human functioning within civilian and military aviation. However, given their shared roots, the history and growth of clinical psychology, aeromedical psychology, and aviation medicine are closely linked, and like many scientific fields of the last century (Kennedy, Hacker Hughes and McNeil 2012), they evolved to meet the needs of various militaries, particularly during World War I (WWI) and World War II (WWII).

World War I (1914–1918)

Preparing for a war which would be fought partially in the air was a daunting task. As Giulio Douhet, an Italian staff officer serving in WWI, correctly predicted, "the sky is about to become another battlefield no less important than the battlefields on land and sea … In order to conquer the air, it is necessary to deprive the enemy of all means of flying, by striking at him in the air, at his bases of operation, or at his production centers. We had better get accustomed to this idea, and prepare ourselves" (Bowen 1980, 24). In the US in April 1917, the Aviation Section of the Air Corps had only 52 trained fliers—this would balloon to 16,000 in only two years (Henmon 1919). Consequently, during WWI, the primary efforts of psychologists were focused on the screening of pilots, observers, and balloonists in order to select those most likely to successfully complete training and avoid aviation mishaps (Henmon 1919; Driskell and Olmstead 1989). Indeed, in the run up to WWI, the Italian psychologist, Agostino Gemelli, became noted for his efforts to investigate the assessment, evaluation, and selection of pilots (Barbarino 2006). However, there were no data available regarding the characteristics of a qualified aviator. Emphasis was placed on sensible physical qualifications, such as vision, audition, balance, and reaction time, with the flight physical serving as the primary means of evaluation (McGuire 1990).

It became clear, due to excessive losses of students from flight training, that the flight physical was not adequately capturing an individual's psychological attributes; elements the aeromedical community would rapidly discover were important predictors of flying proficiency. To address this deficiency, flight surgeons added a psychiatric interview to the basic flight physical. Unfortunately, their lack of experience in psychological assessment produced a non-standardized and unreliable interview process that not only failed at predicting training success, but also rejected 50–60 percent (Henmon 1919) to 90 percent (Hilton and Dolgin 1991) of candidates in some instances. In all fairness, little was known by anyone at the time regarding the "right" psychology of the aviator. In 1914, for example, the assumption was simply that "he must possess an unusual amount of dare-devil spirit" (Dockeray and Isaacs 1921) and that "most militaries wanted pilots who could fly by the seat of their pants; excel at dog fighting, evade attackers, and survive to become aces who would inspire others" (Hilton and Dolgin 1991, 83). Psychologists involved in selection at the beginning of WWI were "instructed to select men of good education and high character, men who were in every way qualified and fitted to be officers" and to "remember that the flying officer was not to be an aerial chauffeur but a twentieth century cavalry officer mounted on Pegasus" (Henmon 1919, 103).

During WWI, Robert Yerkes, the President of the American Psychological Association (APA) in 1917, stated in a letter to the entire membership, "It is obviously desirable that the psychologists of the country act unitedly in the interests of defense. Our knowledge and our methods are of importance to the military service of our country, and it is our duty to cooperate to the fullest extent and immediately toward the increased efficiency of our Army and Navy" (Yerkes 1918). As a result, a psychology committee was organized at the request of the National Research Council (NRC) and a number of subcommittees were formed to address specific areas. Recognizing a gap in existing knowledge of non-physical variables correlated with success as an aviator, the Committee on Psychological Problems in Aviation devoted their efforts to the selection and development of mental and physiological tests designed to predict aptitude for flying. This represented the first US organized effort to apply psychological principles, by trained psychologists, to the screening and selection of aviators. Early work demonstrated significant relationships between predictors such as intelligence, emotional stability, spatial perception, and mental alertness and flying proficiency (Henmon 1919; Koonce 1984).

In Italy, researchers divided pilots into three categories: successful, mediocre, and unsuccessful. Results indicated that pilots needed good perceptual speed, attentional resources, psychomotor coordination, and the ability to inhibit emotions such that they did not interfere with cognitive demands. In France, efforts were focused on the study of reaction time and emotional stability. The British, on the other hand, were most concerned with motor coordination and the effects of altitude on pilots (Dockeray and Isaacs 1921; Koonce 1984). However, attention was also paid to the concept of staleness, felt to be related to an inability to appropriately manage the strain of flying.

Toward the end of the war, information began to emerge about the mental health of combat pilots. Anderson (1919), a surgeon–lieutenant in the British Royal Navy, coined the term aero-neuroses to replace the terms flying stress, flying sickness, and aviator's neurasthenia. He observed that pilots either washed out early in training consequent to the stresses involved in learning to fly or due to a fear of flying, or veteran combat pilots developed problems in reaction to the stress of flying in combat and/or after witnessing or experiencing a crash. With regards to flight students he described the following conditions in which aero-neurosis might arise:

> (a) during the period of dual control instruction; (b) more commonly during the first few solo flights; (c) less commonly later, when the pupil is either transferred to another flight or has to fly faster types of aeroplanes; (d) again, the onset often dates back to a flying accident either without injury or after sustaining shock, concussion, or bodily injury; (e) moreover, it has been found to follow where the pupil has not actually been in a flying accident but has been the witness of one; (f) finally there is the case following a crash, usually with severe bodily injury, in while, very late, Aero-Neurosis supervenes. (99)

While today, training crashes are fairly uncommon, during WWI they were not. Henmon (1919) noted one flight student who was "given 55 hours of flying instruction and wrecked five ships before he was relieved as unadaptable for further training" (104). For a comprehensive discussion of the history and concepts of fear of flying and motivation to fly, please see Chapter 7, this volume.

Post-World War I

During the post-WWI period, aeromedical psychology experienced limited growth, though attrition from military training in the US remained high (that is, 40–73 percent) leading up to WWII (Flanagan 1942). Efforts focused on refining existing paper and pencil measures of aptitude and intelligence and conducting research on physiological adaptability to the aviation environment as the capabilities of aircraft grew, enabling people to fly higher and faster than ever before (Hilton and Dolgin 1991). Dissatisfied with the methods currently available, the US Navy Bureau of Medicine and Surgery directed an examination of psychological predictors of successful completion of flight training, involving the psychological assessment of 628 candidates for flight training in Pensacola, Florida (McGuire 1990). Candidates were interviewed and then rated on qualities hypothesized to be related to flight training, such as courage, stability, aggression, concentration, intelligence, and reaction time. Based upon these ratings, 212 candidates were rated as good aviation material while 416 candidates were rated as poor. Of the 212 rated as good, 70 percent successfully passed training. Of the 416 rated as poor, 83 percent failed training. This work represented one of the first attempts to validate selection measures against aviation training outcomes (for modern military and commercial selection procedures, see Chapters 2 and 3). Building upon case studies of WWI aces conducted by European psychologists, the focus on personality traits by German psychologists specifically, and early studies examining the role of emotional stability, a small segment of American psychologists started to explore the personality traits of successful aviators (Hilton and Dolgin 1991). For example, based upon a survey of Naval flight instructors, Parsons (1918) and Dockeray (1920) described essential qualities for successful aviators, including coolness under strain, quick reaction time, steadiness (or lack of tremor), dependability, persistence, perseverance, quiet and methodical, not given to emotional excitement, and able to inhibit instincts of self-preservation. For the first time, the use of alcohol in conjunction with dangerous flying was broached (Dockeray 1920). For a discussion of substance abuse and aviation, please see Chapter 5, this volume.

In Germany, use of psychological testing was widespread within the military beginning in 1927 (Fitts 1946). All officer candidates were subjected to psychological examination and this soon spread to military specialties. The Air Forces (Luftwaffe) branched off in 1939 and with it grew a specialized, though not standardized, psychological selection program. Eventually, 150 individuals were employed as psychologists within recruiting centers and regimental testing centers. Psychological examinations took between one to three days depending on the aviation specialty (pilots and navigators being the most time intensive) and the phase of the war. These examinations were reported to be highly subjective and based largely on the candidate's behavior during the examination as opposed to the psychological testing scores, with the primary purpose of the psychological evaluation process to clinically assess leadership, character, and personality. However, objective tests were administered in the areas of intelligence, judgment, perceptual ability, sensorimotor coordination, character, leadership, and specialty skills, and a clinical interview was given following test completion in order to conduct a final assessment of personality. In general, effort and motivation were given greater weight during the assessment than the scores obtained; "extreme personality types were generally considered unsatisfactory" (Fitts 1946, 155); "pilots should be alert, intelligent, well coordinated, highly motivated, and have good orientation ability" and gunners "should be physically strong and have normal reaction time and coordination" (155).

The US Army's Sanitary Corps (the precursor to the Medical Service Corps) added psychologically-based biographical questions of aviation candidates related to "parentage, education, business experience, athletic attainments, responsibilities placed upon him by others" and required reference letters from a minimum of three people attesting to moral character (Stratton, McComas, Coover and Bagby 1920). By 1935, the US Army and Navy were exploring more standardized psychological assessments of personality and motivation attributes, such as the Strong Interest Inventory and the Bernreuter Personality Inventory, which measured six factors: neurotic tendency, self-sufficiency, relevant dominance–submissive tendency, relevant introvert–extrovert tendency, confidence, and sociability. However, all of these efforts were related to selection of pilots as opposed to the maintenance of them once fully trained. In 1938, Ferree and Rand expressed a concern that insufficient attention was being placed on the role of the pilot in airplane crashes, and stated "it is strange indeed that so much care is taken to see that the plane is in perfect condition before a flight is undertaken and so little attention is given to the condition of the pilot" (192). They recommended a quick test assessing visual tracking and accuracy, potential age effects, and fatigue prior to and following (to measure strain induced by the flight) each flight for both military and commercial pilots. These concerns were expressed at a time when there were roughly 29,000 aircraft in the US between general aviation (non-airline civil aviation), military, and airliners (Brown 1961).

World War II (1939–1945)

During WWII, aeromedical psychology (and clinical psychology as a whole) experienced tremendous growth, establishing the groundwork for modern aviator assessment and selection (Fiske 1946; Lane 1947; Staff, Psychological Research Project 1946a). This was evident in Britain, when the Air Ministry established the Flying Personnel Research Committee, nine months before the start of the war. Much of their effort was focused on the effective application of psychological methods for pilot selection and training. Their work was seen as successful, particularly by the Admiralty's Advisory Panel on Scientific Research, which looked at the naval applications of psychology and physiology; thus further strengthening the role and contribution of aviation psychology within the British military both during and after the war (Hacker Hughes 2007).

Grinker and Spiegel (1945, 9) differentiated between psychological selection and psychiatric selection. Psychological selection was defined as selection of "those who will be able to learn how to fly with the least difficulty," while psychiatric selection dealt with the selection of "those who will be the most suited to withstand the emotional stresses of flying and of combat." This was at a time where military psychologists had few clinical responsibilities. This remained largely unchanged within the UK military throughout WWII (Hacker Hughes 2007). However, toward the end of the war, the role of US military psychologists dramatically shifted from conducting purely assessment to providing clinical care to the nation's psychiatric war casualties (Kennedy, Hacker Hughes and McNeil 2012). This shift in services had the consequent result of expanding roles for aviation psychologists as well, to include tailored mental health interventions (for example, Muench and Rogers 1946). During the war, clinical psychologists began to apply their knowledge and contributed significantly to US Army Air Force (AAF) research, geared to improve flight training methods, selection of instructors and personnel suited to flying in combat, and "the redistribution and rehabilitation of returnees" (Super 1947). Psychologists began to treat stress reactions related to flying, to include phobias, anxiety states/anxiety reactions, reactive depression, neurasthenic syndrome, conversion phenomena, and psychosomatic disturbances to include airsickness (for example, Anderson 1948). The issue of pilot fatigue was also given significant attention (McFarland 1942; see Chapter 9, this volume for a comprehensive discussion of fatigue in the modern aviation age).

During WWII, the potency of air power was fully recognized, significantly increasing the demand for highly-skilled aviators and necessitating a better understanding of the interface of aviators with the new equipment and systems being developed. To meet this demand, the NRC Committee on Aviation Psychology was established in 1939, followed by the development of the US Army Air Force Aviation Psychology Program (AAAPP) in 1941 (Marquis 1945; Koonce 1984). The AAAPP focused on the refinement of intelligence and flight aptitude tests, emphasizing the development of objective tests that could be economically administered to large groups (Hilton and Dolgin 1991) as well as identifying those who were most likely to be successful in various flight positions (that is, bombardiers, pilots, and navigators; Flanagan 1942; Harrell 1945; Staff, Psychological Research Project 1946b). At the end of the war, the AAAPP published 19 reports on the groundbreaking work they had done during the war, covering topics from classification, qualifications, training, research, and equipment design (Flanagan 1947). The battery of tests developed by the USAAF was adapted for use by the Chinese, French, and Royal Air Forces (Marquis 1945).

In addition, the US Bureau of Navigation (later the Bureau of Naval Personnel), authorized the accession of psychologists to focus on administering, scoring, and interpreting various psychological tests being examined by the Navy for use in the selection of aviators. Navy psychologists contributed to a number of important areas, utilizing their quantitative and analytical training to develop and introduce improved training records and forms, prepare and revise training syllabi and materials, conduct analyses regarding the causes of flight training attrition, and design research projects investigating vision and communication techniques within the aviation environment (Ames and Older 1948).

By 1940, Navy psychologists were directly involved in one of the most significant aviation selection efforts—the Pensacola Project on the Selection of Naval Aviators, funded by the Civil Aeronautics Administration (CAA; formerly the Civil Aeronautics Authority) and the NRC (Fiske 1947). At the time, the CAA was supervising the primary flight training of 36,000 pilots every year, while the military was training approximately 10,000 (Matheny 1941). The Pensacola Project evaluated numerous psychological tests, eventually selecting three of the most predictive for use in the selection of Naval aviators—the Wonderlic Personnel Test, the Bennett Mechanical Comprehension Test, and the Purdue Biographical Inventory (Ames and Older 1948). The Flight Aptitude Rating (FAR), a composite of scores based upon the Bennett Mechanical Comprehension Test, the Purdue Biographical Inventory, and the Aviation Classification Test (ACT), a revision of the Wonderlic, demonstrated good predictive validity for success in flight school, ground school, and overall flight training, and became operational in 1942.

The CAA-NRC project went far beyond the Pensacola Project however, obtaining the cooperation of 39 research centers spread all over the US to look at a multitude of aviation problems (Viteles 1945). Given the ongoing war, most of these research efforts were designed to address specific military problems. However, it was recognized that the research results would enjoy a broad application to civil aviation following the end of WWII. This work resulted in the creation of the National Testing Service in 1942 "designed to provide uniform administration and scoring, on a nation wide basis, of tests for screening candidates for training in the Army phase of the Civilian Pilot Training program" (495).

The research also resulted in the validation of numerous selection instruments (and the subsequent exclusion of unsuccessful and impractical tests and techniques), empirical setting of cut-scores, means to observe elementary flight training in situ and conduct subsequent research on effective training methods, resulting in the development of uniform training manuals and procedures, and the development of new techniques in evaluating pilot performance, to include the Ohio State Flight Inventory and the Purdue Scale for Rating Pilot Competency (Viteles 1945). The latter instrument, when factor analyzed, was thought to measure skill, judgment, and emotional stability. In addition, emotional and physiological reactions were researched, to include sleep, salivary secretions, respiratory changes, muscle potentials, tension, skin temperature, perspiration, muscular set, noise and vibration, and airsickness (for a thorough discussion of airsickness in the modern age, please see Chapter 8, this volume). Last, an important focus of the CAA-NRC project was the prevention of aviation accidents. At the time, 65 percent of all fatal accidents involved stalls, most resulting from level turns at low altitudes, with others resulting from (in order) climbs, glides, spins, slips, and skids. Research resulted in the recommendation for a training shift which would include avoidance and recovery from the stall condition. For a comprehensive discussion of this research, please see Viteles, 1945.

Although the Germans and Italians had been assessing personality as part of their aviation selection programs since WWI, it was not until WWII that personality received serious consideration as a pilot screening tool in the US (Dolgin and Gibb 1988). During this period, the US Army and Navy, in conjunction with the AAAPP, conducted numerous investigations of the ability of commercially available personality measures, such as the new Minnesota Multiphasic Personality Inventory (MMPI), to predict flight school success. The War Department (1940) emphasized the integration of behavioral and personality factors related to aviation and published a technical manual comprising those aspects of neuropsychiatry felt to be most relevant. The War Department (1941) also recommended a thorough neuropsychiatric examination of candidates in order to "eliminate all the mental and nervous weaklings including temperamental and personality handicapped individuals such as eccentrics, disturbers, irritable, unsocial, peculiar, gossipy, arrogant, and other mental twists types, all unsuited to aviation" (230). In order to accomplish this, flight surgeons were expected to develop an almost therapy-like atmosphere in order to make for "relaxation and intimacy" such that applicants/aviators would be truthful and cooperative during the interview (Grow and Armstrong 1942). One estimate during WWII was that 90 percent of pilot training failures were due to emotional instability (Harrower-Erickson 1941).

The magnitude of this task was recognized and Armstrong (1943) noted that flight surgeons required thorough training in psychology and psychiatry, significant patience, and an analytical mind in order to parse out those who may not be emotionally suited to aviation. One emphasis of the neuropsychic component of the flight physical consisted of a detailed biographical history of applicants in order to look for psychopathic personalities, further differentiated into seven categories: inadequate personality, paranoid personality, emotional instability, criminalism, pathologic lying, sexual psychopathy, and nomadism. Emphases were also placed on temperament, intelligence, and volition.

However, while investigation into personality appeared pertinent, results did not support the application of personality testing to pilot selection. These findings were further supported by a review of 94 studies examining the predictive validity of personality inventories completed by Ellis and Conrad in 1948 and the results of a study conducted by Voas, Bair and Ambler in 1957 examining the MMPI, Guilford-Zimmerman, Taylor Manifest Anxiety Scale, Saslow Screening Test, and Hanneman Manifest Anxiety Scale among a sample of 2,000 Naval aviation cadets. Despite a growing body of evidence indicating that personality tests were not useful predictors of pilot training success, psychologists continued to evaluate the role of personality in predicting pilot performance (Hilton and Dolgin 1991). Initiated in 1950, a series of Air Force studies re-evaluated the utility of personality measures for predicting flight school success. Although the results of these studies showed that personality was a poor predictor of training success, the measures used were found to accurately classify psychiatrically unsuitable applicants (Sells 1955) and were more predictive of long-term success in aviation than aptitude and ability measures (Sells 1956). These findings represented a turning point in the role of personality assessment within the context of aviation selection in the US.

In addition to personality characteristics, flight surgeons were directed to focus on other psychological issues. Kafka (1942) recommended that flight surgeons observe their fliers for anxieties, lack of confidence, economic insecurity, emotional insecurities, and sexual difficulties. Fatigue in pilots was studied and in addition to the physically exhausting aspects of flight, one conceptualization included staleness. This was thought to represent operational fatigue and was characterized by anxiety, irritability, and poor judgment (McFarland 1941).

Combat effects were once again brought to the forefront. Grinker and Spiegel (1945) observed that fliers were more likely to experience a neurotic reaction than ground personnel; that enlisted men were three times more likely to experience such a reaction than officers; and that these reactions resolved relatively quickly and often spontaneously. In addition, they noted that those who had the least amount of work to do in the combat environment were "most susceptible to accumulations of anxiety" (213) and that operational fatigue occurred in the following ordered frequency: 1. Radio gunner, 2. Gunner, 3. Engineer gunner, 4. Bombardier navigator, 5. Bombardier, 6. Navigator, 7. Pilot (bomber) and 8. Pilot (fighter). Grinker and Spiegel (1945) recommended a comprehensive assessment over a period of time to address presenting symptoms, precipitating causes, personality, psychosocial background, combat flying experiences, interpersonal relationships, interpersonal stressors, morale, and any source of non-flying stress.

Post-World War II

Aircraft continued to increase in speed, complexity, and numbers in the decades following WWII and commercial airlines began to more consistently apply use of psychological tests in flight crew and other employee selection (Feronte 1949). By 1959, there were approximately 110,000 aircraft in the US alone (70,000 in general aviation, 37,000 in the military, and 1,900 in the airlines) and roughly 700,000 certified pilots (Brown 1961).

During this post-WWII period, many returned to civilian careers and sought to apply their knowledge to peace-time endeavors, including the interface between aviation and psychological science. A number of commercial airlines secured and utilized the knowledge of these specialists in military aviation, within the arena of aircrew selection, education, training, and operational stress, particularly in relation to flight safety. One such example was the utilization of work by Alex Cassie, who worked in the psychological division of the Royal Air Force (RAF) and was associated with the development of pilot selection in the RAF from 1940, which addressed personality factors, motivation, and attitude. Others included

Sipke Fokkema, who worked as Head of the Psychological Research Laboratory of the Free University of Amsterdam. He also served as an advisor to the Royal Netherlands Air Force and researched the use of the Rorschach test as a method of pilot selection. These and others played a critical role in setting high standards for research and practice within the field of aviation psychology. This culminated in 1956 with the establishment of the Western European Association for Aviation Psychology, which dropped the term "Western" at the end of the Cold War (Alsina 2006).

Following WWII, there was a growing realization that the aviation environment presented a number of unique stressors. Age of pilots was considered as a factor for the first time in the 1950s. In 1953, approximately 1 percent of all civilian pilots in the US were age 60 or older (McFarland 1956). Research and concerns related to decreased performance in such areas as age-related effects in memory and altitude, hearing problems, and visual decline, especially as related to night vision, culminated in the "Age 60" rule in 1959. For a comprehensive discussion of aging aviators in the modern context, please see Chapter 11, this volume.

It was increasingly recognized that while an individual may have the cognitive aptitude to perform well, he may not have the necessary temperament to manage these stressors (Bowles 1994). As a result of this, psychologists continued to explore potentially relevant psychological attributes, but along increasingly divergent paths. One path focused on the selection of aviators with the goal of predicting success in flight training (for example, Barry, Sells and Trites 1955) as well as using psychological testing during flight training and comparing findings to actual performance outcomes of combat pilots during the Korean War (Trites and Sells 1957).

Focus continued on the development and implementation of tests measuring intelligence, mechanical and spatial ability, and psychomotor skills to identify individuals with the aptitude to achieve flying proficiency. In addition, encouraged by emerging research on the five-factor model of personality (see Barrick and Mount 1991 for a review), psychologists working in this area moved away from the use of personality inventories designed to detect psychopathology to measures designed to predict variability in job performance within a "normal" population of individuals. Psychologists following the second path adopted a more clinical approach, focusing on those psychological attributes, such as personality traits, temperament and coping mechanisms, related to both safe and effective functioning and maladaptive functioning within the aviation environment as a whole (Bucky, Spielberger and Bale 1972; Holtzman and Sells 1954; Kragh 1960; Trites 1960). This concept, termed aeronautical/aeromedical adaptability, serves as the cornerstone for much of the work that aeromedical psychologists are involved in today to assess the psychological fitness of aviation personnel (see Chapter 6, this volume, for more on the concept of aeronautical/aeromedical adaptability).

Psychological tests were also used to study the effects of a variety of substances, such as dextroamphetamine, on performance in the aircraft (Hauty, Payne and Bauer 1957). Please see Chapter 12, this volume for a discussion of psychopharmacology and aviation, including Go pills. Psychologists continued to apply their unique skills to a number of practical problems in aviation. In the early 1970s for example, after an extreme increase in the number of aircraft hijackings, they were tasked to examine the psychological natures of hijackers and the behaviors of the crew which both positively and negatively impacted hijacking events, advising aviation agencies on both passenger screening policies and flight crew training (Dailey and Pickrel 1975).

In addition to both commercial and military aviation, the space age began in the mid-1950s and with it, the most comprehensive psychological selection programs ever seen. In the US, the first astronaut candidates (Project Mercury) were subjected to 12 motivation and personality tests and 12 cognitive function tests, as well as psychiatric interviews and a variety of stress tests (that is, pressure suit, isolation, complex behavior simulator, acceleration, noise and vibration, and heat). Some of the psychological tests were translated (for example, MMPI) and used in cosmonaut selection in the Soviet Union, which also included psychiatric interviews, and individual and group psychometric and stress tests (Santy 1994). Selection procedures have continued to evolve over time and expand as many other countries send people to space, as space missions change (for example, short duration flights to long duration missions), and as commercial space travel moves closer to reality. See Chapter 2 for an overview of US astronaut selection.

Currently, aeromedical psychologists continue to make significant contributions to both the military and civilian aviation communities through involvement in a wide range of activities including the selection of aviation personnel, assistance in aviation mishap investigations and aviation medical board evaluations, expertise about stress recognition and health promotion, and consultation regarding the assessment, treatment, and psychological disposition of aviation personnel. In the UK, for example, the RAF operates a center of excellence at the RAF Centre for Aviation Medicine, where aviation psychology remains a key component in setting standards for the assessment, selection, training, and retention of aviators across the British military. This is furthermore enhanced by specialist clinical services located at the Aviation Psychiatric Clinic, based within the Department of Community Mental Health at RAF Brize Norton. These organizational assets are at the forefront of conducting research and developing standards in aviation psychology within the UK armed forces. Meanwhile, in Germany, the Department for Aviation and Space Psychology at the country's Institute for Aerospace Medicine regularly conducts research and develops scientific methods for the selection of pilots, astronauts, and air traffic controllers, along with psychological adaptation skills associated with adjusting successfully to the demands of aerospace endeavors. Such work has also helped to raise the standards of aviation psychology in Europe and beyond (Steininger and Stelling 2009). Therefore, as outlined by King (1999), it is ultimately the integration of clinical psychology training and a detailed understanding of the aviation environment, supported by decades of international research, which provides aeromedical psychologists with the unique capabilities to enhance performance and improve safety in the aviation community. The following chapters will provide the reader with a synopsis of the primary roles of aeromedical psychologists and the building blocks needed to conduct specialty evaluations. A variety of international experts will provide information on assessing aviators/pilots, regulations related to mental health and substance conditions and flying, fear of flying and motivation to fly, airsickness, fatigue and aviation, cognitive disorders, aging aviators, psychopharmacology in aviation, mishap investigations, and aviation crisis management. The volume concludes with a chapter on the requirements for psychologists to competently participate in the evaluation, disposition, and/or treatment of aviation personnel.

References

Alsina, M. 2006. Open letter, in *50 Years of EAAP 1956–2006:* EAAP History Booklet. Published by the European Association for Aviation Psychology, September 2006. Available at http://www.eaap.net/history.html [accessed 9 August 2012].

Ames, V.C. and Older, H.J. 1948. Aviation psychology in the United States Navy. *Review of Educational Research*, 18(6), 532–542.

Anderson, G.A. 1919. *The Medical and Surgical Aspects of Aviation*. London: Oxford University Press.

Anderson, R.C. 1948. Neuropsychiatric problems of the flyer. *American Journal of Medicine* 4(5), 637–644.

Armstrong, H.G. 1943. *Principles and Practice of Aviation Medicine.* 2nd Edition. Baltimore, MD: The Williams & Wilkins Company.

Barbarino, M. 2006. 50 years of EAAP *[President's Forword]*, in *50 Years of EAAP 1956-2006:* EAAP History Booklet. Published by the European Association for Aviation Psychology, September 2006. Available at http://www.eaap.net/history.html [accessed 9 August 2012].

Barrick, M.R. and Mount M.K. 1991. The big five personality dimensions and job performance: A meta-analysis. *Personnel Psychology,* 44(1), 1–26.

Barry, J.R., Sells, S.B. and Trites, D.K. 1955. Psychiatric screening of flying personnel with the Cornell Word Form. *Journal of Consulting Psychology*, 19(1), 32.

Bor, R. and Hubbard, T. 2006. Aviation mental health: An introduction, in *Aviation Mental Health,* edited by R. Bor and T. Hubbard. Farnham: Ashgate, 1–9.

Bowen, E. 1980. *Knights of the Air.* Time-Life Books.

Bowles, S.V. 1994. Military aeromedical psychology training. *International Journal of Aviation Psychology,* 4(2), 167–172.

Brown, H.N. 1961. Private flight safety, in *Accident Prevention: The Role of Physicians and Public Health Workers,* edited by M. Halsey. New York: Blakiston Division/McGraw Hill Book Company, 249–277.

Bucky, S.F., Spielberger, C.D. and Bale, R.M. 1972. Effects of instructions on measures of state and trait anxiety in flight students. *Journal of Applied Psychology*, 56 (3), 275–276.

Dailey, J.T. and Pickrel, E.W. 1975. Some psychological contributions to defenses against hijackers. *American Psychologist*, 30(2), 161–165.

Dockeray, F. 1920. Department of psychology, in *Aviation Medicine in the A.E.F,* edited by W. Wilmer. Washington DC: Government Printing Office, 113–132.

Dockeray, F.C. and Isaacs, S. 1921. Psychological research in aviation in Italy, France, England, and the American Expeditionary Forces. *Journal of Comparative Psychology*, 1(2), 115–148.

Dolgin, D.L. and Gibb, G.D. 1988. *A Review of Personality Measurement in Aircrew Selection; NAMRL Monograph (AD-A200392)*. Pensacola, FL: Naval Aerospace Medical Research Laboratory.

Driskell, J.E. and Olmstead, B. 1989. Psychology and the military: Research applications and trends. *American Psychologist*, 44(1), 43–54.

Ellis, A. and Conrad, H.S. 1948. The validity of personality inventories in military practice. *Psychological Bulletin*, 45(5), 385–427.

Feronte, N.C. 1949. Tests used by United States air carriers. *Journal of Applied Psychology*, 33(5), 445–448.

Ferree, C.E. and Rand, G. 1938. Pilot fitness and airplane crashes. *Science*, 87(2252), 189–193.

Fiske, D.W. 1946. Naval aviation psychology. III. The special services group. *American Psychologist*, 1(11), 544–548.

Fiske, D.W. 1947. Naval aviation psychology. IV. The central research groups. *American Psychologist*, 2(2), 67–72.

Fitts, P.M. 1946. German applied psychology during World War II. *American Psychologist*, 1(5), 151–161.

Flanagan, J. C. 1942. The selection and classification program for aviation cadets (aircrew–bombardiers, pilots, and navigators). *Journal of Consulting Psychology*, 6(5), 229–239.

Flanagan, J.C. 1947. Research reports of the AAF aviation psychology program. *American Psychologist*, 2(9), 374–375.

Grinker, R.R. and Spiegel, J.P. 1945. *Men Under Stress*. Philadelphia, PA: The Blakiston Company.

Grow, M.C. and Armstrong, H.G. 1942. *Fit to Fly: A Medical Handbook for Fliers*. New York: D. Appleton-Century Company.

Hacker Hughes, J. 2007. *British Naval Psychology 1937–1947: Round pegs into square holes?* Unpublished Thesis: MS Degree, War and Psychiatry. King's College, University of London, August, 2007.

Harrell, T.W. 1945. Aviation psychology in the Army Air Forces. *Psychological Bulletin*, 42(6), 386–389.

Harrower-Erickson, M.R. 1941. Psychological factors in aviation. *The Canadian Medical Association Journal*, 44(4), 348–352.

Hauty, GT, Payne, RB, and Bauer, RO 1957. Effects of normal air and dextro-amphetamine upon work detriment induced by oxygen impoverishment and fatigue. *Journal of Pharmacology and Experimental Therapeutics*, 119(3) 385–389.

Henmon, V.A.C. 1919. Air service tests of aptitude for flying. *Journal of Applied Psychology*, 3(2), 103–109.

Hilton, T.F., and Dolgin, D.L. 1991. Pilot selection in the military of the free world, in *Handbook of Military Psychology,* edited by R. Gal and A.D. Mangelsdorff. Chichester, England: John Wiley and Sons, Ltd, 81–101.

Holtzman, W.H. and Sells, S.B. 1954. Prediction of flying success by clinical analysis of test protocols. *Journal of Abnormal and Social Psychology*, 49(4), 485–490.

Jones, D.R. and Marsh, R.W. 2001. Psychiatric considerations in military aerospace medicine. *Aviation, Space, and Environmental Medicine*, 72(2), 129–135.

Kafka, M.M. 1942. *Flying Health.* Harrisburg, PA: Military Service Publishing Company.

Kennedy, C.H., Hacker Hughes, J.G.H. and McNeil, J.A. 2012. A history of military psychology, in *Military Psychology: Clinical and Operational Applications*. 2nd Edition. New York: Guilford, 1–24.

King, R. 1999. *Aerospace Clinical Psychology: Studies in Aviation Psychology and Human Factors.* Aldershot: Ashgate.

Koonce, J.M. 1984. A brief history of aviation psychology. *Human Factors*, 26(5), 499–508.

Kragh, U. 1960. The Defense Mechanism Test: A new method for diagnosis and personnel selection. *Journal of Applied Psychology*, 44(5), 303–309.

Lane, G.G. 1947. Studies in pilot selection: Prediction of success in learning to fly light aircraft. *Psychological Monographs*, 61(5), 1–17.

Marquis, D.G. 1945. Psychological activities in the training command, Army Air Forces. *Psychological Bulletin*, 42(1), 37–54.

Matheny, W.G. 1941. The effectiveness of aptitude testing in the civilian pilot training program. *Transactions of the Kansas Academy of Science*, 44, 363–365.

McFarland, R.A. 1941. Fatigue in aircraft pilots. *The New England Journal of Medicine*, 225(22), 845–855.

Mcfarland, R.A. 1942. Fatigue in aircraft pilots, in *War Medicine*, edited by W.S. Pugh. New York: Philosophical Library, 430–448.

McFarland, R.A. 1956. Functional Efficiency, Skills and Employment, in *Psychological Aspects of Aging*, edited by J. Anderson. Washington, DC: American Psychological Association, 227–235.

McGuire, F.L. 1990. *Psychology Aweigh: A History of Clinical Psychology in the United States Navy, 1900-1988*. Washington, DC: American Psychological Association.

Muench, G.A. and Rogers, C.R. 1946. Counseling of emotional blocking in an aviator. *Journal of Abnormal and Social Psychology*, 41(2), 207–215.

Parsons, R.P. 1918. A search for non physical standards for naval aviators. *U.S. Naval Medical Services Bulletin*, 12, 155-172.
Razran, G.H.S., and Brown, H.C. 1941. Aviation. *Psychological Bulletin*, 38(6), 322–330.
Santy, P.A. 1994. *Choosing the Right Stuff: The Psychological Selection of Astronauts and Cosmonauts.* Westport, CT: Praeger.
Sells, S.B. 1955. Development of a personality test battery for psychiatric screening of flying personnel. *Journal of Aviation Medicine*, 26(1), 35–45.
Sells, S.B. 1956. Further developments on adaptability screening of flying personnel. *Journal of Aviation Medicine*, 27(3), 440–451.
Staff, Psychological Research Project. 1946a. History, organization, and research activities, psychological research project (bombardier) Army Air Forces. *American Psychologist*, 1(9), 385–392.
Staff, Psychological Research Project. 1946b. Psychological research on pilot training in the AAF. *American Psychologist*, 1(1), 7–16.
Steininger, K. and Stelling, D. 2009. Operational and clinical aviation psychology, in *World Scientific*, edited by C. Curdt-Christiansen, J. Draeger and J. Kriebel. Hamburg: World Scientific, 611–638.
Stratton, G.M., McComas, H.C., Coover, J.E. and Bagby, E. 1920. Psychological tests for selecting aviators. *Journal of Experimental Psychology*, 3(6), 405–423.
Super, D.E. 1944. Clinical research in the aviation psychology program of the Army Air Forces. *Psychological Bulletin*, 41(8), 551–556.
Trites, D.K. 1960. Adaptability measures as predictors of performance ratings. *Journal of Applied Psychology*, 44(5), 349–353.
Trites, D.K. and Sells, S.B. 1957. Combat performance: Measurement and prediction. *Journal of Applied Psychology*, 41(2), 121–130.
Viteles, M.S. 1945. The aircraft pilot: 5 Years of research: A summary of outcomes. *Psychological Bulletin*, 42(8), 489–526.
Voas, R.B., Blair, J.T. and Ambler, R.K. 1957. *Validity of Personality Inventories in the Naval Aviation Selection Program* (Technical Report No. 13). Pensacola, FL: Naval School of Aviation Medicine.
War Department. 1940. *Technical Manual 8-325: Outline of Neuropsychiatry in Aviation Medicine.* Washington, DC: Author.
War Department. 1941. *Technical Manual 8-320: Notes on Psychology and Personality Studies in Aviation Medicine*. Washington, DC: Author.
Yerkes, R.M. 1918. Psychology in relation to the war. *Psychological Review*, 25(2), 85–115.

Chapter 2
Assessment and Selection of Military Aviators and Astronauts

Brennan D. Cox, Lacey L. Schmidt, Kelley J. Slack and Thomas C. Foster

The introduction of modern aircraft forever changed how the US Armed Forces would determine candidates' fitness-for-duty. Just a few years after the Wright brothers' inaugural flight, which demonstrated that heavier-than-air travel was even possible, the US Army was acquiring and arming aircraft for training military aviators. Such novel and advanced machinery was sure to generate many questions: *How do we fly them? When should we use them, and where?* Not to be lost among this list were two questions in particular: *Who should fly them, and how do we know?*

The assessment and selection of aviators is an evolving science. From the outset of WWI, periods of armed conflict have profoundly influenced the theoretical and technological underpinnings that shape aircrew selection systems. Although the criteria for selecting flight personnel have changed dramatically over time, the impetus remains the same: operating outside of a 1G environment requires a unique set of knowledge, skills, abilities, and other attributes (KSAOs) that demand only the most qualified men and women be permitted to pursue careers in military aviation.

Individuals selected and trained as military aviators have, over the years, been sought after by a variety of aerospace communities ranging from commercial aviation to the astronaut community and more recently the operation of Unmanned Aerial Systems (UAS). In 1992, approximately 90 percent of commercial pilots in the US had a military background (Weber 2009). Although this statistic dropped to around 28 percent by 2008, military aviators still account for a large proportion of commercial pilots today. (See Chapter 3 of this volume for a detailed overview of the assessment and selection of commercial aviators and air traffic controllers.) Similarly, the first astronauts were drawn almost exclusively from the ranks of military aviators, as there is some natural overlap between the two groups. Of course, the National Aeronautics and Space Administration's (NASA's) missions are distinct from those of the military, and these differences manifest in how assessment and selection is handled between the two groups. To illustrate these differences, a section discussing some of the unique aspects of astronaut selection concludes this chapter.

The following pages provide: (1) an overview of the procedures aviation psychologists use to develop and maintain selection systems, characterized by milestones in US military aviation history; (2) a description of the aviation assessment tools used by the US Army, Navy, and Air Force; (3) insight into future selection challenges in military aviation; and (4) an overview of astronaut assessment and selection.

The Psychology of Selection

A selection system is a process of collecting and analyzing data about job candidates to determine who is most likely to perform a job successfully (that is, at or above an established standard). At its core, selection is a matter of prediction. To predict candidates' likelihood of success effectively, the information gathered during the selection process must be linked to job-relevant criteria. Thus, the first step in developing a selection program is conducting a thorough job analysis to determine which work-related elements are appropriate for the candidate assessment battery. Only after this initial step is complete can a selection system be developed, evaluated, and put into operational use.

Job analysis—deciding what to measure

A job analysis is an exhaustive study of the information relevant to a particular job. Job analysis data include descriptions of the work environment, equipment, and materials used, the position's unique tasks, duties, and responsibilities, and the KSAOs required for performing in these areas. In-depth examinations of each work function include its frequency and duration of activity, level of complexity, and acceptable performance standards. This information helps determine which elements are critical to the job, necessary upon entry, and/or highly desired but trainable.

Job analysts collect data from a variety of sources, including site observations, subject matter expert (SME) interviews, questionnaires, reviews of existing documents and databases, and even by participating directly in the work itself. Of course, the preferred method is a combination of approaches. As a starting point, a simple Internet query now makes available basic job analytic information for virtually any profession. O*Net (www.onetonline.org), for instance, provides a searchable library of thousands of occupations, complete with the general tasks, tools, technologies, KSAOs, activities, environments, and education requirements associated with each line of work. In an age of instant information, it is easy to take such readily accessible resources for granted. Naturally, this was not always the case for military aviation. Prior to WWI, not only were there no job analysis databases for aviators—there were hardly any aviators at all. This made the work of the original aviation job analysts a truly revolutionary undertaking.

World War I selection criteria As Armstrong (1948, 1952) described, military selection before WWI focused on candidates' ability to sustain lengthy, physically demanding hardships, as infantry walked to war with their equipment (for example, weapons, ammunition, food, water, shelter) on their backs. The first shift in this paradigm occurred in 1912, when a Medical Corps officer observed that physical stature, strength, and stamina were less important to flying than having an alert mind, quick reactions, and keen sensory perception. That year the US War Department published its initial criteria for selecting aviators, which included assessments of visual and hearing acuity, circulatory system functioning, movement of limbs, and health history. These elements, derived from observations of aviators at work, provided the foundation for the modern flight physical used by all military services today.

The payoff for introducing job-relevant criteria to the pilot selection process was readily apparent. In the first years of WWI, there were more aviator deaths due to physical or medical complications than to enemy opposition (Selz 1919). However, by the war's end, strict adherence to aviation medical standards saw casualties due to physical deficiencies become more the exception than the rule (Armstrong 1948). This considerable improvement highlights the importance of building selection tools based on the unique aspects of a job, as established through job analysis.

Though valuable, the WWI flight physical was by no means a cure-all for aviator selection. Overly-cautious physicians performing these initial screenings disqualified up to 90 percent of potential candidates (Armstrong 1952). Those who passed the physical then entered a training environment characterized by exceptionally high and costly accident and attrition rates, which emphasized the need to modify the aviator selection system (Damos 2007). Thus, in 1917 the Army Air Services introduced Form 609, a revised medical examination that included many non-medical elements, including candidates' parentage, education, business experience, athletic attainments, responsibilities, and military training (Stratton, McComas, Coover and Bagby 1920). Tests of intelligence, including the famed Army Alpha and Beta Tests (Yerkes 1919), were also introduced at this time; however, the degree to which these test were used for aviation personnel is unclear (Damos 2007).

Unfortunately, much of the literature documenting selection work during WWI is unpublished, lost, or difficult to obtain (Damos 2007). Further, though many scientific studies of aviators at work were underway, most of these research programs were phased out when the war ended. (For more aviation psychology history, see Chapter 1 of this volume.)

World War II selection criteria Following a relatively dormant interwar period, aviation selection research experienced a boom in preparation for WWII. In 1939, an unprecedented demand for pilots led the Civil Aeronautics Authority (now the Federal Aviation Administration) to fund the National Research

Council's Committee on Selection and Training of Aircraft Pilots. In cooperation with the Navy, the Council initiated the Pensacola Study of Naval Aviators to evaluate approximately 60 psychological, physiological, and psychomotor tests for predicting success in Navy flight training (McFarland and Franzen 1944). Results of the Pensacola project suggested that psychological and psychomotor tests were better predictors of training success than physiological measures. These findings helped produce the Navy's Aviation Classification Test (ACT) and Flight Aptitude Rating (FAR), which together assessed candidates' general intelligence, mechanical comprehension, spatial apperception, and biographical history. The Navy did not include psychomotor tests in this battery, presumably due to expense (North and Griffin 1977). Of note, many elements of the ACT/FAR are still found in today's Naval Aviation Selection Test Battery (ASTB).

Concurrent with the Navy's aviator analysis efforts, the Army Air Force (AAF) was conducting related work with its personnel. The AAF Aviation Psychology Program examined flight training failure data and conducted interviews with experienced pilots to determine factors important to training success. Their study revealed five KSAO categories critical to training completion: (1) intelligence/judgment; (2) alertness/observation; (3) psychomotor coordination/technique; (4) emotional control/motivation; and (5) ability to divide attention (Wise, Hopkin and Garland 2009). These factors helped produced the AAF's Aviation Cadet Qualifying Examination, for assessing intelligence and aptitude, and the Aircrew Classification Battery, for assessing psychomotor skills. The AAF published this research in a series of 19 reports, including R.L. Thorndike's *Research Problems and Techniques* (1947), which became a popular reference for job analysis and test construction techniques following WWII.

Advances in aviation selection during WWII saw such success that post-war efforts have focused more on refinement than the creation of new tests (Hilton and Dolgin 1991). Nevertheless, enhancements in aircraft components and training technologies continually necessitate modification of the tools used to select flight personnel. Kubisiak and Katz (2006), for example, described a recent US Army effort to replace its existing Flight Aptitude Selection Test (FAST) with a measure focused on future fleet requirements. To initiate the project, the Army developed a Job Analysis Questionnaire (JAQ) using a four-step process: (1) review existing job analysis literature; (2) brainstorm tasks and KSAOs; (3) conduct SME workshops to revise the task/KSAO list; and (4) finalize the JAQ. This process yielded an instrument containing 101 task statements and 92 KSAO statements. Representative samples of incumbent aviators completed the JAQ by rating each item on a five-point importance scale. Across all airframes, emergency procedures and safety were deemed the most important tasks, while situational awareness, operation/maneuvering, psychomotor ability, information processing, and decision-making were the most important KSAOs. These data provided the Army direction in creating its new computer-based Selection Instrument for Flight Training (SIFT).

Modern selection criteria Meta-analytic research suggests that the best predictor of pilot performance is previous training experience (Martinussen 1996). However, the majority of candidates seeking military aviation training have little to no previous flight experience. For these candidates, three assessment areas have withstood the test of time: cognitive ability, aviation job knowledge, and psychomotor ability (Carretta and Ree 2003; Hunter and Burke 1994; Martinussen 1996).

General cognitive ability (*g*) tends to be the best predictor of performance for inexperienced workers, regardless of occupation (Schmidt and Hunter 1998). It is not surprising then that the US Army, Navy, and Air Force aviator selection systems all feature cognitive-based subtests, including tests of mathematic, verbal, spatial, and mechanical abilities (Carretta, Retzlaff, Callister and King 1998). The cross-service Armed Services Vocational Aptitude Battery (ASVAB), used in the selection of air traffic controllers and Army Aviation Warrant Officers, also features tests of specific cognitive abilities. As scores on specific cognitive abilities tests exhibit strong positive correlations with each other, their combined influence is indicative of a *g* factor (Grubb, Whetzel and McDaniel 2004). The use of *g*-based measures during selection not only improves the validity of the overall system, but also serves as a vital baseline for the clinical assessment of flight personnel should they be neurologically compromised by illness or in the evaluation of a decline in flight performance later in their careers.

Aviation job knowledge concerns candidates' familiarity with aviation history, aircraft components, aerodynamic principles, and flight rules and regulations. By obtaining a thorough understanding of these areas prior to selection testing, candidates demonstrate their interest and motivation toward pursuing aviation careers (see Chapter 7, this volume for a discussion of the importance of motivation to fly). In addition, because flight training begins with ground school instruction, candidates with pre-training knowledge of aviation basics improve their likelihood of succeeding academically.

Psychomotor abilities include multi-limb coordination, spatial orientation, perceptual speed, reaction time, dexterity, and control precision. Up until the 1980s, aviation-based psychomotor measurement required the use of complicated, expensive, high-maintenance machinery. Today, the availability of cost efficient and transportable computer-based testing platforms has enabled each of the services to include psychomotor tests in their selection batteries with favorable results.

Test development—deciding how to measure

The job analysis serves as the foundation for selection system development. After identifying all critical work elements, the next step is deciding how to measure them. For military aviation, selection devices have generally fallen into three categories: interviews, paper-and-pencil tests, and apparatus tests (Hilton and Dolgin 1991).

Interviews Aviation candidate interviews were prevalent throughout WWI and WWII, but have since become impractical for selection purposes. Interviews benefit organizations by serving as both recruitment and selection tools, as candidates and interviewers alike are encouraged to ask and answer questions. Best practices in personnel selection suggest that interviews be structured, with each candidate receiving the same set of questions, designed to assess specific KSAOs, and asked by job-familiar interviewers, each of whom should be trained to evaluate candidates along the same verbal and non-verbal dimensions, without bias (Gatewood, Feild and Barrick 2010). However, because most interviews occur on an individual basis, they are among the most time-consuming and expensive selection tools available (Cox, Schlueter, Moore and Sullivan 1989).

In WWI, military aviation interviews were conducted by flight surgeons and line officers, and were intended to evaluate candidates' emotional stability, maturity, aptitudes, and strength of character (Hilton and Doglin 1991). Although these interviews were intended to compensate for the shortfalls of the medical examination, they tended to be unstructured and non-standardized, and therefore did little to improve selection system efficiency (Hubler 1947). As Hubler (1947, 100) described:

> One interviewing officer—and it was then possible for any applicant to be turned down as the result of a single interview—believed that no man could be a good pilot if he were a virgin at the age of 24. Another interviewing officer held that careful, painstaking persons such as stamp-collectors could not become effective fighting pilots; a third discarded all those interviewees who were left handed "because I never knew a left-handed man who made a good aviator."

In the mid-1930s, aviation psychologists introduced standardized questionnaires capable of assessing the characteristics under evaluation in the psychiatric interview (Kellum 1940). With the addition of these more formal instruments, interviews became reserved for making final selection decisions (Jenkins 1946). By the end of WWII, more cost-efficient measures had largely replaced interviews for selecting military aviators.

Paper-and-pencil tests Paper-and-pencil tests are appropriate for assessing candidates' job knowledge, cognitive abilities, and other personal characteristics (for example, personality and biographical data) usually with multiple-choice, true–false, or rating scale items. Although now often featured in electronic format, paper-and-pencil tests have always benefited from being quick and cost-effective to administer and score, which explains their long-standing popularity in large-scale selection programs.

When developing new paper-and-pencil tests, it is critical to utilize item writers who are skilled in creating unbiased, clear, and unambiguous items. Items must cover all relevant content areas, target specific KSAOs, and be of varying difficulty levels. In practice, this requires the drafting of considerably more items than necessary for the initial item pool, with each item subject to SME review. After field-testing the items on individuals from the ultimate group of interest (for example, student or fleet aviators), items that do not provide useful data are discarded, leaving a final measure composed of only the most informative items.

To develop new test forms for the ASTB, for example, the Navy consulted a team of industrial–organizational psychologists to draft an initial item pool of over 1,000 items covering a wide spectrum of difficulty levels (Moclaire, Middleton and Phillips 2006). Several short-form test booklets containing equal representations of each item type were then created and administered to samples of student naval aviators. Analyses of the data identified a much smaller subset of items for use on the final subtests, which together consisted of a mere 80 items. In this case, fewer than 10 percent of the original item pool survived the final cut, which calls attention to the amount of work required for just this one component of paper-and-pencil test development.

Apparatus tests Apparatus testing for military aviation grew out of a need to assess candidates' susceptibility to dizziness, airsickness, and disorientation, but has since expanded to assess a range of candidate KSAOs (Hilton and Dolgin 1991). Experimental apparatus testing by the AAF during WWII demonstrated the potential application of this form of assessment. Using motion picture technology and stick-and-rudder devices, the AAF could capture candidates' complex coordination, two-hand coordination, discrimination reaction time, finger dexterity, steadiness under pressure, aiming stress, divided attention, rudder control, estimation of target speed and direction, pattern memory, and pattern reconstruction, among other abilities (Melton 1947; Thorndike 1947). These early apparatus tests, however, were found to be too confusing, complex, costly, and time-consuming for widespread operational use (Flanagan 1948; Marquis 1944).

Efforts to develop economic and efficient apparatus tests for military aviation selection did not find a foothold until the 1980s when tabletop computer-based simulators became a viable option. During this time, the Air Force introduced the Basic Attributes Test (BAT), a computer-administered test battery designed to assess candidates' psychomotor skills, information processing, and attitude toward risk (Carretta 1987). The portable BAT featured a monochrome monitor, alphanumeric keypad, and two joysticks, which candidates used to complete subtests assessing airplane tracking, mental rotation, immediate/delayed digit memory, encoding speed, and decision-making speed, to name a few. The BAT went operational in 1993 as part of the Air Force Pilot Candidate Selection Method (PCSM, Carretta and Ree 1993), and has since been replaced by the Test of Basic Aviation Skills (TBAS, Carretta 2005).

Off-the-shelf measures As an alternative to developing new selection measures, organizations can also adopt commercially available tests (or adapt existing measures). The use of existing measures saves considerable time and resources that would otherwise be spent writing, reviewing, and field-testing items or test components. The original source should provide documentation of this work in technical reports or research papers. Therefore, electing to use an off-the-shelf measure comes with the responsibility for reviewing its supporting documentation to verify the test's psychometric properties and job-relatedness. Consideration must also be given to how the test has been used in the past, including its availability to the public and, more importantly, potential candidates. Exposure to components of a selection system prior to actual testing can provide an unfair advantage, and may increase Type I error rates if otherwise undeserving candidates are selected over their more qualified counterparts. A final concern for using pre-developed measures is cost (for example, is there a one-time fee or per-administration fee?). The economics of "buying" a selection instrument must be weighed against the time and energy gains to be had by not developing the measure in-house.

The Navy has a history of considering pre-developed tests for selection purposes. The ACT, for instance, was a revision of the Wonderlic Personnel Test, while the FAR was based on the Bennet Mechanical Comprehension Test and Purdue Biographical Inventory (Hilton and Dolgin 1991). For personality assessment, the Navy evaluated the Minnesota Multiphasic Personality Inventory (MMPI), Guilford-Zimmerman Temperament Survey, Taylor Manifest Anxiety Scale, Saslow Screening Test, and the Heineman Manifest Anxiety Scale, though none of these tests saw operational use (Voas, Bair and Ambler 1957). The most recent version of the ASTB, however, features a psychomotor test battery based largely on the Air Force's TBAS, which itself was based on the BAT.

Psychometric evaluation—does it work?

To add value, selection measures must be reliable and valid. Reliability concerns measurement consistency. Two forms of reliability relevant to military aviation selection are retest reliability and alternate form reliability. Reliability is necessary to establish validity, or measurement accuracy. For aviation selection, the two most relevant forms of validity are content and criterion-related validity.

Retest reliability Retest reliability concerns stability in candidates' scores when provided multiple opportunities to complete equivalent tests. With repeated test attempts, any score changes are indicative of either measurement error or changes in the candidate's KSAOs. Repeated test scores are evaluated using correlation analyses, with values of 0.7 or higher commonly considered acceptable. The retest interval, whether immediate or delayed, plays an important role in these analyses. Immediate retest attempts should produce strong, positively correlated scores, as the testing conditions (including the examinee's KSAOs) should be relatively constant between administrations. Weak, negative, or non-correlated scores between immediate retest attempts are evidence of measurement error, which would render the test inappropriate for selection. Immediate retesting, however, tends to produce inflated reliability estimates, as candidates are likely to remember their previous responses. As the retest interval increases, candidates' scores are expected to become less correlated as environmental factors, including exposure to study materials, are likely to affect their subsequent test performance. To increase the likelihood that score gains are the result of actual learning versus practice effects, the services have established retest policies for aviation selection. The Army has a six-month retest window, the Navy has a 90-day retest window, and the Air Force has a 180-day retest window.

Alternate form reliability Alternate form reliability comes into play when a measure is available in more than one version or format. The ASTB-E, for instance, is available in static and computer-adaptive test (CAT) format. A static test features the same set of items for all candidates. ASTB-E has three static test forms, each of which contains unique items designed to assess equivalent candidate characteristics. The ASTB-E static tests are available in print and electronic formats. The CAT version is more complex in that no two candidates (with rare exceptions) receive the same set of items. Based on item response theory, CATs estimate the candidate's ability level based on each item response. Generally, a CAT will begin with a medium difficulty item. If answered correctly, the CAT program re-estimates the candidate's ability level and presents a more difficult item (or vice versa). This process continues until the candidate's standard error of measurement falls below a set threshold confirming measurement precision (see Huelmann and Oubaid 2004 for more on CAT). For fairness, all versions of the ASTB-E should yield comparable scores. The correlation of scores between test forms represents alternate forms reliability, with higher values indicating increased equivalency.

Content validity Content validity refers to whether an assessment device represents all relevant aspects of a work-related KSAO. There is no statistical procedure for estimating content validity directly. Instead, content validity is inferred via inter-rater agreement among SMEs tasked to examine test material. The information collected during the job analysis should provide the key areas on which to focus. A content valid test addresses all critical aspects of the KSAOs under assessment, with more critical aspects receiving increased, but not excessive coverage. For example, the Army's FAST Helicopter Knowledge subtest appropriately features items unique to helicopter flight principles, such as skid design and hovering. If the FAST placed too much attention on autorotation, it would exhibit less content validity, as this is only one of many elements representing basic helicopter knowledge.

Criterion-related validity Criterion-related validity refers to the degree to which selection test scores predict job-relevant performance criteria. In military aviation, most criterion variables are training outcomes. This is the case for several reasons. For one, military aviation training is a lengthy and costly process. It is not uncommon for student aviators to take over a year to complete training, and per-student training costs can reach into the hundreds of thousands (even millions) of dollars. When students fail to complete training, this time and resource-investment is lost. On the other hand, when students complete training, their skills are so advanced that their subsequent performance reflects far fewer instances of poor performance. With minimal performance variability among fleet operators, it is nearly impossible for pre-training selection tests to predict performance distinctions among post-training aviators. Performance criteria for military aviation selection therefore include academic grades, ground school grades, flight performance, flight hours required to complete training, number of failures, class rank, and training completion/attrition status.

When the Air Force revised its PCSM and pilot training program, there was a need to examine impact on the criterion-related validity of its selection battery. Carretta (2011) described this effort, which included trainees' PSCM composite and component scores as predictors of academic grades, daily flying grades, check flight grades, and final training outcome (that is, graduation/elimination). Because the analyses were performed on selected candidates, the data required correction for range restriction (see Thorndike 1949). Results showed statistically significant correlations between each predictor and criterion variable, with one exception (previous flight experience did not predict academic grades). The strongest correlation was found between the PCSM composite and the final training outcome criterion ($r = .53$ in the fully corrected model), which provided sufficient evidence of criterion-related validity for the revised selection system.

Current Measures

Each branch of the US Armed Forces has a unique aviator selection system designed to predict effective performance associated with its particular mission.

Army Flight Aptitude Selection Test (FAST)

The Army's FAST contains 200 multiple-choice items designed to assess candidates' background information, personality characteristics, and special abilities related to successful completion of helicopter flight training. The seven sections of the FAST include:

- Background Information: general information related to candidates' education and work experiences, hobbies, and home conditions.
- Instrument Comprehension: candidates determine the position of an airplane in flight (amount of climb/dive, degree of left/right bank) by reading dials showing an artificial horizon and compass heading.
- Complex Movements: presents an image of a dot outside of a circle and five symbol pairs representing potential direction and distance movements. Candidates select the symbol pair that would place the dot in the center of the circle.

- Helicopter Knowledge: candidates select the response option that best completes a series of incomplete statements representing general helicopter principles.
- Cyclic Orientation: presents three sequential pictures from the pilot's perspective of a helicopter in a climb, dive, bank, or combination of these maneuvers. Candidates indicate the cyclic movement that would enable the helicopter to perform the maneuver displayed in the pictures.
- Mechanical Functions: candidates select between two response options that describe a picture representing a mechanical principle (for example, properties affecting pressure, volume, and velocity; performance of gears, pulleys).
- Self-Description: assesses candidates' interests, likes, dislikes, opinions, and attitudes.

There are two operational versions of the FAST and therefore only one retest opportunity. Candidates, including Army Officer and Aviation Warrant Officer candidates, require a minimum score of 90 on the FAST to be considered for Army aviation. Once candidates meet this qualifying score, they become ineligible for retesting.

Naval Aviation Selection Test Battery (ASTB)

The ASTB is the primary tool for selecting student pilots and flight officers for the Navy, Marine Corps, and Coast Guard. ASTB-E assesses candidates' cognitive abilities, personality traits, personal life experiences, and psychomotor skills associated with success in aviation training.

- Cognitive Abilities: subtests assess math skills (arithmetic, algebra, and geometry), reading skills (extracting meaning from text), mechanical comprehension (properties of physics and mechanical devices), and aviation and nautical information (aviation history, terminology, principles, and practices).
- Personality Traits: the Naval Aviation Trait Facet Inventory (NATFI) presents pairs of favorable or unfavorable trait-based statements matched on desirability. Candidates select the statement that is most descriptive of their own behavior, regardless of whether they would ever perform the behavior itself.
- Personal Life Experiences: the Biographical Inventory with Response Verification (BIRV) assesses candidates' general background, education, skills, interests, and values, with several items requiring detailed explanations used to verify authenticity of the response.
- Psychomotor Skills: the Performance Based Measures (PBM) battery is a series of computer-based tasks requiring the use of headphones and a USB-enabled stick-and-throttle set. The PBM tasks include:

- Direction Orientation: presents paired images of a parking lot and an airborne aircraft, with each image rotated in a different cardinal direction. Candidates orient the aircraft to face a particular side of the parking lot (North, South, East, West) using the computer mouse.
- Dichotic Listening: separate streams of numbers and letters are read aloud in each ear. Candidates attend to a specified ear and depress a button on the stick upon hearing a specified character (for example, an even number).
- Airplane Tracking: candidates use the stick and/or throttle to manipulate a set of crosshairs in an effort to "target" a moving aircraft. Difficulty levels vary based on the direction capabilities of the aircraft (vertical versus multidimensional axis), number of aircrafts (one or two), and aircraft speed (slow, medium, or fast). These tasks are performed individually (vertical, then multidimensional), concurrently (vertical and multidimensional together), and concurrently in conjunction with the dichotic listening task. In the final trial, candidates perform both tracking tasks while also inputting a series of learned control commands using the stick and throttle to address one of three emergency scenarios (for example, responding to a fire light).

Candidates may complete the cognitive-based ASTB-E subtests in paper-and-pencil or electronic format, including CAT format. The NATFI and PBM are available only on computer workstations featuring the Navy's Automated Pilot Examination (APEX) system. The BIRV is an electronic inventory that can be completed and submitted from any Internet-enabled computer.

Air Force Pilot Candidate Selection Method (PCSM)

The PCSM is a weighted score composite derived from three components associated with performance in the Air Force undergraduate pilot training program: experience, knowledge, and aptitude.

- Experience: previous flight hours, coded using an unequal interval scale.
- Knowledge: the Air Force Officer Qualifying Test (AFOQT) contains 11 cognitive subtests and an experimental personality scale. Separate weighted combinations of scores from each subtest yield five scores: Pilot, Combat Systems Operator, Academic Aptitude, Verbal, and Quantitative. The Pilot and Combat Systems Operator scores qualify candidates for aircrew training, but only the Pilot score contributes to the PCSM.
- Aptitude: the TBAS is a computer-based test of psychomotor skills, multitasking ability, and spatial orientation requiring the use of headphones and a USB-enabled stick-and-rudder pedal set. The TBAS tasks include:
  - Direction Orientation: presents paired images of a parking lot and an airborne aircraft, with each image rotated in a different cardinal

direction. Candidates orient the aircraft to face a particular side of the parking lot (North, South, East, West) using the computer mouse.

- Three-Digit and Five-Digit Listening: presents a series of number and letters via headphones. Candidates squeeze the stick trigger when they hear any of the three or five specified numbers.
- Airplane Tracking: candidates use the stick and/or rudder pedals to manipulate a set of crosshairs in an effort to "target" a moving aircraft. Difficulty levels vary based on the direction capabilities of the aircraft (horizontal versus multidimensional axis), number of aircrafts (one or two), and aircraft speed. These tasks are performed individually (horizontal, then multidimensional), concurrently (horizontal and multidimensional together), and concurrently in conjunction with the three-digit and five-digit listening test. In the final trial, candidates perform both tracking tasks while also inputting a series of learned control commands on the keypad to resolve emergency scenarios.

Future Outlook for Military Aviation Selection

The amount of time and effort required to "go live" with a selection system is considerable, even for mere revisions. For this reason, organizations performing large-scale selection efforts must constantly evaluate their assessment devices to determine areas for improvement.

To improve its selection process, the Army initiated the SIFT project. The SIFT project objectives included: (1) computer-based and web-administrated; (2) correct/minimize FAST deficiencies; (3) rapidly modifiable and adaptable to other occupational categories, including Unmanned Aerial Vehicle (UAV) Operators and Special Operations Aviators; and (4) maximize use of existing measures or those under development within the Department of Defense (DoD, Paullin et al. 2006).

The SIFT project calls attention to several issues that are common across all aviation services branches. First, aviation selection is becoming increasingly electronic. Electronic test administration, and web-based test administration in particular, provides for worldwide application, instant scoring, secure data storage, and decreased costs associated with the printing and shipping of test materials. Second, increased use of UAVs in military aviation is a certainty. Although manned and unmanned aircraft require different KSAOs for operation, it would be wise for future selection systems to account for both forms of flight to maximize resource allocation and minimize unnecessary overhead. Third, the SIFT project emphasizes collaboration among the aviation arms of the DoD to address future selection needs. Joint force efforts support the DoD's overarching strategy for unified action and provide the same resource benefits as having a shared system for manned and unmanned aviation selection.

Beyond the objectives listed in the SIFT report, the US Armed Forces are continually evaluating ways to enhance their next generation selection systems. The

Air Force has identified several constructs that may add value to the PCSM, including interpersonal/personality factors (integrity, assuming responsibility, cooperativeness, decisiveness), communication skills (listening and reading comprehension, oral and written expression), and prioritization/task management (Carretta 2011). Alternative measurement techniques (for example, situational judgment tests, simulated team tasks, behavioral personality assessments) and tools for informing advanced training classifications (for example, the Navy's helicopter, jet, or multi-engine pipeline assignment) also warrant consideration. Suffice to say, considerable time, effort, and resources are necessary to build or update a selection system. Therefore, it is best practice (indeed, a necessity) for large-scale selection programs to maintain a constant state of internal revision and development.

United States Astronaut Assessment and Selection

Astronaut selection occurs as needed to replace astronauts who have left the space program and to meet the requirements for the myriad technical assignments for which the Astronaut Office is responsible, one of which is manned space exploration. Candidates include military pilots, scientists, academicians, physicians, and educators.

Selecting astronauts is an incredibly involved process, given the demand for high performance coupled with high risks and the stressors inherent in space travel. As such, astronaut selection requires the coordination of many NASA organizations, one of which is the Behavioral Health and Performance Group (BHP) within the Space and Clinical Operations Division at NASA's Johnson Space Center. This section focuses on the role of BHP as it relates to astronaut assessment and selection. Medical and psychiatric assessments run concurrently during the astronaut selection cycle and for maximum crew safety, each crewmember must be free of medical conditions that would either impair the person's ability to participate in, or be aggravated by, space flight. BHP is mandated to assist in conducting psychiatric evaluations of the applicants to determine whether each meets NASA Psychiatric Medical Standard criteria for qualification and to report this determination to the Aerospace Medicine Board (AMB). BHP determinations of psychiatric qualification as provided to the AMB are binding and cannot be overturned by the Astronaut Selection Board (ASB). Additionally, BHP is asked to provide information on applicants' psychological suitability for long-duration space flight to the ASB and to provide consulting on psychological factors as needed to inform the ASB's interviews and decisions. Suitability information provided by BHP is non-binding and the ASB uses the information at their discretion. Direct BHP involvement in astronaut selection commences once the applicant pool has been narrowed from thousands to approximately 120 to 150 applicants.

Psychological and psychiatric screening for illness that might jeopardize a space flight mission has been an integral part of astronaut selection at NASA since 1959. However, until 1989, astronaut applicant psychiatric evaluations were conducted by individuals with varying styles and methods. In 1989, a method of providing well-grounded, standardized evaluations for psychiatrically qualifying or disqualifying astronaut applicants was initiated. A battery of appropriate psychometric tests (including measures of intelligence, personality, and psychopathology) was chosen and administered by a group of qualified professionals. A semi-structured interview method based on the latest version of the Diagnostic and Statistical Manual was also adapted to the NASA context and standardized across professionals engaged in the psychiatric screening of astronaut candidates. All tests and interviews used by BHP to select astronauts for short-duration space missions that occurred prior to and during the Space Shuttle Program emphasized the identification of psychopathology.

Long-duration mission focus guides selection changes

NASA's focus began to shift toward long-duration missions as experience was gained with the NASA-Mir Program in the 1990s and crews were being selected for the first International Space Station (ISS) missions. The unique demands of missions longer than the typical two-week shuttle mission placed a greater emphasis on the qualities and skills that allow astronauts to adapt to living and working in space for long periods of time. Current missions on the ISS are typically six months.

Identifying new selection criteria Galarza and Holland (1999) conducted a preliminary job analysis to identify the skills necessary for long-duration versus short-duration missions in order to inform the initial astronaut candidate selection process. Twenty SMEs, including astronauts and psychologists, rated 47 relevant skills on criticality and rated an additional 42 environmental and work demands in terms of their probability of occurrence. The environmental and work demands for long-duration space missions included group dynamics within a heterogeneous crew and with external groups such as ground control. The SMEs' ratings resulted in ten broad factors important for long-duration missions, including performance under stressful conditions, judgment/decision-making, and teamwork skills. These ten factors overlap somewhat with those identified in previous peer rating studies that suggested both a job competence dimension and an interpersonal dimension for astronaut performance (McFadden, Helmreich, Rose and Fogg 1994; Santy 1994). Identification of these ten factors important for successful adaptability to long-duration missions led BHP to update the tools and procedures used for astronaut applicant psychological screening in 1998 (Galarza and Holland 1999; Galarza et al. 1999).

Building the tools BHP collects information regarding the clinical indicators of suitability for living and working in space flight conditions for months at a time. Selection efforts focused on psychological suitability began with the astronaut class of 1992 and included two groups of tools still presently in use: personality inventories and structured interviews. Each grouping of tools was chosen in order to support best practices exhibited in similar industries or in validated selection programs for jobs similarly involving public safety and trust (for example, government agencies, police, special forces, firefighters, Antarctic expeditioners, flight controllers). The tools have evolved since their initial inclusion to reflect changes in the science of selection and changes in the astronaut job demands and job context.

The first group of tools, the personality measures, was incorporated as a secondary source of information to support clinical judgments made by the clinical/operational psychologists regarding suitability and the aerospace psychiatrists regarding medical qualification. Personality tests were chosen by BHP to meet a specific set of criteria. First, a preliminary list of personality tests was identified. From that list, five psychologists, with expertise in using personality tests to assess individuals in work settings, examined the psychometric evidence of the personality tests, such as face validity, content validity, construct validity, test reliability, and other psychometric properties. Practical concerns such as test duration, computer-based administration, scoring, and reporting logistics were also taken into consideration. The extent to which the test could identify a pattern of response-faking and socially desirable responding was also weighed in the review of available personality tests. This information and the experts' ratings and recommendations resulted in the selection of the final personality tests to include in the current astronaut selection test battery. Tests that did not assist clinical or psychiatric evaluations and were not correlated with suitability factors were dropped from the overall test battery whenever sufficient information was available to determine test utility.

The second tool intended to help examine critical skills for long-duration space flight was a structured interview based on those psychological skills identified to be critical to long-duration space flight (Galarza and Holland 1999). This psychological interview is in addition to the standardized psychiatric interview already used to screen for pre-existing illnesses. Research shows interviews that are structured and based on job analyses yield more valid judgments about normal psychological and behavioral tendencies than interviews that are unstructured (Murphy, Haynes and Heiby 2004; Schmidt and Hunter 1998). BHP's structured interview consists of interview questions from each of the ten critical proficiencies or suitability factors outlined by Galarza and Holland as necessary to adapt and cope successfully with short- and long-duration space missions. The structured psychological interview also provides rating forms to guide the psychologist interviewers in the assignment of ratings of the applicant's suitability to perform and adapt safely to short- and long-duration space missions.

Recent assessment and selection changes

In 2008, NASA created a new training flow in order to reduce the inexperienced crewmember training sequence for ISS missions, from assignment to launch, from four years to around two to two and a half years. One consequence of the shortened training flow for the ISS is a reduced opportunity for crewmembers to train together as a team, which impacts the crews' ability to establish shared mental models or interpersonal experiences prior to flight. Unlike Space Shuttle crews and three-person ISS crews that were the normal size prior to 2010, the ISS six-person crews of today may have little contact with one another prior to their mission. Another issue is that since the three crewmembers are rotated every three months on the ISS, one crewmember may interact with five to eight other astronauts or cosmonauts during one six-month mission. Although the training flow for the ISS now spans two and a half years, rarely do all ISS crewmembers train together; even more rarely have they lived together prior to launch. These changes in the training flow have led to questions regarding how best to select candidates for the training flow.

A series of team experiential exercises were added to the 2009 astronaut selection cycle by BHP to augment psychological testing and interviews. An additional benefit was the ability to observe applicants' behaviors in a newly created team setting similar to what they may experience during their space mission. The mix of exercises allowed factors more essential to effective teamwork to be assessed under this low-fidelity field setting.

Future outlook for astronaut selection

As NASA looks toward moving out of low Earth orbit and into a return to the Moon or travel to an asteroid or Mars, the selection and psychological screening of astronauts will again need to adapt. The new missions will require the crew to become more autonomous. The ISS was never developed to be an autonomous space platform, but instead to have Earth-based control. Longer flights mean that crewmembers will be required to take greater responsibility for training as they will need to remember technical information for longer periods and potentially will need to complete just-in-time training while en route. Other challenges to be addressed in selection and training include communications delay with Earth, limited access to mission support, friends and family support, the lack of widely varied entertainment, an increased isolation, limited confinement, and an extremely hostile environment. Not only are partnerships with other countries expected to continue, but NASA has begun to partner with commercial space companies. Future challenges on selection will be impacted by decisions that have yet to be made and include issues like crew composition, single or multinational explorers, commercial explorers, and multi-space agency involvement. To address these future needs, NASA continues to evaluate the changing job requirements of the astronauts and seek improved methods of selection.

Summary

The dynamic nature of aerospace operations requires that the systems used to assess and select personnel be capable of adaptation. Whether it is expansion of air warfare to include unmanned aviation or a mission to Mars, success will depend on the ability of aeromedical practitioners involved in the design and development of selection systems to make advancements concurrent with any growth in the theoretical and technological environment. The fundamentals of selection system design (for example, job analysis, test development, validation), however, are well established and unlikely to change in the near future. Therefore, it is essential that researchers, practitioners, and vested decision-makers take into account lessons-learned from aeromedical history to better prepare the selection programs of years to come.

References

Armstrong, H.G. 1948. USAF developments in the selection and classification of fliers. *The Military Surgeon*, 102, 469–473.

Armstrong, H.G. 1952. *Principles and Practice of Aviation Medicine*. 3rd Edition. Baltimore: Williams and Wilkins Co.

Carretta, T.R. 1987. *Basic Attributes Test (BAT) System: The Development of an Automated Test Battery for Pilot Selection,* (Technical Report No. 649). Brooks AFB: Manpower and Personnel Division, Air Force Human Resources Laboratory.

Carretta, T.R. 2005. *Development and Validation of the Test of Basic Aviation Skills (TBAS)*, (Technical Report No. 0172). Wright-Patterson AFB: Air Force Research Laboratory, Human Effectiveness Directorate, Warfighter Interface Division.

Carretta, T.R. 2011. Pilot candidate selection method: Still an effective predictor of US Air Force pilot training performance. *Aviation Psychology and Applied Human Factors,* 1(1), 3–8.

Carretta, T.R. and Ree, M.J. 1993. Basic Attributes Test (BAT): Psychometric equating of a computer-based test. *International Journal of Aviation Psychology*, 3(3), 189–201.

Carretta, T.R. and Ree, M.J. 2003. Pilot selection methods, in *Human Factors in Transportation: Principles and Practices in Aviation Psychology,* edited by P.S. Tsang and M A. Vidulich. Mahwah: Erlbaum, 357–396.

Carretta, T.R., Retzlaff, P.D., Callister, J.D. and King, R.E. (1998). A comparison of two US Air Force pilot aptitude tests. *Aviation, Space, and Environmental Medicine*, 69(10), 931–935.

Cox, J.A., Schlueter, D.W., Moore, K.K. and Sullivan, D. 1989. A look behind corporate doors. *Personnel Administrator*, 34(3), 56–59.

Damos, D.L. 2007. *Foundations of Military Pilot Selection Systems: World War I*, (Technical Report No. 1210). Arlington: US Army Research Institute for the Behavioral and Social Sciences.

Flanagan, J.C. 1948. *The Aviation Psychology Program of the Army Air Forces,* (Report No. 1). Washington, DC: US Government Printing Office.

Galarza, L. and Holland, A. 1999. *Critical astronaut proficiencies required for long-duration space missions*. Paper presented at the International Conference on Environmental Systems, Denver, CO, July 19, 1999.

Galarza, L., Holland, A., Hysong, S., Lugg, D.J., Palinkas, L.A., Stuster, J., Witham, L. and Wood, J.A. 1999. Preparing for long-duration space missions: Discussion and resource guide for astronauts [Online: NASA]. Available at: http://ntrs.nasa.gov/search.jsp. [accessed: 7 December 2007].

Gatewood, R., Feild, H.S. and Barrick, M. 2010. *Human Resource Selection*. 7th Edition. Mason: South-Western.

Grubb, W.L., Whetzel, D.L. and McDaniel, M.A. 2004. General mental ability tests in industry, in *Comprehensive Handbook of Psychological Assessment: Vol. IV. Industrial and Organizational Assessment*, edited by J.C. Thomas. Hoboken: John Wiley and Sons, 7–20.

Hilton, T.F. and Dolgin, D.L. 1991. Pilot selection in the military of the free world, in *Handbook of Military Psychology*, edited by R. Gal and A.D. Mangelsdorff. New York: John Wiley and Sons, 81–101.

Hubler, R.G. 1945. The psychologist picks a pilot. *Flying*, 36(5), 54–55, 100–107.

Huelmann, G. and Oubaid, V. 2004. Computer Assisted Testing (CAT) in aviation psychology, in *Aviation Psychology: Practice and Research*, edited by K. Goeters. Burlington, VT: Ashgate Publishing Company, 123–134.

Hunter, D.R. and Burke, E.F. 1994. Predicting aircraft pilot training success: A meta-analysis of published research. *International Journal of Aviation Psychology*, 4(4), 297–313.

Jenkins, J.G. 1946. Naval aviation psychology (II): The procurement and selection organization. *American Psychologist*, 1(2), 45–49

Kellum, W. 1940. *Preliminary Report of Psychological Research: 2 January to 15 February 1940*, unpublished.

Kubisiak, C. and Katz, L. 2006. *US Army Aviator Job Analysis,* (Technical Report No. 1189). Arlington: US Army Research Institute for the Behavioral and Social Sciences.

Martinussen, M. 1996. Psychological measures as predictors of pilot performance: A meta-analysis. *International Journal of Aviation Psychology*, 6(1), 1–20.

Marquis, D.G. 1944. Psychology and the war. *Psychological Bulletin*, 41(9), 103–115.

McFadden, T.J., Helmreich, R.L., Rose, R.M. and Fogg, L.F. 1994. Psychological predictors of astronaut effectiveness: A multivariate approach. *Aviation, Space, and Environmental Medicine*, 65(10: Pt 1), 904–909.

McFarland, R.A. and Franzen, R. 1944. *The Pensacola Study of Naval Aviators. Final Summary Report,* (Report No. 38). Washington, DC: Civil Aeronautics Administration, Division of Research.

Melton, A.W. 1947. *Apparatus Tests*, (Research Report No. 4). Washington, DC: Government Printing Office.

Moclaire, C.M., Middleton, E.D. and Phillips, H.L. 2006. *3PL Parameter Estimation on Prototype Items for Multiple-Choice Subtests of the Aviation*

Selection Test, (Technical Report dated 2006). NAS Pensacola: Naval Aerospace Medicine Institute.

Murphy, K.R., Haynes, S.N. and Heiby, E.M. 2004. Assessment in work settings, in *Comprehensive Handbook of Psychological Assessment, Vol. 3: Behavioral Assessment*, edited by S.N. Haynes and E.M. Heiby. Hoboken, NJ: John Wiley and Sons, 346–364.

North, R.A. and Griffin, G.R. 1977. *Aviator Selection 1919–1977* (Technical Report dated 04 October 1977). Pensacola: Naval Aerospace Medical Research Laboratory.

Paullin, C., Katz, L., Bruskiewicz, K.T., Houston, J. and Damos, D. 2006. *Review of Aviator Selection,* (Technical Report No. 1183). Arlington: United States Army Research Institute for the Behavioral and Social Sciences.

Santy, P.A. 1994. *Choosing the Right Stuff: The Psychological Selection of Astronauts and Cosmonauts*. Westport, CT: Praeger Scientific.

Schmidt, F.L. and Hunter, J.E. 1998. The validity and utility of selection methods in personnel psychology: Practical and theoretical implications of 85 years of research findings. *Psychological Bulletin*, 124(2), 262–274.

Selz, O. 1919. Über den anteil der individuellen eigenschaften der flugzeugführer und beobachter an flugunfällen (On the influence of individual qualities in pilots and navigators in air accidents). *Zeitschrift für Angewandte Psychologie*, 15, 254-300.

Stratton, G.M., McComas, H.C., Coover, J.E. and Bagby, E. 1920. Psychological tests for selecting aviators. *Journal of Experimental Psychology*, 3(6), 405–423.

Thorndike, R.L. 1947. *Research Problems and Techniques,* (Report No. 3). Washington, DC: US Government Printing Office.

Thorndike, R.L. 1949. *Personnel Selection*. New York: John Wiley and Sons.

Voas, R.B., Bair, J.T. and Ambler, R.K. 1957. *Validity of Personality Inventories in the Naval Aviation Selection Program,* (Technical Report No. 13). NAS Pensacola: Naval School of Aviation Medicine.

Weber, H.R. 2009. As pilots age, airlines hire fewer from military. *USA Today* [Online]. Available at: http://www.usatoday.com/travel/flights/2009-01-16-pilots-age_N.htm [accessed: 26 July 2012].

Wise, J.A., Hopkin, V. D. and Garland, D. J. 2009. *Handbook of Aviation Human Factors*. 2nd Edition. Boca Raton: CRC Press.

Yerkes, R.M. 1919. Report of the Psychology Committee of the National Research Council. *Psychological Review*, 26(2), 83–149.

Chapter 3

Commercial Airline Pilot and Air Traffic Controller Selection[1]

Gary G. Kay, Andrew J. Thurston and Chris M. Front

On 1 January 1914, Tony Jannus flew the first scheduled commercial airline flight from Tampa, Florida to Saint Petersburg, Florida in his Benoist XIV flying boat. The flight lasted only 23 minutes and passed between 50 and a mere five feet in altitude above the clear waters of Tampa Bay. Some 3,000 spectators, a parade band, and reporters from the *Saint Petersburg Times*, watched as Jannus and his passenger left for the peninsular side of the Bay. The passenger paid $400 for the inaugural voyage, and the first cargo transported were the pictures taken during the flight.

Six months later, with the advent of WWI, planes were needed as well as pilots. As reported by Damos (2007) in approximately 20 months, the Army went from two flying schools to 25 and from 26 fully-qualified pilots to over 16,000. At the time, Thorndike (1919) reported that scores on the Mental Alertness Test, a measure of intelligence, correlated 0.50 with success in ground school. In contrast, the correlation between years of schooling and success in ground school was reported to be 0.25.

By the time WWII began, the need for pilots was even more manifest. Aerial combat, strategic bombing, flight deck complexity, and dogfights meant pilots required more training than their predecessors. At the end of the war, veteran pilots were eager to return home to use their training and expertise in the burgeoning commercial aviation industry. The newly forming airlines were able to fill their cockpits with these military pilots, selecting the best of these candidates for the esteemed job of airline pilot. From the 1950s to 1970s the job of airline pilot was portrayed as prestigious and glamorous (for example, the character of Frank Abagnale, Jr. played by Leonardo DiCaprio in the 2002 film *Catch Me If You Can*).

Since then, a litany of changes to commercial aviation has made selection of qualified aviators far more complicated. The Airline Deregulation Act of 1978 ended government control over fare prices. Discount airlines began to emerge, challenging the status quo. The lower cost of flying meant that those who previously would have travelled across country via bus, car, or rail could now afford to fly.

---

1 Author note: Special thanks to Arianna Hoffmann, Carl Hoffmann, Carlos Porges, and Sally Spetz for their expert insight into the current state of commercial aviation selection. Additional thanks to Kristin Saboe for her review and comments.

However, for the airlines, and their pilots, what followed were bankruptcies, furloughs, mergers, and loss of promised pensions.

Over the next 20 years there is expected to be a serious shortage in qualified pilots. According to Boeing, 460,000 new pilots will be needed worldwide by 2031, with 69,000 needed for North America, and 185,000 in the Asia–Pacific region. The problem is already acute in Asia and the Middle East. In the US the shortage is related to a combination of factors including a wave of pilot retirements, a reduction in the number of military pilots entering civil aviation, and tougher qualification standards that become effective in the US in 2013. John Allen, Federal Aviation Administration (FAA) Director of Flight Services, expressed concern that this looming shortage may have an impact on flight safety. His comments made in an interview with the Associated Press, directly address the need for pilot selection: "If the industry is stretched pretty thin … that can result in someone getting into the system that maybe isn't really the right person to be a pilot. Not everybody is supposed to be a pilot" (as cited in Lowy 2012).

How are the airlines to select future pilots? What factors are impacting the pool of applicants? What selection tools have been found effective in predicting success in training and subsequent performance in the job? In 1986 Helmreich wrote:

> The prototypical pilot continues to be drawn from the era of white scarves and open cockpits: a man with lightning reflexes, high intelligence, extraordinary courage, X-ray vision, and a legendary capacity for alcohol and sleep loss, an individual who can be characterized as having "The Right Stuff" (Wolfe 1979) or being a "Macho Pilot" (Helmreich 1979) … In the less florid language of experimental psychology, we could describe the best of this genre as having extremely high aptitude and motivation for the task at hand … How do we know when we have selected an individual with the requisite qualities? (274–275)

The following chapter addresses the issues of pilot selection from an updated perspective. In addition, the chapter briefly reviews the selection process used with air traffic controllers, a related aerospace occupation which also faces shortages, and which shares the need to find individuals with the capability to perform a job demanding of exceptionally high levels of reliability.

Issues of the Current Era

Pilot compensation

In order to remain competitive and reduce costs, airlines have significantly reduced pilot compensation, especially for those in entry-level positions. The marked reduction in pilot pay would be expected to have an impact on those considering a career as a pilot. In 2009 the average starting pay at major airlines was $36,283

(McCartney 2009), about 36 percent below the median US income level in 2010 (US Census Bureau 2011). Similarly, according to O*NET (Occupational Information Network) the median wages for commercial pilots, co-pilots, and flight engineers for 2011 was $105,580 (O*NET Online 2010). This is a considerable decline from the relative income level of pilots from 20 years ago.

The prospective pilot not only faces lower wages but also job instability (due to bankruptcies, mergers, and consequences of terrorism) and risk of their pension disappearing. Individuals who previously would have considered flying as an occupation may find the occupation (and the decline in prestige) to be less than desirable. In spite of their having a passion for flight, highly intelligent individuals who may have pursued a career in aviation are more likely to choose careers in fields where they are more likely to earn an income more deserving of their abilities. Normative data collected at a major US airline in the late 1980s showed that the average IQ was close to 125 for the pilots over the age of 45. In contrast, the average IQ found for the younger pilots at a large regional airline was closer to 105 during the same time period. This suggests that the profession is attracting fewer individuals with higher levels of intelligence. How will this impact training and flight safety?

Education

Military (fixed wing) pilots are required to be officers and therefore, at a minimum, have a four-year college degree. There is no such global educational requirement in commercial aviation. Education requirements are set by the airlines. Pilots currently filling the ranks at regional carriers often have less than a traditional college degree. For those with a college degree it is increasingly the case that their degree has been obtained at a college specializing in aviation. The academic rigor of these programs can range from being a glorified flight school to being a top-notch university. Some of the less rigorous programs offer substantial college credit for work experience. Under these circumstances the meaning of a Bachelor's degree is questionable.

Training

With fewer pilots being trained by the military, the training of aviators is mostly accomplished at civilian flight training programs ranging from flight schools where training is offered by Certified Flight Instructors (CFIs) at community airports operated by Fixed Base Operators (FBOs), to major university-based aviation training programs (for example, University of North Dakota Aerospace), to airline-based *ab initio* (see below) programs.

Those seeking a career as a pilot via the civilian route generally begin by earning their single-engine license. They subsequently acquire multi-engine and commercial certificates. They typically become a CFI, and complete their instrument ratings. The aspiring pilot then generally works as a CFI to build up their single-engine flight hours while seeking opportunities to obtain multi-engine hours. Prior to passage of new regulations in August 2010 (PL 111-216; Airline Safety and Federal Aviation Administration Extension Act 2010), an aspiring pilot with 350 to 700 flight hours was considered qualified to submit an application for a job with a regional airline. The number of hours required has depended upon the quality of the applicant pool. However, in the wake of a tragic regional airline crash, a new law in the US mandates a minimum of 1,500 hours, with exceptions to be determined (for example, prior military and other as of yet specified training programs). Once hired by a regional carrier the pilot commonly begins the application process anew seeking a position with a major carrier. The typical number of flight hours for an individual pursuing the civilian path to be considered by a major airline has been in the range of 2,000 to 5,000 hours (again with fewer hours required of military applicants).

The new PL 111-216 regulations require Part 121 Air Carriers (that is, scheduled air carriers) "to develop and implement means and methods for ensuring that flight crewmembers have proper qualifications and experience" (Airline Safety and Federal Aviation Administration Extension Act, 2010, Section 216. Stat. 2367). Relevant to this chapter is the requirement that prospective flight crewmembers shall "undergo comprehensive pre-employment screening, including an assessment of the skills, aptitudes, airmanship, and suitability of each applicant for a position as a flight crewmember in terms of functioning effectively in the air carrier's operational environment" (Airline Safety and Federal Aviation Administration Extension Act 2010, Section 216. Stat. 2367). Under this regulation air crewmembers are required to hold an Air Transport Pilot (ATP) license and have "appropriate multi-engine aircraft flight experience" (Airline Safety and Federal Aviation Administration Extension Act 2010, Section 216. Stat. 2367). Furthermore, in Section 217 of the new law, there is a requirement for the ATP certificate holder to have a minimum of 1,500 flight hours, including sufficient flight hours, "in difficult operational conditions that may be encountered by an air carrier to enable a pilot to operate safely in such conditions" (Stat. 2368). These regulations apparently don't apply to Part 135 operations (that is, commuter and non-scheduled commercial operations) which has historically been another opportunity for civilian pilots to build the hours needed to become an airline pilot.

The *ab initio* approach to pilot development is an alternative which aims to have the trainee begin flying as an airline's first officer upon course completion with 350 hours of total flight time. How this will work under PL 111-216 is unclear, though the law states that "specific academic training courses" may be credited toward the total required flight hours. The *ab initio* approach to pilot development is increasing in popularity, particularly outside the US, where carriers prefer to provide their own training. Training is focused from the beginning on preparing the student pilot to be a first officer (often for a specific aircraft). International carriers have developed sophisticated two-year training programs (for example, Lufthansa). These programs make extensive use of simulator-based training.

The manner in which pilots are trained is changing as well. Training that was primarily comprised of logging hundreds (or thousands) of hours in the cockpit may become obsolete. Additionally, pilots may no longer be required to sit and watch other pilots log hours while serving in the role of flight instructor. While time spent as a CFI probably has provided valuable training experience, at some point it is likely of diminishing value, other than for the purpose of building up flight hours for the candidate's resume. In *ab initio* programs the student pilot logs maneuvers in a simulator rather than spending the hours in the cockpit. Pilots will train to execute responses to hundreds of stalls and engine failures. This is believed to be an efficient and effective way of preparing the student pilot to assume the role of airline pilot. This approach to training is expected to reduce training time and to allow pilots to be better prepared for emergencies. However, opinions are mixed on this type of training. The conventional wisdom is that logging long hours in the cockpit is a rite of passage which provides good training for the types of unplanned events that the pilot may face. Others feel that the new approach provides direct training in handling safety-critical low-frequency events.

Licensure of pilots is likely to change as well. For the first time in several decades, the International Civil Aviation Organization (ICAO) reviewed the standards of pilot licensure and developed a new license, the Multi-Crew Pilot License (MPL). The MPL is a license that qualifies the aviator to serve as a first officer (a co-pilot). MPL pilots must be at least 18 years old, have a minimum of 240 hours of flying training, and 750 hours of theoretical knowledge instruction. This license allows a pilot to have the same privileges as a first officer in commercial air transportation on multi-crew airplanes. The MPL is a variant of the *ab initio* training described above. It is referred to as a competency-based curriculum integrating Crew Resource Management (see Chapter 14, this volume) as well as Threat and Error Management skills throughout training. As with other *ab initio* training, training for the MPL typically makes heavy use of flight simulators to enable scenario-based training. Once in the cockpit the MPL pilot gains further experience and training by working with seasoned captains.

Some have argued the MPL was designed to reduce training time at the cost of safety. The dissenters argue that the MPL was designed to meet the needs of countries in Asia; especially China and India. Although empirical data is hard to come by, early reports indicate that new training methods and certification requirements (for example, MPL) are producing pilots as safe as those trained using traditional methods. This view is challenged by the Air France Flight 447 (AF447) disaster. "What doomed the flight was a series of almost incomprehensible mistakes on the part of the flight crew. Two were particularly astounding. First, a co-pilot at the controls pulled the plane up into a climb—simply leveling out the aircraft could have saved it. Second, he and the two other pilots on the flight deck failed to realize that, as a result of climbing, the plane had entered an aerodynamic stall and began plummeting toward the ocean" (Wise 2012, para. 3). The report of the Bureau d'Enquete et d'Analyse (BEA), the investigative body, reported:

"When crew action is expected, it is always supposed that they will be capable of initial control of the flight path and of a rapid diagnosis that will allow them to identify the correct entry in the dictionary of procedures … During this event, the initial inability to master the flight path also made it impossible to understand the situation and to access the planned solution" (Wise 2012, para. 8). The report offered the following summary: "As AF447 demonstrates, if a flight crew lacks the training and the cognitive resources to figure out what the problem is, they not only won't be able to do anything useful, but they could also turn a minor crisis into a catastrophe" (Wise 2012, para. 9). At the time that the plane entered its fatal descent the plane was under the control of the most junior member of the flight crew (with fewer than 3,000 total flight hours) and was being aided by another first officer. The captain, a veteran with 11,000 flight hours, was taking a routine short break (or nap) and was not on the flight deck.

Theoretical Basis of Selection

Knowledge, skills, abilities and other attributes (KSAOs)

KSAOs refer to the knowledge, skills, abilities, and other attributes deemed as either necessary or desirable for a candidate to possess (see also Chapter 2, this volume). *Knowledge* refers to the "specific types of information people need in order to perform a job," *skills* are "the proficiencies needed to perform a task," *abilities* are "relatively enduring attributes that are relatively stable over time," and *other characteristics* "are all other personal attributes, most often personality factors … or capacities" (Muchinsky 2012, 69). Abilities can be trained into skills. Individuals who are weak or deficient in abilities require more training resources and are more likely to fail to achieve the requisite level of skill; even with extensive training.

For commercial aviation selection, KSAOs are determined by job analysis. While some job requirements are applicable across airlines others are specific to a given airline. For this reason it is often recommended that each carrier conduct their own job analysis to determine company specific requirements. Some KSAOs however are considered to be universal for pilots. It has been suggested that deficiencies in job analyses have undermined efforts to develop more effective pilot selection programs (Damos 1996).

Knowledge Pilots must possess knowledge of aircraft and aircraft systems. They must possess a minimum level of job-related knowledge. This is typically assessed by job knowledge tests, by quizzing the candidate during an interview, and by questions posed during simulated flight. Topics generally include basic aspects of aviation engineering, navigation, aerodynamics, aviation regulations, and meteorology. Some argue that sophisticated computer systems reduce the job knowledge requirements for pilots. Others argue that the reverse is true. More complicated systems require the pilot to possess a more sophisticated understanding of flight systems, not less. Experience is generally considered a proxy for knowledge. However, not all flight hours are equal. Spending 1,000 hours watching a general aviation student make approaches in a single-engine airplane under Visual Flight Rules (VFR) conditions is not equivalent to 100 hours of making challenging maneuvers under instrument conditions in a regional jet or airliner. As described above, at the global level, training models and licensing standards are moving in the direction of requiring fewer hours, while in the US the move is in the opposite direction, requiring more flight hours.

Skills Flight skills are not innate; except for birds. They are developed by training and practice. There are some flight skills that apply across all aircraft and others which are specific to a particular aircraft (and even to a particular model/version of the aircraft). Pilots who have attained an ATP license meet one of the FAA's minimum requirements to become a Part 121 pilot. PL 111-216 required the establishment of a working group to improve the ATP examination. Weekend courses are available that provide training on answering the questions that appear on the current version of the ATP written test. Some argue that the ATP is a better demonstration of the ability to memorize (test answers) than a demonstration of flight skill. Possession of an ATP license does not establish that the candidate has the ability to develop new skills or to learn and apply new procedures at a reasonable pace. Because airlines cannot assume that an ATP license holder possesses the level of flight skills demanded by their airline, they generally perform a careful review of the candidate's work history and training experience. They attempt to determine if the candidate has a history of training failures or other performance problems. As discussed later in this chapter, a candidate's flight skills are often assessed using a flight simulator.

Abilities Job analyses, in spite of their limitations, have helped to identify the required abilities for a pilot. These include cognitive, perceptual, and motor abilities. In the post stick-and-rudder era, flying has become much less dependent upon psychomotor abilities and far more dependent upon cognitive and perceptual abilities. This is not to say that psychomotor skills or manual flight skills have become obsolete. Recent mishaps, such as US Airways Flight 1549, which had to make an emergency landing on the Hudson River, demonstrate the need to be able to fly without reliance upon autopilot systems. Through the superior abilities of the pilot and crew, Flight 1549 did not result in any loss of life. Autopilot systems can fail due to mechanical or software issues or may become inoperable under adverse weather conditions. Under these conditions the aviator must have sufficient psychomotor skills to fly the aircraft. Furthermore, pilots must have the ability to multitask. They must be able to rapidly analyze information and make decisions. They must be able to communicate effectively in the cockpit and with air traffic control. They need to have situational awareness and attentional abilities including good working memory, vigilance, and processing speed. Abilities may be improved by training and practice; though at a considerable cost of training resources. Furthermore, if a pilot is truly deficient in an ability it may not be possible to sufficiently improve that ability to meet the requirements of commercial aviation.

Damos (1996) was critical of pilot selection tests for their failure to be based on legitimate task analysis. Early pilot tests were based upon face validity rather than task analyses. Tests developed in the 1970s and 1980s (for example, the Basic Attributes Test; BAT) utilized what Damos refers to as the theoretical approach, such as the work of Fleishman and Reilly (1992), which identified what were believed to be essential job abilities. According to Damos, the "…primary limitation of this approach seems to be that many psychological theories are concerned with basic elements of human cognition or personality. These elements may not play a significant role in the performance of a task as complex as flying" (1996, 204).

Other attributes Other attributes considered essential include personality characteristics. Leadership is one of these characteristics. Pilots are expected to become captains. Eventually they will serve as the pilot in command (PIC). Therefore, it is critical that candidates have the capacity to lead others. They must be able to guide other members of their crew and keep them focused on safety. In the role of first officer, they must have integrity and sufficient assertiveness to express themselves when their commanding pilot has made an error.

Although organizational ability could be considered an aspect of cognitive ability (that is, executive functions), it is also considered an important aspect of personality; namely, conscientiousness (for example, the Orderliness facet of the NEO-PI-R; Costa and McCrae 1992, 18). Pilots must be organized, punctual, and disciplined. They need to be methodical and detail-oriented in following procedures. Additionally, they must be able to plan ahead and anticipate problems long before they arise. Furthermore, they need to have the social skills to get along with and work effectively with fellow pilots and other crewmembers, air traffic controllers, flight operations personnel, passengers, and other airport personnel (for example, Transportation Security Administration).

The Screening Process

After completing the application process and being found eligible for hire, the candidate is typically offered an opportunity to visit the airline for an interview and undergo selection testing (depending on the carrier). This portion of the chapter will focus on the interview and other testing procedures. Some components of the hiring and screening process, such as the physical examination and drug testing, will not be covered in this chapter.

The interview

Interviews are the most commonly used method of selection. In some airlines (and other corporate cultures) the interview (along with the application form) is the *only* selection method used (Muchinsky 2012, 115). Typically, at these airlines you will find a retired, former airline captain who conducts the interview of candidates. If asked, these captains will usually state that their job is to determine if the candidate is a "good fit" for their airline. Their opinion is that testing is unnecessary. They typically utilize an unstructured interview with the goal of getting to know if the candidate, "is the kind of person you'd want to be stuck with in the cockpit for eight hours, three to four days in a row?"

In contrast, structured interviews are those in which each candidate is asked roughly the same set of questions, and responses are evaluated using specified criteria. These interviews have been shown to have greater criterion-related validity than unstructured interviews. One study conducted with US Air Force pilot trainees found that interviews provided no incremental validity beyond assessments of aptitude, cognitive ability, and personality (Walters, Miller and Ree 1993). However, the criterion used in this study was related to success in undergraduate pilot training. In contrast, more recent research conducted at a regional airline demonstrates a strong relationship between interview scores and success in initial airline pilot training (Carl Hoffman, personal communication, 6 July 6 2012).

In spite of limited empirical evidence to support their use, interviews play a key role in pilot selection. They provide an opportunity for social interaction between the airline and the candidate. There is a widely held belief that interviews allow the airline to determine if the candidate will be a good fit within the culture of the airline. At the same time, the interaction between candidate and airline also provides the candidate, who almost always has applied to multiple airlines, the opportunity to shop around and choose the airline with which they feel most comfortable. While this may be seen to some as reinforcing in-group homogeneity, and at an extreme could be construed as discriminatory, selectors and candidates characteristically prefer to work with others that share their interests and values.

Interviews yield information about the candidate that is not readily gleaned from the results of testing. Is the candidate's motivation realistic or unrealistic? Does the applicant have a passion for flying? Has the applicant always dreamed of becoming an aviator? How has this motivation been demonstrated? The interview provides the airline with an opportunity to further explore the applicant's education, training, and work history. The skilled interviewer is capable of evaluating the applicant's ability to work as a team member and their ability to express themselves clearly, succinctly, and accurately. Lastly, the interview, to the extent that novel, unexpected questions are asked, provides an opportunity to observe how the candidate thinks on their feet.

Simulator testing

Although simulators play an essential role in flight training, for the skilled aviator they appear to provide little information with respect to their success in training or adapting to a new airline. The flight scenarios are typically well known to the aviator. The candidate generally has the opportunity to purchase flight simulator time from vendors who will provide them with the training and practice necessary to pass the screening scenario. Simulator-based testing is also costly for the airline and difficult to schedule. Airline simulators are employed around-the-clock for training and checkrides. Several airlines which have examined the incremental validity of simulator testing in their selection program have abandoned simulator testing in order to focus more on other selection methods.

Job Knowledge tests

There is a range of opinion regarding the value of job knowledge tests in pilot selection. Job knowledge tests evaluate a candidate's understanding of aviation concepts such as avionics, navigation, and engineering. It is important to recognize that job knowledge tests used in selection testing vary considerably in their construction and demonstrated validity. There is empirical data demonstrating that some job knowledge tests contribute significant variance to predicting training success in pilot applicants. Martinussen and Torjussen (1998) demonstrated that knowledge of instruments, mechanical principles, and aviation information, are significant predictors of pilot performance. Similarly, studies by Hoffman et al. (Hoffman, Hoffman and Kay 1998; Hoffman, Spetz and Hoffman 2011) have demonstrated that job-knowledge tests not only predict success in training, but are also predictors of long-term performance.

However, a considerable problem with job-knowledge tests is test security. As will be discussed below in the section on "Gouge," a candidate can often find the actual test items online and memorize the answers prior to taking the test. In response to this, test authors now generate very large pools of items from which the questions appearing at a particular test session will be selected. At a minimum, performance on these tests provides a measure of the candidate's preparation (and motivation).

Cognitive ability testing

Cognitive ability testing is conducted to determine if the candidate has the cognitive abilities required for the job. The tests assess the likelihood that the candidate can complete the airline's training syllabus in the expected amount of time. Cognitive assessments have been found to provide predictive validity with respect to identifying aviator success in training. They also identify individuals who have the information processing speed that plays a key role in cockpit resource management (CRM) (Hoffman, Hoffman and Kay 1998). The aviator without sufficient cognitive capacity is less able to handle novel or abnormal flight situations, or other situations in which they are task-saturated and overloaded. The following section provides a brief description of some of the tests commonly used to assess cognitive ability in pilot selection.

WOMBAT-CS (Aero Innovation Inc. 2012)

The WOMBAT-CS (Complex System Operators) is described as a test of situational awareness and stress tolerance and includes measures of psychomotor ability. The examinee is required to operate joysticks and enter responses on a keyboard. The primary function of the assessment is to evaluate the candidate's situational awareness and stress tolerance.

The WOMBAT-CS consists of primary and secondary tasks. The primary task is a target tracking task. For this task the examinee is required to monitor and adjust the position of two figures on a computer screen using a set of two joysticks. The secondary tasks are as follows:

- Figure Rotation Task: a fixed figure is displayed with a matching rotatable figure. The examinee rotates the second figure to match the orientation of the first.
- Quadrant–Location Task: the numbers 1–32 are displayed in a tabular format. The examinee's task is to find each number in each quadrant in ascending order.
- Two-Back Cancelling Task: the examinee is shown a sequence of three digits from one to eight. After the third digit is displayed, the examinee is tasked with recalling the digit which was displayed "two-back" in the sequence.

These are classic and empirically validated cognitive psychology tasks. The figure rotation test is a variation of the Shepard and Metzler figure rotation task (Shepard and Metzler 1971). The Quadrant–Location task appears to be similar to the Trail Making Test (Lezak, Howieson and Loring 2004) and the Two-Back Task is a variant of the n-Back test (Owen, McMillan, Laird and Bullmore 2005). Results are combined to generate a Final Score which is used as a cut-score in selection programs. The cut-score reportedly determines if the candidate has the necessary ability to maintain situational awareness and tolerate stress. The WOMBAT-CS was designed to be culture-independent, and to not require computer literacy. Scores from the tasks are reported to generalize beyond pilot selection to other careers which require monitoring of complex systems.

CogScreen-Aeromedical Edition (Kay 1995)

CogScreen-AE is a computer-administered and scored cognitive screening instrument designed to assess aviation-related aspects of attention, immediate and short-term memory (that is, working memory), visual-perceptual speed, sequencing functions, information processing speed, logical problem solving, calculation skills, reaction time, multitasking ability, and executive functions. The test was developed for the FAA as a measure to detect changes in cognitive functioning that can impact operational performance in the cockpit.

CogScreen-AE is comprised of 11 subtests:

- Backward Digit Span: recall of a sequence of visually presented digits in reverse order.
- Math: traditional word problems with multiple-choice answer format.
- Visual Sequence Comparison: comparison of two simultaneously presented series of letters and numbers.
- Symbol Digit Coding: substitution of digits for symbols followed by immediate and delayed recall of the symbol digit pairs.
- Matching to Sample: discrimination of a checkerboard pattern.
- Manikin: mental rotation task requiring examinee to identify the hand in which a rotated human figure is holding a flag.
- Divided Attention: visual monitoring task, which is presented alone and in combination with the Visual Sequence Comparison subtest.
- Auditory Sequence Comparison: comparison of two series of tone patterns.
- Pathfinder: visual sequencing and scanning task that requires the examinee to sequence numbers, letters, and an alternating set of numbers and letters.
- Shifting Attention: rule-acquisition and rule-application under time pressure. The task requires mental flexibility and conceptual reasoning.
- Dual Task: consists of two tasks, each of which is performed alone and then together as a simultaneous test. One task is a visual-motor tracking test. The other task is a continuous memory task involving serial digit recall.

CogScreen-AE provides measures of Speed (median reaction time on correct trials of tasks), Accuracy (percentage of correct responses), and Thruput (a measure of response efficiency based on the number of correct responses per minute). The reliability and validity of CogScreen-AE has been established in a variety of settings and populations (for example, Taylor, O'Hara, Mumenthaler and Yesavage, 2000). The test–retest reliability coefficients for each of the measures ranges from .63-.91 (Kay 1995). Predictive validity of the instrument has been evaluated against flight performance errors made by commercial pilots as detected by analysis of the cockpit data recorder (Yakimovich et al. 1994). Other studies have compared CogScreen-AE performance to flight simulator performance (for example, Taylor, O'Hara, Mumenthaler and Yesavage, 2000). Relevant to selection testing was the work of Bettes, Lehenbauer and Wolbrink (2007), which evaluated the validity of the test for identifying commercial pilots who had problems completing a regional airline's training program, and the recent work of Hoffmann, Spetz and Hoffman (2011) which investigated the long-term value of selection testing.

IPAS (Initial Pilot Aptitude Screening System; APST Holding 2003)
The Initial Pilot Aptitude Screening System (IPAS) is designed to provide an initial screening of aptitudes considered necessary for a pilot. The test is administered online. Individuals with an interest in aviation can take the test before they start training, allowing the candidate to evaluate their own level of ability. Based on their performance, potential candidates can then assess whether or not paying for flight training will be a good investment. The IPAS screening program is intended to be especially useful for *ab initio* pilots.

The IPAS assesses aptitude in the following five areas (Initial Pilot Aptitude Screening System 2003):

- Ball Game: test of hand–eye coordination.
- Total Recall: test of short-term memory.
- Math Class: test of mathematic ability.
- Wings And Things: an orientation test.
- Bingo: a measure involving multitasking.

COMPASS (European Pilot Selection and Training 2012)
The Computerized Pilot Aptitude Screening System (COMPASS) assesses many of the same areas of aptitude as the IPAS (for example, hand–eye coordination), but also assesses verbal reasoning and comprehension. The COMPASS battery is comprised of the following subtests:

- Control: a task assessing basic hand/foot/eye coordination.
- Slalom: a tracking task assessing hand–eye coordination.
- Mathematics: a test of basic applied mathematical understanding and speed.
- Memory: a measure of short-term memory recall.
- Task Manager: a measure of multitasking.
- Orientation: a task which evaluates instrument interpretation, comprehension, and spatial orientation.
- Tech-Test: a measure of physics knowledge.

Additionally, candidates are assessed on their understanding of the English language, including their ability to understand air traffic control messages. There is also a measure designed to evaluate verbal reasoning ability.

DLR-Test; ready entry The DLR-Test is administered by the German Aerospace Center. It provides a basic skills assessment and a company assessment. The basic skills assessment measures the following abilities: ability to absorb audio and visual information, three-dimensional spatial sense, and the ability to cope with multiple tasks and stresses. There are also measures of mental arithmetic and English. The test includes a personality component and a portion where the examinee provides handwritten answers to ten questions. The test is described as having been based upon the typical demands found in the working environment of a pilot. A passing score on the basic skills assessment leads to an invitation to take the company assessment. This is described as an assessment of the applicant's psychological profile, including the ability to work as a team member, work under stress, and handle conflict. There is also a psychological interview and a screening conducted in a simple simulator (Airline Test Training Center 2012).

Personality assessment: identifying psychopathology

While the base rate of psychopathology in commercial airline pilots is historically low, screening out persons with mental disorders is considered essential. In 2012, a captain for a major US airline had a psychotic break during a flight, becoming floridly delusional. At some point he left the cockpit which gave the first officer an opportunity to lock him out. The disordered pilot started to bang on the cockpit door and had to be restrained by flight crew and passengers (Mouawad 2012). The pilot's erratic behavior was displayed on the 24-hour news networks as videos of the incident from passenger cell phones began to surface on the Internet. Subsequently, ten of the passengers from that flight filed suit against the airline for gross negligence because they had allowed the pilot to fly. As this example demonstrates, airlines can be held responsible for failure to hire mentally healthy and stable pilots.

Assessment designed to detect psychopathology has little predictive value as a predictor of flight performance. Martinussen (1996) reported a correlation of 0.17 between personality inventories and measures of pilot success. Results also indicated that only 3 percent of the variance in pilot success was accounted for by clinical personality inventories. This confirmed findings from an earlier study showing that personality measures designed to assess psychopathology fail to predict training attrition (Griffin and Mosko 1977). While measures of psychopathology are not predictors of flight performance or training success, such instruments are relied upon in selection testing to screen out psychopathology after a conditional offer of employment has been made to the candidate. The most commonly used instruments are the Minnesota Multiphasic Personality Inventory-2 (MMPI-2) and the Personality Assessment Inventory (PAI).

Assessment of personality traits (or factors)

In contrast to assessment aimed at evaluating the presence and/or severity of mental disorders, this type of assessment is aimed at characterizing the examinee with respect to basic personality traits or factors which are known to be predictors of flight performance. Helmreich (1982) reported that trait constellations of instrumentality, expressiveness, and achievement motivation were predictive of check airmen's ratings of a pilot's line performance. Investigators (Chidester and Foushee 1991) found that these personality characteristics were predictors of pilot performance. Instrumentality is a personality trait associated with the ability to be focused in a competitive way, to be objective, independent, and to be able to make decisions easily. Other descriptions associated with instrumentality include striving for independence, task orientation, and goal seeking. Expressiveness pertains to interpersonal behaviors, including sensitivity and warmth. Other descriptions for expressiveness include cooperativeness and emotional openness.

The Five-Factor Model (FFM; Costa and McRae 1992) According to the Five-Factor Model, personality can be generally classified into five broad factors. Although a number of psychometric instruments provide assessment of personality based directly or indirectly on this model, the test most commonly used in aviation has been the NEO Personality Inventory-Revised (NEO-PI-R; Costa, & McRae, 1992). The following description of the model is based on the work of Costa and McRae (1992). Street and Helton (1993) reported that numerous researchers have found the FFM to have considerable potential for pilot selection and training. The five factors are as follows:

Openness-to-experience The construct of openness-to-experience is designed to assess elements such as imagination, creativity, and curiosity. For potential aviators, a moderate level of openness-to-experience is considered beneficial. Candidates should be curious and want to learn new things. However, extreme levels of openness are considered undesirable. Based upon a job analysis an airline can identify the level of openness-to-experience most appropriate for employment in their organization.

Conscientiousness Among the five personality factors, conscientiousness is considered to have the greatest predictive validity with respect to job performance (Barrick and Mount 1991; Hurtz and Donovan 2000). Conscientiousness refers to an individual's propensity toward organization, thoroughness, and self-discipline. Facets of the construct of conscientiousness include competence, dutifulness, and self-discipline (Costa and McCrae 1992). Commercial airlines need candidates who can be trusted to operate the aircraft autonomously. Conscientiousness is also related to integrity. Airlines depend upon their pilots to follow company flight policies, to conduct their flight checks properly, scan the console diligently, and follow aviation regulations to the letter, even when no one else is watching. They also expect their pilots to be prepared for training events. The importance of conscientiousness as a predictor of pilot performance is well-established.

Extraversion Extraversion is most easily understood as an individual's propensity toward sociability. The extraversion component of the FFM is designed to assess factors such as warmth, gregariousness, assertiveness, excitement-seeking, and activity (Costa and McCrae 1992). Pilots range from one end of the spectrum to the other with respect to extraversion. Extraversion is not synonymous with social skills. A highly introverted individual may have excellent social skills and be reasonably friendly in the context of the cockpit. However there are aspects of extraversion considered undesirable such as excessive excitement-seeking, risk taking, and hyperactivity. When in a training environment, trainees need to have at least moderate levels of extraversion in order to succeed but, if their extraversion is excessively high their need for interaction may distract from training (Barrick and Mount 1991).

Agreeableness Agreeableness refers to an individual's propensity to be interpersonally cooperative, and to be accommodating and accepting of others. Agreeable people can be altruistic or sympathetic to the needs of others. In situations of disagreement, agreeable persons tend to acquiesce to the demands of others. Agreeableness measures facets such as trust, straightforwardness, modesty, and compliance (Costa and McCrae 1992). One might assume that it would be desirable to have as high a level of agreeableness as possible in a pilot. Actually, the extreme of agreeableness reflects a level of dependency on others that would be problematic in the cockpit (Costa and McCrae 1992). Pilots need to critically evaluate the responses of their fellow pilot in the cockpit, and express their opinion or observations. At the same time they need to be sufficiently agreeable to work as a team member and to demonstrate good CRM.

Neuroticism (resilience) The factor generally referred to as Neuroticism, is not a clinical diagnosis, but rather is a factor related to the extent to which individuals are prone to psychological distress (Costa and McCrae 1992). In earlier work with the NEO-PI-R the inverse of the scale was considered to be a measure of resilience. This factor refers to the ease with which an individual reacts to negative stimuli. Some texts have referred to the construct of neuroticism under the title of emotional stability. In this respect, neuroticism is a measure of coping ability. The factor also indicates the individual's vulnerability to stress and their ability to respond and recover from adversity (Costa and McCrae 1992).

For the purpose of selecting pilots, commercial airlines prefer candidates who can handle stress and remain calm, and who can do so without becoming overly anxious or depressed. Additionally, if a pilot is susceptible to experiencing stress from outside the cockpit (for example, work–family conflict) he/she may experience deleterious effects within the cockpit as well. Candidates need to be emotionally stable so that they can perform under pressure.

The resilience/neuroticism factor is a robust predictor of job performance. This was reported in a meta-analytic review examining the correlation between neuroticism and military pilot training success (Campbell, Castaneda and Pulos 2010). The neuroticism factor and the facet of anxiety were both significantly correlated (in the negative direction) with performance. There is a body of research showing that neuroticism is a predictor of poor job performance (for example, Barrick and Mount 1991; Barrick, Mount and Judge 2001).

The pilot profile Fitzgibbons, Davis and Schutte (2004) found evidence for a pilot personality based upon the FFM, as demonstrated by the NEO-PI-R results of 93 commercial pilots (75 percent captains; mean age 42). The researchers described the pilot profile as consistent with an "emotionally stable individual who is low in anxiety, vulnerability, angry hostility, impulsiveness, and depression. This person also tends to be very conscientious; being high in deliberation, achievement-striving, competence, and dutifulness. He also tends to be trusting and straightforward. Finally, he is an active individual with a high level of assertiveness" (Fitzgibbons, Davis and Schutte 2004, 5).

Other personality constructs Although measures of leadership have been developed, and are utilized in industrial–organizational psychology, they are not typically used in pilot selection testing. As stated earlier in this chapter, motivation is integral to success in pilot training, but is generally not assessed by personality instruments. Another personality characteristic, integrity, is critical for a pilot. Unfortunately, most of the measures designed to assess integrity are inappropriate for pilot selection testing (for example, focus on employee theft). Additionally, these measures are particularly subject to response distortion and social desirability bias. To some extent measures of conscientiousness provide a measure of integrity. It can also be argued that validity scales that assess honesty and openness in responding to personality test items (for example, the MMPI-2 and PAI validity scales) constitute indirect measures of integrity.

Humor has also been identified as an essential trait for pilots (Carl Hoffmann, personal communication, July 6, 2012). Sense of humor is not directly assessed by personality testing, but can be evaluated in the interview. For better or worse, selection testing may have to rely upon interviewers to elicit the information needed to assess factors such as humor, motivation, integrity, and leadership.

The Decision Process

Once testing and interviews are complete a decision needs to be made regarding making a job offer to the candidate. There are a range of decision-making approaches taken in the industry. These range from what could be referred to as the clinical model to the actuarial model. In the clinical model, the expert combines or processes information in their head (or according to some, in their "gut"). In the actuarial (or statistical) model the human judge is eliminated. Decisions are based solely on empirically derived relations between the predictor variables and the criterion. All of the data, including results from the interview, are amenable to actuarial interpretation. A third approach is a combined clinical and actuarial method. This approach becomes difficult to apply when there is a lack of agreement between the clinical and the actuarial approaches. Meehl (1954) is credited with initiating research on clinical and actuarial judgment. Comparative studies generally favor the actuarial method as providing more accurate judgments (Dawes, Faust and Meehl 1989). This may only be the case where the relationships between predictors and criterion are known and account for a substantial degree of the variance: "When properly derived, the mathematical features of actuarial methods ensure that variables contribute to conclusions based on their actual predictive power and relations to the criterion of interest" (1671). Certainly, the actuarial approach is the most legally defensible.

In pilot selection programs where a strictly actuarial approach is employed, the airline uses empirically derived cut-scores to determine whether or not a candidate is to be offered a job. These cut-scores are based upon established relationships between performance on selection metrics and job criteria. In practice, a lack of consensus may exist between an interviewer's impression and the findings obtained from testing. For example, if a cut-score for a measure of cognitive ability is set at 170.00 and a candidate obtains a score of 169.98, the strictly actuarial approach would lead to the candidate not being hired. A mixed clinical–actuarial approach might "bump" the candidate up to the hired category based on favorable comments from the interviewer. The interviewer may be considering factors not otherwise included in the equation. The danger is that the (impressionistic) factors not in the equation may not truly be predictors of the criterion.

Criterion-related validity

Criterion-related validity is the extent to which the predictor variable (for example, a test score, or an interview rating) predicts the score on a criterion variable (for example, completion of pilot training or successful job performance). From such information, it is possible to model the likelihood of an individual achieving a particular score on the criterion variable based on their performance on the predictor variable. Organizations use this information to establish a cut-score. The cut-score is the score on a predictor variable that a candidate must reach before he/she can be hired. Failure to reach this score, in a strictly actuarial model, will lead to the individual not being given a job offer. Candidates who are selected are those judged to be likely to succeed in the position.

The criterion (that is, outcome variable) which is generally the focus of pilot selection testing is some aspect of training success. Helmreich (1986) wisely challenged the idea that success in training should be the criterion of selection. His view was that "the ultimate performance criterion … is how the individual performs over time in the operational setting." He shared research findings showing that personality and motivational factors measured prior to employment were only found to be good predictors of job performance after the pilots had been out of training and on the job for more than three months. The predictors were described as unrelated to performance in training and to performance during the initial honeymoon period immediately following training. Furthermore, he pointed out that for multicrew aircraft, overall performance is as much a result of the ability of the team to work effectively as it is a function of individual proficiency.

Damos (1996) provided further criticism of these issues in her review of pilot selection batteries. She pointed out that the purpose of the batteries is to select individuals for the job of flying an airplane. She noted that "no examination of concurrent validity or of post-undergraduate training performance has been made for transport pilots" (Damos 1996, 200). She concluded that the existing selection batteries were not concerned with operational performance but rather with training.

In spite of these caveats, the primary criterion for evaluating the validity of selection batteries continues to be training outcomes. There is no argument that success in training is critically important to the airlines. Training is costly and uses up limited resources (that is, flight simulators). Readers may be surprised to learn that close to 20 percent of the hired pilots at one regional airline were reported to experience significant training problems; and a substantial number failed to graduate. Frequency of training problems (for example, during checkrides, or while completing an Initial Operating Experience (IOE)) is one of the most commonly used validation criteria. Completion of training (versus failure to complete training) has also been used to validate selection programs, though dismissal from training occurs with relatively low frequency and therefore is not a particularly good criterion. If the criterion is too rare of an event (that is, the base rate is too low), it is unlikely that it can be predicted with selection instruments.

Other measures of pilot operational success are difficult to obtain or define. Damos (1996) pointed out that "job analyses for operational pilots are surprisingly difficult to find" (201). The situation is complicated by the difficulty in obtaining reliable measures of job performance. Damos stated that "for legal reasons or because of agreements with the pilot's union" data from simulator scenarios used in checkrides and other company flight check data are typically destroyed and unavailable to be used as a measure of performance (1996, 202). Furthermore, even if checkride ratings were available, the reliability of these ratings is also questionable.

These barriers to evaluating the performance of pilots "flying the line" have not disappeared. However, there have been some notable exceptions. Research was conducted with airline pilots in the former Soviet Union. During the late 1980s in the USSR it was possible to assess the relationship between flight performance violations recorded by the cockpit data recorder on Aeroflot flights and performance on CogScreen-AE (for example, Yakimovich et al. 1994). The test was found to be a good predictor of the rate of flight performance violations. Around the same time, Hoffmann, Hoffman and Kay (1998) developed and validated an observational tool to assess line pilot performance of 300 line pilots at a major US airline. The pilots also completed CogScreen-AE and personality testing. One of the interesting findings from that research was that CRM was better predicted by performance on CogScreen-AE than by results from personality testing. Recent studies conducted by Hoffmann and his associates have investigated results from performance evaluations collected over a 12-year period with line pilots. The findings from this longitudinal research show that CogScreen-AE test findings predict training and non-training outcomes such as loss of productivity (for example, scheduled rotation hours not flown due to sick leave usage).

Gouge

Gouge is military jargon referring to the information required to pass a test. With respect to pilot selection testing, gouge is readily available on the Internet. Not only can the candidate find study guides with sample questions, they can now purchase complete versions of the tests. Interview questions and test items are posted online by former candidates who have been examined.

A recent search engine query for "aviation job knowledge test gouge" yielded numerous websites offering access to test items and even structured interview questions. One website offered, "peace of mind knowing you are prepared!" (Cathay Pacific 2012). The site was a repository of hundreds of interview experiences and test questions. Needless to say, test security is lacking. Candidates can download entire tests with a single click of the mouse (and by providing their credit card information).

Gouge has become so rampant there is often a delay referred by those providing selection services to the industry as a "gouge-lag" (Carl Hoffmann, personal communication, 6 July 2012). Gouge-lag is the time that it takes for candidates to disregard an outdated gouge when a new version of the test has been deployed.

Is this cheating? Or is it akin to taking a prep course for a college entry examination (for example, SAT or GRE)? Is the candidate trying to find a shortcut to being truly prepared? Or does the use of the gouge demonstrate their desire to be prepared … and their desire for the job? The mere fact that the candidate is researching the test is considered by some to be an indication of motivation. On the other hand, pursuit of some types of gouge could indicate an integrity problem. For example, obtaining a bank of test questions and answers published by the FAA for public consumption is fundamentally different from attempting to obtain the unpublished "correct" responses to MMPI-2 items in order to avoid disclosing psychopathology.

On the positive side, gouge can help to reduce test anxiety. Alternatively, this can be accomplished by providing test examples for candidates to review. Test anxious candidates may have difficulty demonstrating their true knowledge and abilities due to their poor test taking ability. The gouge may be helpful in reducing anxiety, allowing the examinee better access to the test answers they know.

Air Traffic Control Specialists

Selection of Air Traffic Control Specialists (ATCS) has a number of similarities but also some key differences with commercial pilot selection. First of all, unlike commercial pilots, controllers are employed by either the military or by the FAA. The various branches of the military are responsible for the selection, training, and medical certification of the controllers used in military flight operations. The requirements and duties across the branches are very similar, with selection based upon entry-level aptitude testing as well as the ability to meet the physical and psychological standards associated with flight duties. Ongoing medical examination and review of medical qualification is conducted by military flight surgeons.

In the US, the vast majority of controllers are employed by the FAA. This group includes former military controllers, graduates of civilian colleges and universities with aviation programs that provide training in air traffic control, as well as *ab initio* applicants with no prior air traffic control education, training, or experience. The FAA conducts a stringent selection process that is fully compliant with statutes that apply to pre-employment testing. Applicants undergo aptitude testing, interviews with Air Traffic Control Managers, a security background investigation, testing for illicit drug use, and physical examination and psychological screening in order to determine medical qualifications.

The psychological selection process for ATCSs employed by the FAA is divided into a select-in phase that is aimed at ensuring a goodness of fit between the applicant and the job duties and a select-out phase that is aimed at identifying any disqualifying psychological conditions that would pose a hazard to safety in the conduct of ATCS duties. Requirements imposed by the Americans with Disabilities Act (ADA 1990) permit the "select-in" aptitude testing phase of the selection process to occur as part of the initial application process. In contrast, the select-out phase results are considered to be a medical and psychological examination and therefore must not occur until the applicant has received a tentative job offer.

The aptitude testing for FAA ATCS applicants is conducted with a test battery developed by FAA research psychologists at the Civil Aerospace Medical Institute (CAMI) in Oklahoma City, Oklahoma. The resulting Air Traffic–Selection and Training (AT-SAT) test battery is a computer-administered battery of tests that assesses the fund of knowledge, skills, and abilities that were previously empirically identified as being central to the effective performance of ATCS duties (Ramos, Heil and Manning 2001). Domains assessed include: prioritization, planning, situation awareness, execution, thinking ahead, projection, tolerance for high intensity, composure, recall from interruption, reasoning, time sharing, decisiveness, taking charge, short term memory, visual scanning, flexibility, and problem identification. Subtests used to assess these domains range from traditional reasoning tasks such as Analogies to highly interactive, dynamic visual tasks such as Air Traffic Scenarios. High scores on the AT-SAT have been strongly correlated with subsequent performance on practical testing at the FAA's Air Traffic Control Academy. Applicants for FAA ATCS positions must obtain an acceptable score on the AT-SAT in order to proceed in the application process. Those applicants who have already demonstrated their aptitude by virtue of prior military air traffic control qualifications are not required to take the AT-SAT.

After the applicant has successfully completed the application process and received a tentative job offer, the medical examination is conducted to ensure that the applicant meets the medical qualifications specified by the FAA. The psychological screening aspect of the medical examination is conducted using the MMPI-2, in accordance with FAA policy. The resulting MMPI-2 profiles are reviewed by the Licensed Clinical Psychologist in the Office of Aerospace Medicine at FAA Headquarters in Washington, DC, who then makes recommendations regarding psychological clearance to the Regional Flight Surgeons offices responsible for the medical examinations.

Applicants who produce satisfactory results on the MMPI-2 are immediately cleared from a psychological standpoint. If no other disqualifying medical conditions have been identified, then the Regional Flight Surgeon responsible for the medical examination of the applicant medically clears the applicant for continuation in the hiring process. If, on the other hand, the MMPI-2 profile contains scores of concern to the reviewing Licensed Clinical Psychologist, then the applicant is referred by the Regional Flight Surgeon for a more complete psychological assessment that the FAA refers to as a Tier 2 psychological evaluation (Front 2012). These Tier 2 psychological evaluations are conducted by licensed clinical psychology consultants with expertise in personality assessment who have been trained by the FAA Licensed Clinical Psychologist. The resulting Tier 2 evaluation report and psychological test data are reviewed by the FAA Licensed Clinical Psychologist, and a final recommendation regarding the psychological aspect of the medical clearance is then made to the Regional Flight Surgeon with medical clearance authority. If a disqualifying psychological condition is identified, then medical clearance is denied and the applicant is afforded an opportunity for appeal. If no disqualifying psychological condition is identified, and no other disqualifying medical conditions have been identified, then the Regional Flight Surgeon responsible for the medical examination clears the applicant for continuation in the hiring process.

Summary

Selection of pilots and air traffic controllers is referred to as high-stakes testing. The costs associated with errors in selection are tremendous. The training of pilots and air traffic controllers is expensive. Poor students cost more to train, are more likely to fail out of the training program, and are often poor performers following training. Failure to select individuals with the right cognitive abilities and personality traits for these professions impacts the safety of the airways.

Airlines utilize pilot selection for a number of reasons including; minimizing training costs, reducing liability from claims of discrimination from disgruntled applicants, and reducing liability from passengers claiming negligent hiring. Screening is also performed to find candidates who are the best match for the airline's culture. From the passenger's perspective they would simply be delighted if airlines could clone Captain Sully, the US Airways captain (Sullenberger) who safely landed an Airbus A320 in the Hudson River, after a bird strike shut down both engines on takeoff. Captain Sully and his crew saved the lives of all 155 passengers.

Current selection systems for pilots and controllers have been proven to be legally defensible and effective at reducing training costs. In the future, there will undoubtedly be further development of metrics to assess the performance of line pilots (that is, not just students or new hires). Unobtrusive measures will hopefully be employed to measure CRM, compliance with company policies and procedures, and adherence to checklists. These measures could be obtained and related to pre-hire test findings. Airlines that continue to rely on unstructured, unstandardized interviews need to seriously consider the value of a bona fide selection program. In the US, airlines will also need to consider the role of their selection program in complying with PL 111-216 which requires "comprehensive pre-employment screening, including an assessment of the skills, aptitudes, airmanship, and suitability of each applicant for a position as a flight crewmember in terms of functioning effectively in the air carrier's operational environment" (Airline Safety and Federal Aviation Administration Extension Act 2010, Section 216. Stat. 2367).

Furthermore, once a selection program has been put into operation it requires monitoring and updating. Prediction equations need to be monitored over time. New versions of the tests may need to be introduced as a result of a failure in test security. Changes to the size and qualifications in the pool of available candidates and to the methods used to train new hires (for example, *ab initio* or more in-depth training in flight skills) may necessitate a review and revision of current selection methods (and cut-scores). Consultants to the airlines need to remain on the lookout for newer and better tools to optimize the results of selection.

References

Aero Innovation Inc. 2012. What is Wombat? [Online]. Available at: http://www.aero.ca/what_is_WOMBAT.html [accessed 8 August 2012].

Airline Safety and Federal Aviation Administration Extension Act of 2010, Pub L. No. 111–216, 124 Stat. 2348. 2010. Available at: http://www.gpo.gov/fdsys/pkg/PLAW-111publ216/pdf/PLAW-111publ216.pdf [accessed 6 April 2013].

Airline Test Training Center. 2012. DLR-Test Training. [Online] Available at: http://www.dlr-test.com/dlrtesttraining/index.html [accessed 8 August 2012].

Americans with Disabilities Act of 1990, Pubic Law No.101-336, §2, 104 Stat. 328. 1991. Available at www.ada.gov/pubs/adastatute08.htm [accessed 6 April 6, 2013]

Barrick, M.R., and Mount, M.K. 1991. The Big Five personality dimensions and job performance: A meta-analysis. *Personnel Psychology*, 44(1), 1–26.

Barrick, M.R., Mount, M.K. and Judge, T.A. 2001. Personality and performance at the beginning of the new millennium: What do we know and where do we go next? *International Journal of Selection and Assessment*, 9(1–2), 9–30.

Bettes T., Lehenbauer, L. and Wolbrink, A. 2007. *Relationship between CogScreen-AE and Regional Airline Pilot Training Scores*. Proceedings of the XXth

Annual Conference of the Aerospace Medical Association, New Orleans, Louisiana, May 2007.
Campbell, J.S., Castaneda, M. and Pulos, S. 2010. Meta-analysis of personality assessments as predictors of military aviation training success. *The International Journal of Aviation Psychology*, 20(1), 92–109.
Cathay Pacific. 2012. Pilot interview information gouge, flight attendants and dispatchers. Available at http://www.aviationinterviews.com/pilot gougedisplay/Cathay-Pacific-69.html [accessed 28 August 2012].
Chidester, T.R. and Foushee, H.C. 1991. Leader personality and crew effectiveness: A full-mission simulation experiment, in *Proceedings of the Fifth International Symposium on Aviation Psychology, Vol.11* edited by R.S. Jensen. Columbus, OH: The Ohio State University.
Costa, P.T., Jr. and McCrae, R.R. 1992. *NEO PI-R Professional Manual*. Odessa, FL: Psychological Assessment Resources, Inc.
Damos, D.L. 1996. Pilot selection batteries: Shortcomings and perspectives. *The International Journal of Aviation Psychology*, 6(2), 199–209.
Damos, D.L. 2007. *Foundations of Military Pilot Selection Systems: World War I*. Technical Report 1210 submitted to US Army Research Institute, September 2007, [US Army Project Number 633007A792], Alexandria, Virginia.
Dawes, R.M., Faust, D. and Meehl, P.E. 1989. Clinical versus actuarial judgment. *Science*, 243(4899), 1668–1674.
European Pilot Selection & Training. 2012. Computerized pilot aptitude screening system. Available online at: http://www.epst.nl/com/compass.htm [accessed 8 August 2012].
Fitzgibbons A., Davis, D. and Schutte P.C. 2004. Pilot personality profile using the NEO-PI-R. *NASA/TM-2004-213237. National Aeronautics and Space Administration.* Langley Resaerch Center, Hampton, VA. November 2004.
Fleishman, E.A. and Reilly, M.E. 1992. *Handbook of Human Abilities: Definitions, Measurement, and Job Task Requirement*. Palo Alto, CA: Consulting Psychologists Press.
Front, C. 2012. *Personality Assessment Consultation Opportunities with the FAA: An Orientation to FAA Practices and Standards*. Workshop presented at the annual meeting of the Society for Personality Assessment, March 2012, Chicago, IL.
Griffin, G.R. and Mosko, J.D. 1977. *A Review of Naval Aviation Attrition Research (1950–1976): A Base for the Development of Future Research and Evaluation (NAMRL 1237)*. Pensacola Naval Air Station, FL: Naval Aeroapce Medical Research Laboratory.
Helmreich, R.L. 1979. Social psychology on the flight deck, in *Resource Management on the Flight Deck*, edited by G.E. Cooper, M.D. White and J.K. Lauder. NASA Report No. CP-2120.
Helmreich, R.L. 1982. *Pilot Selection and Training*. Paper presented at the Annual Meeting of the American Psychological Association, Meeting, August 1982. Washington, D.C.

Helmreich, R.L. 1986. *Pilot Selection and Performance Evaluation: A New Look at an Old Problem*. Proceedings Psychology in the Department of Defense 10th Symposium, Colorado Springs, Colorado, 16–18 April 1986. [USAFA-TR-86-1], 274–278.

Hoffmann, C.C., Hoffmann, K.P. and Kay, G.G. 1998. *The Role that Cognitive Ability Plays in CRM*. Paper presented at the RTO Human Factors & Medicine Panel (HFM) Symposium on Collaborative Crew Performance in Complex Operational Systems, Edinburgh, Scotland, April 1998. North Atlantic Treaty Organization (NATO) Research and Technology Organization (RTO) Meeting Proceedings 4.

Hoffmann, C.C., Spetz, S.H. and Hoffmann, A.K. 2011. *Long-term Study of Pilot Selection at Delta Airlines*. Presented at the 16th International Symposium on Aviation Psychology, Dayton, Ohio, May 2011.

Hurtz, G.M. and Donovan, J.J. 2000. Personality and job performance: The big five revisited. *Journal of Applied Psychology*, 85(6), 869–879.

Initial Pilot Aptitude Screening System. 2003. *European Pilot Selection and Training*. Available at http://www.nitex.nl/apst/ [accessed 19 July 2012].

Kay, G.G. (1995). *CogScreen Aeromedical Edition: Professional Manual*. Odessa, FL: Psychological Assessment Resources.

Lezak, M.D., Howieson, D.B. and Loring, D.W. 2004. *Neuropsychological Assessment*. 4th Edition. New York: Oxford University Press.

Lowy, J. 2012. Safety concerns raised by possible pilot shortage. *Star Tribune*, July 13, 2012. Available at: http://www.startribune.com/printarticle/?id=162275406 [accessed August 28, 2012].

Martinussen, M. 1996. Psychological measures as predictors of pilot performance: A meta-analysis. *The International Journal of Aviation Psychology*, 6(1), 1–20.

Martinussen, M. and Torjussen, T. 1998. Pilot selection in the Norwegian Air Force: A validation and meta-analysis of the test battery. *The International Journal of Aviation Psychology*, 8, 33–45.

McCartney, S. 2009. Airline pilot salaries: How much does your captain earn? *The Wall Street Journal*, June 16. Available at: http://blogs.wsj.com/middleseat/2009/06/16/pilot-pay-want-to-know-how-much-your-captain-earns/ [accessed 19 July 2012].

Meehl, P. 1954. *Clinical versus Statistical Prediction*. Northvale, NJ and London: Jason Aronson Inc., 1996 (first published in 1954).

Mouawad, J. 2012. Fracas on JetBlue flight shows gap in screening. *The New York Times*, March 28. Available at: http://www.nytimes.com/2012/03/29/business/jetblue-incident-raises-questions-about-screening-pilots.html?pagewanted=all [accessed 18 July 2012].

Muchinsky, P.M. 2012. *Psychology Applied to Work: An Introduction to Industrial and Organizational Psychology*.10th Edition. Summerfield, NC: Hypergraphic Press.

O*NET Online. 2010. *Summary Report for: 53-2011.00—Airline Pilots, Copilots, and Flight Engineers.* In O*Net online. Available at: http://www.onetonline.org/link/summary/53-2011.00 [accessed 18 July 2012].

Owen, A.M., McMillan, K.M., Laird, A.R. and Bullmore, E. 2005. N-back working memory paradigm: A meta-analysis of normative functional neuroimaging studies. *Human Brain Mapping*, 25(1), 46–59.

Ramos, R.A., Heil, M.C. and Manning, C.A. 2001. *Documentation of Validity for the AT-SAT Computerized Test Battery, Volume II.* (DOT/FAA/AM-01/6). Washington, DC: Federal Aviation Administration Office of Aviation Medicine.

Shepard, R.N. and Metzler, J. 1971. Mental rotation of three-dimensional objects. *Science*, 171(3972), 701–703. Available at: http://www.cs.virginia.edu/cs150/ps/ps3/mental-rotation.pdf [accessed 28 August 2012].

Street, D. and Helton, K. 1993. The 'right stuff': Personality tests and the five factor model in landing craft air cushion crew training. *Proceedings of the Human Factors and Ergonomics Society 371th Annual Meeting*, 920–924.

Taylor, J.L., O'Hara, R., Mumenthaler, M.S. and Yesavage, J.A. 2000. Relationship of CogScreen-AE to flight simulator performance and pilot age. *Aviation, Space and Environmental Medicine*, 71(4), 373–380.

Thorndike, E.L. 1919. Scientific personnel work in the Army. *Science*, XLIX (1255), 53-61.

United States Census Bureau. 2011. *Income, Poverty, and Health Insurance Coverage in the United States: 2010 (Publication No. P60-239).* Washington, DC: US Government Printing Office. Available at: http://www.census.gov/prod/2011pubs/p60-239.pdf [accessed 18 July 2012].

United States Public Law 111-216. 2010, August. Washington, DC. Available at http://www.gpo.gov/fdsys/pkg/PLAW-111publ216/pdf/PLAW-111publ216.pdf [accessed 8 August 2012].

Walters, L.C., Miller, M.R. and Ree, M.J. 1993. Structured interviews for pilot selection: No incremental validity. *The International Journal of Aviation Psychology*, 3(1), 25–38.

Wise, J. 2012. Air France 447 and the limits of aviation safety. *Popular Mechanics*, July 2012. Available at: http://www.popularmechanics.com/technology/aviation/crashes/air-france-447-and-the-limits-of-aviation-safety-10487501?click=main_sr [accessed 27 August 2012].

Wolfe, T. 1979. *The Right Stuff.* New York, NY: Bantam.

Yakimovich, N.V., Strongin, G.L., Govorushenko, V.V., Schroeder, D.J. and Kay, G.G. 1994. *CogScreen as a Predictor of Flight Performance in Russian Pilots.* San Antonio, TX: 65th Annual Scientific Meeting of the Aerospace Medical Association.

Chapter 4
Aviation Mental Health and the Psychological Examination

Robert W. Elliott

Introduction

Aviation psychology is a specialty field (Jensen 1991; Roscoe 1980) within the discipline of psychology. The practice of aviation psychology represents an integration of clinical and counseling psychology, human factors psychology, and aviation medicine involved in the application of the principles, methods, and techniques applied within the aviation community. During the early days of aviation the focus of aviation psychology was on the selection of pilots (see Chapter 1, this volume). However, with changing technologies and the expansion of flight in the commercial, military, and general aviation arenas the role of the aviation psychologist has expanded. Today, aviation or aeromedical psychologists deal with issues involving accident investigations (see Chapter 14, this volume), mental disorders, the selection and training of flight personnel (see Chapters 2 and 3, this volume), maintenance of mental health, fitness-for-duty evaluations, and medical certification of pilots. While the mental health and personality characteristics of military aviators were recognized as important variables to success in military aviation, mental health concerns involving commercial pilots was seldom the focus of research or general articles, although working as a commercial pilot is recognized as a significantly stressful job (Brienza 2011; Shappell et al. 2007). In a survey by Career Cast (Brienza 2011) it was determined that coping with irregular work hours, jet lag, faulty equipment, battling weather systems, and encountering terrorist threats while ensuring the safety of passengers places a great deal of stress on this occupational group. It has been well-documented for decades that there is a strong relationship between stress and the development of chronic physical conditions, mental illness, and psychological symptoms (Dewa, Lin, Kooehoorn and Goldner 2007; Kivimäki, Hotopf and Henderson 2010; Mazure 1995; Vinokur and Selzer 1975) including depression, paranoid thinking, suicidality, anxiety, panic attacks, and behavioral indicators including alcoholism, reduced social contact, and traffic accidents correlated with stress.

The assessment of the mental health of pilots is deemed critical because of safety-related concerns. Seventy percent of flight accidents are caused by pilot error and it is believed that the biggest risk factors involved in general aviation accidents are the “psychology of pilots” and decision-making (McClellan 2010).

For example, at-risk pilots may ignore warnings about deteriorating weather, stretch fuel reserves, make impulsive decisions, and try to do several things at one time. Under such conditions pilots may be unable to tune out passengers or controllers and become easily distracted. Such pilots take on added risk rather than minimizing or making conservative decisions to divert a flight.

The UK Civil Aviation Authority conducted a review of commercial pilots who reported incapacitations either in flight or off duty in 2004 (Evans and Radcliffe 2012). Impairment was defined as a partial incapacitation with symptoms that could have resulted in a reduction of function or distraction from a flight crew task. Musculoskeletal disorders were the most common incapacitating condition (18 percent), cardiovascular illnesses were second at 14 percent, and psychiatric disorders were rated as the third most common incapacitating medical condition (10 percent). The authors point out that while none of these incapacitating conditions resulted in fatal accidents in commercial aviation, if these conditions had been encountered in single pilot operations, fatal accidents would potentially have occurred. The authors concluded that a high proportion of psychiatric disorders associated with incapacitation "reflects the need to monitor the mental health of pilots" (49).

A review of the content of the *International Journal of Aviation Psychology*, for the years 2001 through 2005, revealed that only 19 percent of articles in the journal included content relevant to personality or cognitive topics associated with flight performance. Eighty percent of the articles in the journal were focused on human–machine interactions, equipment displays, or design systems (Elliott 2011). This is in spite of the fact that human error is involved in the great majority of civilian and military aviation accidents (McClellan 2010; Boeing 2009; Shappell and Wiegmann 2001).

Aeromedical psychologists and psychiatrists are in unique positions in that their training and background affords them the ability to assess fitness to fly from a mental health perspective within the commercial and general aviation environment. Within the military environment aeromedical (also termed aerospace) psychologists and psychiatrists evaluate pilots to determine if they are physically qualified (PQ)/not physically qualified (NPQ) for flight status (see Chapter 6, this volume). Aeromedical psychologists, whose clinical assessment tasks and responsibilities are the focus of this chapter, conduct evaluations which involve review of records, clinical interviews, and administration of standardized measures which assess cognitive functioning and mental health status to establish an airman's fitness to fly. Some airmen may manifest significant and obvious deficits in neurocognitive functioning or symptoms associated with mental illness, while other airmen may manifest only subtle deficits in specific areas of functioning and/or mental health. A crucial issue centers on whether the deficit and/or mental illness or symptom might adversely impact flight safety. The aeromedical psychologist's job is to render an opinion about the airman's fitness to engage in flying based upon the totality of the mental health evaluation process.

The framework for conducting a mental health evaluation of a pilot must include a review of relevant records, in-depth interview of the referred airman, comprehensive psychodiagnostic and/or neuropsychological assessment (see Chapter 10, this volume for more on neuropsychology), determination of an appropriate neurocognitive or psychiatric diagnosis, description of the individual's cognitive and affective strengths and weaknesses, identification and clarification of behavioral-related issues, identification of any limitations in the evaluation that was conducted, and recommendations for treatment and/or follow-up. It is advisable that collateral sources of information, such as spouses, flight surgeons, employers, treatment providers, and/or other individuals who have engaged with the pilot/aviator in or out of the job setting be interviewed. Collateral interviews with individuals who have flown with the pilot/aviator may be informative but require the interviewer to be sensitive to issues related to confidentiality. Documentation of the information reviewed and interviews conducted, assessment findings, conclusions, and recommendations are then summarized in a mental health report. The structure for an aviator mental health report is detailed in Appendix A.

Mental Health Considerations

When conducting an aviation mental health examination of a commercial pilot or military aviator, the primary consideration is ultimately flight safety. Within the commercial realm, concerns are for crewmembers, including the pilots, and the passengers, depending on whether the flight is operating as a commercial or non-commercial venture. Within the military, flight safety encompasses a different set of flight variables, such as flying in a combat/hostile zone with weapons and ammunition on board, and types of aircraft, including high-performance tactical aircraft which may include sensitive or classified equipment. Despite this, the military evaluation clinical components are essentially the same as for a commercial pilot, though rules regarding qualifications to fly are different both between the commercial guidelines and even between branches of service. For an in-depth discussion of mental health diagnoses and the military, to include contact information for the pertinent US medical waiver/disqualification guidelines and authorities, refer to Chapter 6, this volume. Psychologists or psychiatrists conducting a military aviation evaluation should contact the pertinent branch of service to ensure the provision of a complete evaluation. The remainder of this discussion focuses on the system for commercial pilots in the US, however, it is important to note that the clinical aspects of this chapter may be applied to military evaluations.

When mental health issues arise involving commercial pilots the referring agent, whether it be the pilot him/herself, the pilot's employer, the Federal Aviation Administration (FAA), the pilot's union, Aviation Medical Examiner (AME), or another source is ultimately interested in knowing whether the referred pilot is fit for duty and/or whether the pilot needs to suspend his/her flight responsibilities. Such a referral may work its way through the mental health, legal, or an aviation-related system resulting in a meeting between the pilot and an aviation psychologist or psychiatrist who will be responsible for conducting a mental health examination. The evaluator will be responsible for conducting an assessment within the context of flight safety, aviation standards, and, in many cases, in compliance with regulations that apply to the pilot (Federal Aviation Administration 2013b).

When a referral is made to an aeromedical psychologist, mental health testing may be a component of the evaluation process. Such an evaluation must be a valid and reliable assessment of the airman's behavior and emotional status at the time. A goal of the assessment process will be to identify personality traits and/or psychopathology that may adversely impact the referred airman's fitness to fly. Although psychological testing cannot predict specific behavior with certainty, mental health should be assessed since it has been established that there is a correlation between flight safety and an airman's emotional status (Bor and Hubbard 2006).

The aviation mental health examination is likely to be a challenge both for the airman and also for the mental health professional who is responsible for conducting the examination. Since the early days of flying, pilots have distrusted doctors, especially psychologists and psychiatrists, because such mental health professionals can affect the pilot's career or eligibility for medical certification required for flying (Musson 2006). It is common lore that the doctor is the natural enemy of the aviator.

In the *Aeronautical Information Manual* (*FAR/AIM*) (Federal Aviation Administration 2013a) the association between flight safety and emotionality and mental health is addressed in FAA regulations. The regulations (Chapter 8, Medical Facts for Pilots, Section 1, Fitness for Flight, subsection g [Emotion], p. 938) state that "stress from the pressures of everyday living can impair pilot performance, often in very subtle ways. Difficulties, particularly at work, can occupy thought processes enough to markedly decrease alertness. Distraction can so interfere with judgment that unwanted risks are taken, such as flying into deteriorating weather conditions to keep on schedule. Stress and fatigue can be an extremely hazardous combination" (938). The *FAR/AIM* (Federal Aviation Administration 2013a) states that:

> Certain emotionally upsetting events, including a serious argument, death of a family member, separation or divorce, loss of job, and financial catastrophe can render a pilot unable to fly an aircraft safely. The emotions of anger, depression, and anxiety from such events not only decrease alertness but also lead to taking risks that border on self-destruction. Any pilot who experiences an emotionally upsetting event should not fly until satisfactorily recovered from it (938).

The World Health Organization (WHO 2005) defines mental health as "a state of well being in which the individual realizes his or her own abilities, can cope with the normal stresses of life, can work productively and fruitfully, and is able to make a contribution to his or her community" (18). Although it is apparent that there is not an agreed upon single definition for the concept of mental health (Jones and Marsh 2001), mental health issues can have a significant impact on an individual's functioning within the work environment, family, and in the community (Vinokur and Selzer 1975).

Mental health is a critical component within the flying environment and pilots are part of that environment. A pilot's mental health status is briefly evaluated by the pilot's AME during the mandated flight physical examination which takes place every six months to as infrequently as once every five years depending on the medical certificate applied for. A number of questions involving mental health issues are listed on the FAA Form 8500-8 (Department of Transportation 2013). Such issues may also be identified by the AME during an interview process. When possible mental health issues are identified by the AME, the pilot may be referred to a mental health specialist for further evaluation or to the FAA to review.

The aviation mental health practitioner who plans to conduct a mental health evaluation of a private, commercial, or military pilot must be prepared to recognize that issues involving mental health for an airman may be manifested at a pre-clinical level that nonetheless may have an adverse impact on flight performance. The pre-clinical symptoms may not have reached a level which supports a diagnosis although such symptoms may have implications for fitness and flying safety. There are many mental health warning signs that the aeromedical practitioner should be alert for. Such signs include reckless behavior, mood changes, mistakes in the cockpit, distractions, hopeless feelings, inability to feel happy, disturbed sleep, unexplained weight changes, diminished interest in sex, increased use of alcohol or sleep medications, appearance, mood, behavior, memory, and/or cognitive changes or limitations. Mental health-related problems for the airman may emerge from a number of sources including those associated with: (1) coping, safety, and survival; (2) work load or work organization; (3) personal problems; (4) worry about loss of medical certification because of a potentially disqualifying medical condition; and/or (5) normal psychological challenges that one encounters on a day-to-day basis (Bor and Hubbard 2006).

Components of Mental Health Assessment

An understanding of the nature of pilots in the context of conducting airmen mental health examinations requires not only an understanding of the pilot personality and the environment in which airmen operate, but also a comprehensive understanding of the pertinent agency standards relevant to the aviator (for example, Federal Aviation Administration 2013b) that apply to mental health aspects required for medical certification (see also Chapter 15, this volume for a discussion of the necessary qualifications of an aeromedical psychologist). There are a number of reasons why a psychologist would be asked to conduct an aviation mental health examination. Beyond the technical skills required for functioning as a pilot, there are also mandates that the pilot be psychologically fit and attest to such fitness during the periodic flight physical examinations conducted by AMEs. Any evidence during a flight physical examination process that raises questions about the psychological fitness of a pilot is of concern. If there is any concern regarding the mental health fitness of a pilot, or if the pilot acknowledges a mental health issue or condition, the pilot can be grounded, at least temporarily, while the mental health issue is evaluated. It is under these circumstances that the assessment of the pilot, which includes a comprehensive mental health examination, is conducted.

US medical certification standards for pilots are described in the Federal Aviation Regulations (FARs) (Federal Aviation Administration 2013b) and in the *Guide for Aviation Medical Examiners* (Federal Aviation Administration 2013a) both of which are edited annually. While the standards described in the FARs are relied upon by the FAA when considering medical certification for a pilot, other diagnostic criteria can be helpful in the decision process.

The most relied upon descriptions of mental disorders in the US are embodied in the *Diagnostic and Statistical Manual of Mental Disorders, Fourth Edition, Text Revision* (*DSM-IV-TR*), published by the American Psychiatric Association (2000). The *DSM-IV-TR* works within a framework that supports a diagnosis of a mental health condition by identifying a combination of symptoms. To qualify for a diagnosis an individual must meet a set of criteria that are characteristic of the disorder. The *DSM-IV-TR* classification system includes five different Axes which are used to designate groups of mental disorders or mental health conditions, personality disorders, intellectual disabilities, concurrent medical disorders, stressors impacting the individual, and a rating scale that identifies the individual's overall level of functioning (Federal Judicial Center 2011). Other diagnostic criteria resources, such as the *International Classification of Diseases, Ninth Revision, Clinical Modification* (*ICD-9-CM*), published by the Centers for Disease Control and Prevention (2011), provide diagnostic criteria and listings of all disease entities including all mental disorders.

Reasons for referral

Pilots are referred for mental health and neuropsychological evaluations for a variety of reasons. Common reasons for such referrals, which may be initiated by an airline or aviation-related operation (that is, flight school, corporate aviation department, aircraft manufacturer's flight testing department, military command, and so on), the FAA, AME, pilot union, employee assistance program (EAP), flight training department, human resources department, disability insurance provider, or by the pilot him/herself, include:

- identification of a mental condition that involves neurocognitive and/or mental health implications;
- flight-related difficulties;

- training difficulties;
- behavior on or off the job; or
- a number of specific conditions including substance abuse, HIV+ status, Attention Deficit/Hyperactivity Disorder (ADHD), development of a neurological condition, or use of an approved antidepressant (SSRI) medication. Common mental health/psychiatric conditions which are specifically identified and may be disqualifying in the FARs (Part 67.107, 67.207, 67.307) include a medical history or clinical diagnosis of a personality disorder, neurosis, bipolar disorder, or psychosis.

In addition to such potentiality disqualifying mental health conditions pilots may be referred for a mental health assessment to:

- determine the presence, nature, and severity of emotional issues;
- establish or confirm a diagnosis;
- assist in the development of a treatment program in association with a mental health condition;
- establish differential diagnosis involving a complex clinical presentation;
- provide baseline information and/or monitor progress to track cognitive and/or emotional/affective changes over time;
- identify subclinical mental health symptoms which may lead to a mental disorder;
- provide information to assist training departments in their efforts to support pilot skill development or enhancement; or
- describe functional limitations.

When conducting mental health assessments there exists a specific need for identification or determination whether an identifiable mental health condition exists. If a mental health condition does exist, does the condition have aeromedical significance? While difficult to ascertain, if a mental health practitioner is able to establish whether an identifiable mental health condition has aeromedical significance, such information should be provided in the mental health assessment. A mental health assessment should provide information on the rate of progression of any identified disorder when such information can be determined with reasonable certainty. Ultimately, an airman mental health assessment needs to provide insights about the commercial or general aviation airman's fitness to fly or the military aviator's PQ status. If the mental health practitioner is of the opinion that his/her background and knowledge is insufficient to render an opinion regarding such information the mental health practitioner should identify the issues and consult with a trusted colleague before offering any opinion regarding fitness-for-duty to the FAA or to the appropriate military authorities. Aviators participating in either civilian or military evaluations should not be informed during the evaluation whether the mental health specialist believes they are medically qualified or not as this determination can only be made at the agency/administrative level.

Review of records

When conducting a mental health examination of an airman it is critically important that the examiner conduct a review of relevant records. The airman's FAA medical file is available upon request (with the airman's authorization) through the FAA Aviation Medical Certification Division (AMCD) in Oklahoma City. Records that are not likely to be included in the FAA medical file, which should be obtained from other sources, include flight performance evaluations from an airline training department, documentation of any work-related disciplinary action, airline or flight department ordered fitness-for-duty psychological or psychiatric evaluation reports (including test scores), or counseling records.

Obtaining informed consent

During the initial meeting the mental health practitioner needs to begin by providing detailed information to the airman about the nature of the mental health examination. The informed consent process, which should be consistent with the American Psychological Association (APA) Ethical Principles of Psychologists and Code of Conduct (American Psychological Association 2002) or other applicable ethics code, requires that the client be identified and fees, access to records, release of information, limits of confidentiality, and the distribution of any reports that are generated be discussed. Bennett et al. (2006) emphasize that the critical information that needs to be provided is information that the average person would want to know under the circumstances. Although the airman may agree to the evaluation, the examination should proceed only after the examinee has been provided with a written informed consent document and a copy of all authorizations that have been signed. It is a good practice to ask the airman to state back the purpose of the evaluation and the risks and benefits derived from the meetings.

Clinical interview

Although the mental health evaluation includes several different components, including mental health testing, the clinical interview with the referred airman remains an essential part of any mental health evaluation. The clinical interview will assist in determining if the symptoms indicate a medical, psychiatric, or neurological condition impacting behavior, affect, or cognition. The interview should include a history of the presenting issue from the airman's perspective, course of development of any symptoms, medical history, psychiatric history, use/abuse of any substance including recreational drugs and/or alcohol, and review of medications. In some cases it will be appropriate to include a significant other in the clinical interview of the airman in order to verify the information the airman presents. On occasion, the interview may need to take place over a series of meetings.

The format of the interview should include, initially, open questions concerning the nature of the referring problem which are then followed by more focused questions to clarify sequence of events and the evolution of key symptoms. Open questions will encourage the airman to speak openly and to help establish rapport. Closed questions then develop as there is an understanding of the underlying disorder. Symptom checklist instruments are available in computer scored form, such as the Symptom Checklist-90-Revised (Derrogatis and Unger 2010), or hand scored format, such as the Personal Problems Checklist Adult (Piedmont, Sherman and Barrickman 2000), which will discern differences between different mental health conditions which can be used during the interview process. Although pilots tend to be very task-oriented individuals and typically want to rapidly focus on the problem during the clinical interview, it will be important that the pilot fully explain their perspective about the circumstances of the referral as well as what led up to the referral.

The examiner needs to be prepared to allow time for the airman's emotions to calm and to offer reassurance as appropriate. In many cases appointments should be scheduled for two consecutive days to ensure that the referred pilot does not feel rushed and that all planned testing can be completed. Approximately 20 percent of the pilots referred to the Aerospace Health Institute in Los Angeles require more than one day to complete the interview and assessment process (Elliott 2011). For high-stakes evaluations, such as at the Naval Aerospace Medical Institute (NAMI) for Navy/Marine Corps aviators, evaluations are scheduled for a week to accommodate serial interviews with multiple providers, in-depth record review, psychological and cognitive testing, time for collateral interviews and debriefing of the aviator.

When conducting an airman mental health interview the pilot's appearance, attitude, behavior, mood and affect, speech, thought processes, thought content, perceptions, cognition, insight, and judgment should be documented. Commentary about any difficulties the pilot had locating the assessment site and whether the pilot was on time should be noted.

Collaboration

Professional pilots, as a general rule, tend to be very defensive and guarded individuals who minimize psychological distress and conflicts (Kay 2002). Because they are anxious to return to their aviation responsibilities they will minimize any factors that interfere with return to flying. On occasion a referral of a pilot, who self-declares he or she is not prepared to return to work, will be received but such pilots tend to be in the minority of referrals.

To overcome the minimization of symptoms that underlie the referral, the mental health practitioner needs to rely upon collateral sources of information. In addition to a review of the records provided by the FAA, employer, or other medical practitioners, it is advisable to work with the commercial pilot's airline flight department, the employer's human resources and/or EAP, the pilot's AME, treatment providers, spouse, and/or significant others who may have information which is critical for an understanding of the referred pilot's condition.

Psychological testing

Psychological testing, as differentiated from the assessment process, involves administration of standardized measures which provide information about the psychological and/or cognitive status of the referred pilot. The assessment process includes not only the clinical interview and psychological testing but also summarization of information compiled from reviewed records, information from collateral sources, and behavioral observations made during the interview and testing segments of the evaluation process. Psychological tests that are administered during a mental health examination should be reliable, validated as measures of the characteristics and traits that are under study and meet the *Standards for Educational and Psychological Testing* (American Psychological Association 1999).

The aviation psychologist should include tests which have norms available for aviators (Kay 2002) when such norms are available (see Chapter 10, this volume). The FAA has drafted minimal specifications for psychological evaluations for a variety of mental health and related conditions which are available online to the examiner in the FAA *Guide for Aviation Medical Examiners* (Federal Administration 2013a). The psychological assessment measures required by the FAA include measures of neurocognitive functioning involving intellect, general cognitive functioning, personality measures, and other measures of mental health functioning (Elliott 2011, 2007).

The primary purpose of psychological testing is to evaluate personality traits, affective functioning, behavior, and cognitive abilities to assist in making judgments, predictions, and decisions about the status of individual pilots who are participating in the mental health testing. Psychological test results are used to assist in the diagnosis of psychological and neurocognitive conditions, prescribe psychological and mental health treatments, identify changes while involved in or following discharge from intervention programs, measure emotional and/or behavioral progress, and determine whether a referred pilot is fit to fly.

Personality and measures of affective functioning are designed to assess "normal and abnormal dimensions of personality and psychopathology" (Kay 2002, 235). Psychological tests which have been used in research and relied upon in clinical applications with commercial and military pilots are described below.

Written report

The mental health report, whether intended for an airline, the FAA, a pilot union, a military service, or another aviation-related agency, should clarify the referral question and related issues and list all psychological tests administered, norms relied upon for interpretation of the psychological test results, and collateral sources of information. A summary score sheet with all relevant test information should be attached to the report. Reports submitted to the FAA and any military service branch should include the raw test score data in the likelihood that the report will be evaluated by designated aeromedical neuropsychology/psychology consultants (Kay 2002). All of the information reviewed should be considered when formulating an opinion, not just the test scores. Any testing modifications that were implemented (for example, within session repetition of the same test) and which may have a potential impact on interpretation should be described. Recommendations should be clearly articulated.

Effort and malingering determination

The vast majority of airmen who are undergoing mental health evaluations are motivated to return to work. Airmen will almost always attempt to maximize their ability to return to flying. Effort and malingering testing, also referred to as symptom validity testing, is important when there is a possibility that the examinee is not putting forth good effort or when the examinee is trying to perform poorly. *DSM-IV-TR* (American Psychiatric Association, 2000) defines malingering as the "intentional production of false or grossly exaggerated physical or psychological symptoms, motivated by external incentives" (739). *DSM-IV-TR* guidelines for suspecting malingering specify any combination of the following including: (1) referral by an attorney; (2) marked discrepancies between objective data and claimed stress or disability; (3) failure to cooperate with the evaluation process or with treatment; and/or (4) a diagnosis of antisocial personality disorder (American Psychiatric Association 2000).

In unusual cases where the aviation psychologist "becomes suspicious of insufficient effort or inaccurate or incomplete reporting" (Bush et al. 2005, 425–426) symptom validity testing and measures of effort must be included to "assist in the determination of the validity of the information and test data obtained" (426). Effort testing should be included in a mental health testing battery when litigation is involved (that is, assigning blame, assessing financial responsibility, or assessing disability status). Assessment of effort relies upon clinical decision-making which is based upon multiple factors, and psychometric testing which can improve the decision-making process (Greiffenstein 2010; Johnson and Lovell 2011). Collateral interviews structured to examine the nature of the individual complaints, can provide information about behavioral inconsistencies, establish timelines for symptom onset, symptom progression, and course (Pella et al. 2012).

Because effort can vary over the course of an evaluation, symptom validity testing should be assessed throughout the evaluation process in those cases in which validity of the test results is of concern. Commonly relied upon measures used in mental health examinations include the Miller Forensic Assessment of Symptoms Test (MFAST) (Miller 2001), Minnesota Multiphasic Personality Inventory-2 (MMPI-2) F-K ratio (Greene 2011), MMPI-2 Fake Bad Scale (Greene 2011), Millon Clinical Multiaxial Inventory-III (MCMI-III) X and Z scales (Strack

2008), and Trauma Symptom Inventory-2 (TSI-2) (Briere 2010). Pella et al (2012) provides a comprehensive listing of self-report, informant rating, interview scales, and performance measures designed to detect malingering and suboptimal performance. Slick, Tan, Strauss and Hultsch (2004) published a survey of the opinion of neuropsychologists on the subject of malingering. The respondents in the survey indicated that the prevalence of malingering which occurred in their patient population ranged between 10 and 20 percent of cases. No survey has been done of the rate of malingering in airmen but it is estimated that it would be significantly less than the rate in the non-aviation population (Elliott 2011).

Norms

The primary goal of the mental health assessment of a pilot is the determination of whether a referred pilot is manifesting evidence of a psychiatric and/or behavioral condition or psychological features that could compromise flight safety. Examiner decision-making is based, in part, upon norms that have been established for airmen, a defined population group (that is, those diagnosed with depression, alcoholism, bipolar disorder, and so on), general population norms, age and/or education corrected norms, ethnicity corrected norms, or some other combination of norms that include specific features applicable to the referred pilot.

The practitioner conducting a mental health assessment needs to determine whether the airman referred for a mental health assessment is evidencing healthy or unhealthy personality and/or behavioral characteristics and whether those characteristics are likely to compromise flight safety. To address this issue the mental health practitioner must select appropriate norms to assist in making decisions about whether the referred airman is manifesting a disqualifying mental health condition that might jeopardize flight safety. Most test publishers today provide normative information reporting the range of performances for their tests for healthy individuals and individuals who have been diagnosed with defined mental health conditions.

Since different score values may lead to different conclusions the practitioner needs to consider the appropriateness of the norms relied upon for conclusions. Norms exist for some neuropsychological/neurocognitive tests for aviators (Hess, Kennedy, Hardin and Kupke 2010; Kay 2002) but only a few mental health or affective psychological tests report norms for airmen. The few mental health measures which have published profiles for military and commercial pilots include the MMPI-2, the NEO PI-R, Sixteen Personality Factor Questionnaire (16PF), and the MCMI-III. The use of pilot norms is a more critical consideration involving the assessment of neurocognitive functioning. Use of appropriate (for example, age, gender, ethnicity, and so on) general population norms is recommended in the assessment of the emotional/personality functioning of commercial, general aviation, and military pilots. Ultimately, the decision of the selection of norms that will be relied upon will require clinical judgment (Bush 2010).

The selection of norms has been enhanced with access to performance information which is based upon the population on which the test was standardized. Strauss, Sherman and Spreen (2006), Lezak, Howieson, Bigler and Tranel (2012), and Mitrushina, Boone, Razani and D'Elia (2005) have compiled manuals with extensive normative data associated with most published neuropsychological tests as well as a few psychological/mental health measures. In these publications the neurocognitive and psychological tests are described, the normative data listed, and commentary on issues involving the validity and reliability of the tests is provided (Grant and Adams 2009).

The process of deciding what is normal or abnormal or what features would contribute to concerns about flight safety is a challenge for the mental health practitioner, especially when the outcome on psychological mental health measures does not fall within an extreme range. It is not uncommon for practitioners to assume that a relationship exists between observed symptoms and abnormality when no such relationship exists or when there is an overestimation of the relationship between observed symptoms and psychopathology. As pointed out by Faust (2012) a variety of studies have shown that behaviors that indicate pathology may in fact be common among normal individuals (McCaffrey, Palav, O'Bryant and Labarge 2003).

To increase the accuracy of diagnostic conclusions the mental health practitioner should be aware and knowledgeable about the base rates of signs and symptoms within a normative population. The diagnostic capability of a test is "relevant to the base rate of the diagnosis in the population of interest … The importance of base rates in the clinical sciences is fundamental" (McCaffrey, Palav, O'Bryant and Labarge 2003, 1).

Labeling and Decision-making

When considering the outcome of the test results, a determination needs to be made about how the test scores translate into decision-making regarding fitness for flying and use of descriptors which provide information on the level of fitness of the airman referred. According to Faust (2012), "there is no generally accepted system for converting levels of test performance into descriptive labels. What one psychologist or neuropsychologist calls low average, another might call average, or what one calls borderline, another might call abnormal" (398). In the assignment of a descriptor to an airman's test performance the practitioner compares an airman's performance or test profile against a population of other similarly-matched individuals on defined mental health criteria. Whatever system of labeling is used to describe the pilot's level of performance or profile, the system should be consistent across the mental health measures administered and across different points in time when either different or the same mental health tests have been administered.

Federal Aviation Administration Mental Health Standards

DSM-IV-TR and *ICD-9-CM* diagnostic systems list all current psychiatric disorders and provide information on criteria needed to establish a diagnosis. *DSM-IV-TR* and *ICD-9-CM* are relied upon for determination of diagnostic criteria by most of the medical community in the US, to include all branches of the military, as well as most insurance providers. The FAA on the other hand relies upon the Title 14, *Code of Federal Regulations* (CFR) descriptions (Federal Aviation Administration 2013b) under Part 67 (Medical Standards and Certification), subparts 67.107, 67.207, and 67.307, for the definitions for mental standards for all class pilots to qualify for medical certification. The FAA mental standards for medical certification include, "(a) no established medical history or clinical diagnosis of any of the following:

1. a personality disorder that is severe enough to have repeatedly manifested itself by overt acts;"
2. a psychosis which has been manifested by "delusions, hallucinations, or grossly bizarre or disorganized behavior, or other commonly accepted symptoms of this condition;" or
3. bipolar disorder.

Other possible disqualifying conditions include a "personality disorder, neurosis, or other mental condition … which makes the person unable to safely perform the duties or exercise the privileges of the airman certificate applied for or held; or may reasonably be expected, for the maximum duration of the airman certificate applied for or held, to make the person unable to perform those duties or exercise those privileges" (131). In addition to the FAR mental health standards, when an airman undergoes periodic flight physical examinations an FAA Form 8500-8 (Federal Aviation Administration MedXPress Application for Airman Medical Certification) is completed. Questions under Section 18 (Medical History) ask for affirmation or denial of the existence during the applicant's life of any of a number of medical, behavioral, and/or mental health conditions, medical treatment, or other conditions. The conditions which are mandated to be reported include, among others, "a mental disorder of any sort," depression and anxiety, substance dependence or a failed drug test, or "substance abuse or use of an illegal substance in the last two years," suicide attempt, and "other illness, disability, or surgery."

Common Psychiatric Disorders

This section will focus on the mental health and related conditions listed in the FARs (Federal Aviation Administration 2013b), *Guide for Aviation Medical Examiners* (Federal Aviation Administration 2013a), and the FAA Form 8500-8 (Federal Aviation Administration 2008), although other psychiatric conditions may arise which have implications for flight safety and which are critical for maintenance of aviation safety. Any pilot/aviator with a current mental health disorder may be grounded until fully recovered and, in some cases, will be required to undergo a period of observation prior to return to active flight status. It is notable that within the military, any health provider may ground anyone with flight status (that is, aviators, flight officers/navigators, air traffic controllers, and unmanned aerial vehicle operators), though only that individual's flight surgeon may reinstate the military aviator (that is, put back up).

Mental health disorders are fairly common among the general population. While medically certified pilots generally are considered to be at less risk for psychiatric disorders and have been monitored and prepared for dealing with stress situations in an adaptive manner, mental health breakdowns do occur. A recent example involved a JetBlue Airways captain who became delusional, incoherent, manic, and physically threatening during Flight 191 (Blaney 2012). The efficient handling of the crisis involved physical restraint of the captain by a number of passengers and management of the flight by the flight attendants, first officer, and air traffic controller. They coordinated their efforts during the emergency flight situation in an efficient manner resulting in a safe landing in Amarillo, Texas, in part because of the training they had undergone (Jansen 2012). Other mental health episodes have taken place during flights, though such breakdowns are rare. Examples include:

- In 2008, an Air Canada co-pilot was forcibly removed from the cockpit and restrained after experiencing a breakdown during an international flight.
- In 1999, the relief co-pilot of Egypt Air Flight 990 deliberately crashed into the Atlantic according to National Transportation Safety Board (NTSB) investigators. The co-pilot's motive was never determined according to the NTSB (Flight Safety Foundation 2012). All 217 flight crewmembers and passengers were killed.
- In 1996, a co-pilot of a Maersk Airlines flight broke into a sweat during a European flight and told the captain he was afraid of heights. The jet, with 49 passengers, made an emergency landing.
- In 1982, a Japan Airlines flight crashed into Tokyo Bay after the captain, who had previously been grounded for mental illness, reversed some engines. The co-pilot and engineer struggled but were unable to regain control of the aircraft resulting in the death of 24 of the 174 people on board (Musa and Mayer 2012).

In the case of airmen the emergence of mental health symptoms, relevant to aviation safety, may become evident at a subclinical level which may have aeromedical relevance. For example, in a situation involving a pilot who is going through a divorce, the pilot may become so preoccupied with personal events that he/she is unable to attend to a radio call, a cockpit cue or indicator light, or evaluate deteriorating weather conditions. The military refers to such a distraction as being unable to remain within the flight box, or in other words to compartmentalize personal problems in order to perform safely and effectively in the flight environment. Although the symptoms in such a case may not reach a level which supports a clinical diagnosis, the symptoms may significantly impact the pilot's situational awareness and flight performance resulting in a degradation of flight safety and qualify as a disqualifying condition under Title 14, Part 67 of the FARs.

Personality disorder

The FAA (2013a) states in the FARs, under Title 14 CFR 67.107 (a)(1), 67.207 (a)(1), and 67.307(a)(1), that to qualify for an airman medical certificate, an airman must have:

> No established medical history or clinical diagnosis of any of the following: a personality disorder that is severe enough to repeatedly have manifested itself by overt acts … which makes the person unable to safely perform the duties or exercise the privileges of the airman certificate applied for or held; or may reasonably be expected for the maximum duration of the airman certificate applied for or held, to make the person unable to safely perform the duties or exercise those privileges.

There is also a specification that there must be no other medical history or clinical diagnosis of "other personality disorder, neurosis, or other mental health conditions" (130–131).

The *Guide for Aviation Medical Examiners* (Federal Aviation Administration 2013b) indicates that "overt acts" in reference to personality disorders involves acting out behavior. Such personality problems are apt to be manifested by:

> … poor social judgment, impulsivity, and disregard or antagonism towards authority, especially rules and regulations. A history of long-standing behavioral problems, whether major (criminal) or relatively minor (truancy, military misbehavior, petty criminal and civil indiscretions, and social instability) usually occurs with these disorders. Driving infractions and previous failures to follow aviation regulations are critical examples of these acts (135).

The *DSM-IV-TR* lists ten specific personality disorders including paranoid, schizoid, schizotypal, antisocial, borderline, histrionic, narcissistic, avoidant, dependent, and obsessive–compulsive diagnostic conditions. McClellan and Bernstein (2006) note that the existence of personality traits "…does not necessarily lead to the development of a personality disorder. Successful aviators, for example, are likely to have varying degrees of narcissism (i.e., healthy sense of self-confidence) and obsessive–compulsive traits (attention to detail and conscientiousness)." They point out that when such characteristics become "inflexible, maladaptive, and cause significant functional impairment or subjective distress" then they become problematic in the aviation environment (1188).

In an NTSB hearing involving a pilot (petitioner) who was determined to meet FAR criteria for a personality disorder, a central question was whether there were identified criteria that supported an FAA diagnosis of a personality disorder and the impact of such a disorder on eligibility for medical certification. During the hearing the aviation psychologist testified that the airman manifested multiple overt acts, some dating back more than 20 years, including drug abuse during his adolescent years, and a history of multiple arrests and criminal convictions during his adult years. There was also a history of poor adherence to FAA rules, flight violations, physical fights, reckless behaviors, impulsivity, interpersonal conflicts, and alcohol abuse. The overt behaviors indicated a lack of personal responsibility/remorse, poor social judgment, disregard for rules and regulations, and disregard for flight safety. Psychological test results, including the MMPI-2 and MCMI-III summary profiles and diagnoses, did not specify a specific personality disorder but did identify many personality traits consistent with a diagnosis of a personality disorder including: difficulties with anger control, argumentativeness, recklessness, shallow relationships, impulsivity, poor judgment, use of anger to control others, defensive vigilance, proneness to rule infractions, and high-risk behaviors. His history of maladaptive behaviors and clinical evidence (psychological test results) revealed enduring behavior patterns which were traceable back to his adolescent years and which were pervasive across a broad range of personal, social, vocational, and community situations. The Administrative Law Judge ruled that the petitioner "does suffer from a personality disorder serious enough to be manifested by overt acts … his personality is inflexible and maladaptive and causes him significant functional impairment, pervasive across a broad range of personal and social situations" (Richardson v NTSB 2010, 28).

McClellan and Bernstein (2006) point out that personality will promote "crew coordination, mission completion, and safety of flight…" (1188), and when "maladaptive, inappropriate, and inflexible across the broad range of demanding circumstances inherent in aviation" (1188) can be a contributing factor in aviation accidents. In a subsequent effort to firm up the definition of a personality disorder which would be consistent with the FARs, Elliott and Gitlow (Personal communication, Stuart Gitlow 31 March 2012) expanded this definition to include a specification that the problematic behavior(s) should be manifested outside the range of societal expectations and the behavior(s) were not better explained by a metabolic, addiction, or other psychiatric diagnosis. It is noteworthy that in the changes to the criteria for personality disorder that have been proposed for the *DSM-V* (American Psychiatric Association 2013) these additional characteristics are included.

Diagnosing an aviator as having a personality disorder is a challenging task. While other psychiatric conditions, such as psychosis or severe depression, may be easy to recognize and diagnose, the process of identifying whether a personality disorder exists in an airman manifested by difficulties in interpersonal relationships, chronic involvement in work-related disputes, and/or problematic behaviors within the cockpit involving other crewmembers, represents a challenge. A central issue for the referring agency is deciding whether such behaviors should be considered within the context of a mental health condition or treated as unacceptable behaviors subject to administrative intervention.

The decision by an agency to make a referral for mental health examination due to concerns about maladaptive personality traits may be complicated by the individual's behavior. As Jones and Patterson (1989) point out, "The aviator may lack judgment and insight into the situation, not seeing what the problem is, citing personality conflicts or other perceived failings of those around him. Lawsuits, formal complaints, union grievances, or other appeals to authority are mentioned, all aimed at getting other people to change" (86). Noteworthy features of a personality disorder involving the airman is the absence of personal distress and a belief that he or she is being mistreated and is misunderstood. The airman with a personality disorder believes that others need to change and the behaviors that are of concern to the referring agency were totally justified or were misinterpreted.

When evaluating the existence of a personality disorder in an airman it is important that the diagnosis take into consideration the airman's job-related history. The mental health examiner will need to compile documentation and interview significant others to establish the existence of "overt acts" which are required under FARs to support a diagnosis of a personality disorder. Concerns should be focused on repeated acts (generally three or more) that could impact flight safety. Psychological testing, while helpful in evaluating a personality disorder, particularly by identifying maladaptive traits and other evidence of psychopathology is not diagnostic. Diagnosing a personality disorder requires that the practitioner review documentation, interview collateral sources, interview the airman/pilot, and administer tests that are valid for use in assessing personality disorders. The instruments most commonly used include the MMPI-2 and MCMI-III. In particular, the aviation psychologist needs to be attentive to DSM-IV-TR Cluster B personality traits which are likely to be particularly disruptive in an aviation environment.

Psychosis

Under FARs (Federal Aviation Administration 2013b) 14 CFR 67.107(a)(2), 67.207(a)(2), and 67.307(a)(2) there shall be, "No established medical history or clinical diagnosis of … a psychosis. As used in this section, 'psychosis' refers to a mental disorder in which: 'the individual has manifested delusions, hallucinations, grossly bizarre or disorganized behavior, or other commonly accepted symptoms of this condition'" (131–132). The FAA *Guide for Airman Medical Examiners* (Federal Aviation Administration 2013a) specifies that, "Psychotic disorders are characterized by a loss of reality testing in the form of delusions, hallucinations, or disorganized thoughts. They may be chronic, intermittent, or occur in a single episode. They may also occur as accompanying symptoms in other psychiatric conditions including but not limited to bipolar disorder…major depression… borderline personality disorder" (136).

Psychosis and psychotic disorders are disqualifying for airmen except in those rare cases where the psychosis was brief and was found to have resulted from either a general medical condition, was secondary to an adverse drug reaction, was due to alcohol and/or drug withdrawal, or was found to have resulted from sleep deprivation.

The role of the aviation psychologist in evaluating an airman with a history of a psychotic episode is to determine if there is a residual psychosis or other psychopathology. The most useful instruments include the MMPI-2, Personality Assessment Inventory (PAI) and Rorschach Inkblot Method. In using the Rorschach the psychologist should pay particular attention to the Perceptual Thinking Impairment (PTI) score which has been validated as sensitive to distortions of perceptions and thinking. When interviewing the airman, it is critical to include a thorough review of the "positive" and "negative" symptoms associated with psychosis.

Bipolar disorder

Bipolar disorder, or bipolar affective disorder, historically known as manic-depressive disorder, is one of a spectrum of mood disorders which involves significant alterations in mood state and at least one episode of manic or hypomanic behavior. Even in cases where the bipolar disorder is not manifested by symptoms that support a diagnosis of psychosis, the disorder can be so disruptive as to impair judgment and functioning and pose a significant risk to aviation safety, even in milder forms of the disease (Federal Aviation Administration 2013b).

It is not uncommon that individuals with bipolar disorder will manifest abrupt onset of symptoms (Elliott 2011). Ninety percent of individuals experiencing a manic episode will have a recurrence of such an episode within five years. It has been reported that 25 to 50 percent of patients diagnosed with bipolar disorder attempt suicide at least once (Jamison 2000), with 15 percent dying by suicide (American Society of Aerospace and Medicine Specialists 2008).

Under *DSM-IV-TR* there are two types of bipolar disorder, Bipolar I and Bipolar II. The distinction between the two types is that in Bipolar I there has been one or more manic episodes (American Psychiatric Association 2000). Bipolar II patients have recurrent major depressive episodes interspersed with one or more hypomanic episodes (Morrison 1995) but no manic episodes. Pilots with a diagnosis of Bipolar I are disqualified from medical certification. Those with a diagnosis of Bipolar II require thorough psychological and psychiatric evaluation. The evaluation relies upon accurate and thorough history taking and record review in order to make an accurate diagnosis. Psychological testing has been found to be of limited use for evaluating the disorder.

Neurosis

Neurosis, which is listed in 14 CFR 67.107(c), 67.207(c), and 67.103(c) (Federal Aviation Administration 2013a), is a potentially disqualifying mental disorder. Neurosis includes a maladaptive adjustment that includes mood symptoms such as anxiety and fear or depressive symptoms but does not include conditions that are manifested with delusions or hallucinations. Common symptoms may include complaints of physical symptoms that do not appear to have an underlying medical cause. The term neurosis is not in common use today. The general diagnosis of neurosis and many subtypes were included in the earliest editions of the *DSM*. Over time, with changes in the conceptualization of mental illness there has been a shift away from psychodynamically derived categories such as neurosis (Weiner and Craighead 2010). Evaluation of neurosis is likely to include the MMPI-2, PAI, and Rorschach Inkblot Method.

Depression

Depression is a specifically identified mental disorder listed on the FAA Form 8500-8. Major depression within the general population is a common mood disorder. Sixteen percent of the US population reports having experienced a major depressive episode during their lifetime. Within the professional aviator population depression may be compounded by the stress of flight responsibilities and disruption to family relationships that develop with absences from home (Morse and Bor 2006). Approximately 50 percent of individuals who experience a major depressive episode will re-experience such an episode within five years. With a history of two episodes, the probability increases to 70 percent and after three episodes the probability of recurrence increases to approximately 90 percent (Lezak et al, 2012).

Symptom development in depressive disorders may develop slowly over a period of several weeks or months making the identification of the mental health condition challenging. Occasionally, a mental health practitioner will be asked by the referred airman or treatment provider to allow the airman to continue to fly as a form of therapy. There are no studies which recognize or support flying as a safe or effective form of psychotherapy (Jones 2008).

Medication waiver

In April 2010 the FAA announced that under certain conditions and on a case-by-case basis pilots were going to be allowed to fly while taking specific antidepressant medications (see Chapter 12, this volume for a comprehensive discussion of these policies). After years of consideration about such an authorization the FAA decided to provide a mechanism that would allow pilots to take approved Selective Serotonin Reuptake Inhibitor (SSRI) medications, under Special Issuance waiver provisions, and resume flying (Federal Aviation Administration 2013b). The FAA determined that a pilot could apply for the Special Issuance authorization if the pilot was taking one of four medications for treatment of depression (that is, Fluoxetine [Prozac], Sertraline [Zoloft], Citalopram [Celexa], and Escitalopram [Lexapro]. These four SSRIs are non-sedating, commonly prescribed, and result in minimal side-effects.

Because of the impact of depression on both cognitive and affective functioning, psychological/neuropsychological testing that is conducted should address both the cognitive and affective issues associated with depression. Airmen with depression may demonstrate cognitive difficulties including slowed mental processing, psychomotor retardation, mild attentional deficits, decreased drive and initiation, impairment in short-term recall and learning for visual and verbal material, memory problems, and/or slowed psychomotor speed. There may also be impairments in language, perception, and spatial abilities which are secondary to difficulties in attention, motivation, or organization (Johnson and Lovell 2011; Kay 2002).

Specific mental health/affective psychological test measures that are appropriate for use with pilots who are manifesting depression include, but are not restricted to, the MMPI-2, Personality Assessment Inventory (PAI), and the Beck Depression Inventory-II (BDI-II) (Beck, Steer and Brown 1996). A sample of major neurocognitive measures that assess reasoning, information processing speed, memory, and sustained attention include: CogScreen-Aeromedical Edition (Kay 1999), Category Test (DeFilippis 1993), Trail Making Test (Lezak 2012), Paced Auditory Serial Addition Test (PASAT) (Lezak, et al. 2012), California Verbal Learning Test-II, (CVLT-II) (Delis, Kramer, Kaplan and Ober 2000), Connor's Continuance Performance Test-II (CCPT-II) (Conners 2008), and the Test of Variable Attention (TOVA-8) (Greenberg et al. 2012).

Anxiety

Anxiety is a specifically listed mental disorder on FAA Form 8500-8 (Medical History) and is a disorder that may be disabling within the flight environment. Anxiety disorders collectively are among the most common mental disorders experienced by Americans (National Institute of Mental Health 2012). The category includes the following conditions: panic disorder, agoraphobia, PTSD, acute stress disorder, obsessive–compulsive disorder, generalized anxiety disorder, social and specific phobias, anxiety due to general medical condition, substance-induced anxiety disorder, and anxiety disorder not otherwise specified (Morse and Bor 2006). In the US the one-year prevalence rate for all anxiety disorders among adults 18 to 54 years of age exceeds 16 percent (US Department of Health and Human Services 1999). Individuals experiencing an anxiety disorder constantly worry and feel uneasy, are often depressed, and manifest difficulties in decision-making and concentration. While each of the types of anxiety disorders are manifested with unique characteristics, extreme or pathological anxiety and apprehension are the principal disturbances evident in all of the anxiety disorders (Edwards 1990; US Department of Health and Human Services 1999).

Anxiety can interfere with an individual's ability to attend, learn, and remember new information. Mental efficiency problems, such as slowing, scrambled or blocked thoughts or words, memory failure, and increased distractibility (Johnson and Lovell 2011) are common in severe cases. Such features pose significant concerns for aeromedical safety.

Suicide

A history of suicide attempt is specifically listed as a condition which must be reported on FAA Form 8500-8. Suicide by aircraft may be accompanied by a history of mental disorder (for example, depression in most cases). Bills, Grabowski and Li (2005) conducted a review of aviation accidents reported by the NTSB between 1983 and 2003 involving suicide as a probable cause. During this 21-year period 37 pilots committed or attempted suicide by aircraft with 36 resulting in at least one fatality. All of the cases involved younger men who were engaged in general aviation (14 CFR Part 91) flights. Toxicological testing revealed that 24 percent of the cases involved use of alcohol and 14 percent involved illicit use of drugs (for example, cocaine and marijuana). Underlying factors included pre-existing domestic and social problems (46 percent), legal troubles (41 percent), and psychiatric conditions (38 percent). For one of every completed suicide there are eight to 25 attempts. The National Institute of Mental Health (NIMH) reported suicide as the tenth leading cause of death in the US in 2007 with more than 90 percent of suicides committed by individuals who had been identified as having a mental illness (National Institute of Mental Health 2010).

An important clue for the mental health practitioner who is evaluating an airman when there is concern about suicide is determining how the aviator is coping with stress. In the assessment of risk factors, the impact of anticipated adverse events may be preceded by a suicide attempt by days or months. It is therefore critical that information about such an event in the airman's life be discussed and evaluated. The psychological tests most often employed in such an evaluation include the MMPI-2, PAI (SUI scale in particular), and the Suicide Constellation score from the Rorschach Inkblot Method.

Psychological Testing Measures

A referral of a pilot for a fitness-for-duty or mental health assessment is conducted in an effort to determine if the pilot is fit to fly. In most cases the primary referral issue involves numerous factors that become increasingly apparent as information is compiled on the case. Contributing to the primary concern may be stress factors that pilots encounter such as long flights that involve fatigue, periodic flight check-rides, communication problems, interpersonal lifestyle factors, job insecurity and concerns over layoffs, labor unrest, marital and family problems, terrorist attacks, security challenges, and even monitoring by fellow crewmembers (Butcher 2002).

When a mental health evaluation is requested by a referring party because of a behavioral or flight performance-related issue the referral will include meetings between the airman and an aviation psychologist and/or psychiatrist and psychological testing will more likely than not be an integral part of the mental health assessment and evaluation process. If the FAA has become involved in a case it is likely that the airman has received a letter from the FAA. The reasons for the FAA's concerns and minimal requirements for a mental health assessment will be detailed in the letter. The FAA recently revised the specifications for conducting psychological/psychiatric and neuropsychological evaluations of pilots for a number of mental health and neurocognitive conditions. The FAA Specification sheets provide instructions for the pilot who is applying for airman medical certification. The FAA letter specifies the qualifications of the examiner, the required content in the evaluation report and a list of required psychological and/or neuropsychological tests to be administered (at a minimum) for the condition under review (Federal Aviation Administration 2013a).

Both pilots and controllers tend to engage in some degree of response distortion on self-report measures, particularly when being scrutinized for medical certification. These forms of response distortion were referred to by Front (2013) as "socially desirable responding, positive malingering, or positive impression management." While these terms refer to conscious efforts to dissimulate or minimize psychopathology it is also important to consider unconscious efforts and or distortions related to a grandiose perceptual style consistent with Paulhus' self-deceptive enhancement (Paulhus 1984).

Since these patterns of response distortion are so frequently exhibited by pilots it is important to attend carefully to the validity measures of any and all tests administered. Additionally, it is helpful to administer the Paulhus Deception Scales (PDS 1998) which quickly and efficiently assess the presence and type of response distortion tendency for an individual examinee.

The following is a list of psychological tests that are commonly used to evaluate emotional stability, mental health, and personality factors in military, commercial and general aviation pilots.

Minnesota Multiphasic Personality Inventory-2 (MMPI-2)

The MMPI and its successors (that is, MMPI-2 and the MMPI-2-RF) are the most widely used and researched measures of psychopathology (Greene 2011). A unique feature of the MMPI is that Butcher (2002) collected airline pilot norms that can be used in assessing the emotional and personality characteristics of airline pilots. The MMPI-2 is a paper-and-pencil or computer-administered personality questionnaire with 567 true/false statements that is used by clinicians for multiple purposes including personnel selection, identification of mental disorders, and selection of appropriate treatment methods. The MMPI-2 is designed to assess major symptoms of social and personal maladjustment in a wide variety of different populations including professional pilots (Butcher 2002). It was designed with a large normative sample and was last revised in 2001. The test takes between 60 and 90 minutes on average and the adult form is appropriate for individuals age 18 and older.

Butcher (2002) has presented information on MMPI-2 results for 437 commercial airline pilot candidates. He found that the applicants for pilot positions were an "unusually well-functioning group psychologically" but tended to minimize adjustment problems and present themselves in an unrealistically favorable manner. The aviation psychologist is strongly advised to become familiar with the procedures for evaluating response distortion on the MMPI-2 explicated in Nichols and Greene (1997). Pilots were generally low across the board on the clinical scales which measure psychopathology. They tend to exaggerate positive traits and a positive self-image with high self-esteem. In an effort to adjust and increase the effectiveness of the MMPI-2, Butcher and Han (1995) developed a Superlative Self-Presentation Scale (S) which contrasts the MMPI-2 responses of 274 male pilot applicants with those of the 1,123 males in the MMPI-2 standardization sample. The S scale is comprised of five subsets of items which allow for appraising the areas in which the applicant is likely to be projecting an overly positive image.

Personality Assessment Inventory (PAI)

The PAI is a 344-item self-report inventory of adult (18 years or older) personality features designed to assess various domains of personality and psychopathology and provide information for use in the diagnosis of mental health disorders, development of treatment plans, and screening for psychopathology (Morey 2003; Murphy, Conoley and Impara 1994). The PAI includes 22 scales covering constructs relevant to the assessment of mental disorders. The scales include four validity scales, 11 clinical scales, five treatment consideration scales, and two interpersonal scales. Given the tendency towards elevated Positive Impression Management (PIM) profiles, the aviation psychologist must become adept at utilizing the supplemental validity scales, in particular the Defensiveness Index (DEF) and Cashel's Discriminant Function (CDF). The 11 clinical scales can be divided into clusters within the neurotic spectrum, psychotic spectrum, and behavior disorder or impulse control spectrum. A supplemental critical item listing is provided that suggests behavior or psychopathology that may demand immediate attention. Profiles can be compared with both normal and clinical populations.

Millon Clinical Multiaxial Inventory-III (MCMI-III)

The MCMI-III is a self-report personality questionnaire consisting of 175 items. The test is based upon Theodore Millon's model of personality and psychopathology (Strack 2008). According to Weiner and Craighead (2010), "...the MCMI-III has become one of the most frequently used instruments for the identification of personality disorders and clinical syndromes" (999). The MCMI-III was standardized on a sample of psychiatric patients and thus may result in artificial elevations of three scales (that is, Histrionic, Narcissitic, and/or Compulsive) on the profiles of normal individuals and those engaging in impression management. Therefore, the aviation psychologist should be cautious to not diagnose an individual based only upon elevations seen on one or more of these three scales. Originally published in 1994, the norms have been updated.

The MCMI-III takes 30 minutes to administer and is appropriate for adults 18 years of age and older. The computerized report corresponds to the *DSM-IV* classification system, provides treatment information, and includes validity scales as well as facet scales that identify personality processes that underlie overall scale elevations on the personality pattern scales.

NEO Personality Inventory-Revised (NEO-PI-R)

The NEO-PI-R, is a revised version of the NEO Personality Inventory with updated normative data. The NEO-PI-R is the most widely used measure of the Five-Factor Model of adult personality. The test is suitable for individuals 17 years of age and older and covers the domains of neuroticism, extraversion, openness, agreeableness, and conscientiousness (Costa and McCrae 2005).

The NEO-PI-R is frequently used for the assessment of pilots. The test has been shown useful in evaluating aeronautical adaptability, and is considered to be a measure of the "right stuff," the desired personality characteristics for successful aviators and astronauts (Kay 2002). There has been substantial research on the differences between aeronautically adaptable and non-aeronautically adaptable (NAA) pilots within the military with the NEO-PI-R (Campbell, Moore, Poythress and Kennedy 2009, see also Chapter 6, this volume). Researchers have found that with the NEO-PI-R they were able to determine that 19 percent of the NAA pilots had a diagnosed personality disorder and 81 percent had maladaptive personality traits. Non-adaptable pilots were described as inefficient, undependable, self-effacing, pessimistic, easily overwhelmed, sober, and sedentary compared to the aeronautically-adaptable group of pilots (Ellis, Moore and Dolgin 2001).

In a study conducted to determine if general personality measures could identify a personality profile for commercial aviation pilots the researchers determined, based upon their research with 93 pilots, that a "pilot personality" did exist. Commercial aviation pilots scored low on NEO measures of neuroticism, anxiety, angry hostility, depression, self-consciousness, impulsiveness, and vulnerability. The pilot group was high on measures of conscientiousness, competence, dutifulness, achievement, striving, self-discipline, deliberation, assertiveness, activity, and positive emotions. They were average on measures of extroversion, gregariousness, openness, agreeableness, trust, straightforwardness, honesty, and tender mindedness (Fitzgibbons, Davis and Schutte 2004).

Performance based personality measures

Personality tests vary with respect to the content, structure, scoring, and interpretation methodology. At one end of the spectrum are highly structured and objectively administered and scored psychological tests such as the MMPI-2, MCMI-III, PAI, and NEO-PI-R. At the other end of the spectrum are the performance based (previously known as projective) personality tests such as the Rorschach Inkblot Method (Rose, Kaser-Boyd and Maloney 2001), Thematic

Apperception Test (TAT) (Kaplan and Saccuzzo 2009), and various sentence completion tasks (Kaplan and Saccuzzo 2009). The FAA specifications for the evaluation of aviators list the Rorschach Inkblot Method as an assessment measure for conducting mental health evaluations where there are concerns about reality testing, a thought disorder and/or a defensive invalid profile on other psychological test results (Federal Aviation Administration 2013a). The specifications indicate that the Rorschach Performance Assessment System (R-PAS) scoring system is preferred, however either the R-PAS or scoring based upon the Exner Comprehensive Systems are considered acceptable.

In spite of decades of controversy surrounding the use of performance based personality measures, especially the Rorschach Inkblot Method, the use of such psychological tests remains popular with psychologists (Davis 2001). The Society for Personality Assessment (SPA) published an "Official statement by the Board of Trustees of the Society for Personality Assessment" (219) specifying that "the Rorschach possesses reliability and validity similar to that of other generally accepted personality assessment instruments, and its responsible use in personality assessment is appropriate and justified" (221). The society noted that studies had shown that the Rorschach possesses adequate psychometric properties, can be scored reliably, and provides measures of psychological functions that cannot be obtained from other instruments or clinical interviews. Validity was determined to be comparable to other psychological tests (SPA 2005). Reliability was reported to range from the mid-80s to greater than 90 for specified scores (Davis 2001).

When conducting a personality assessment of an aviator, the administration of the Rorschach may be especially appropriate for the identification of impaired reality testing, a thought disorder, stress intolerance, emotional dyscontrol, suicide potential, lack of coping resources, hostility, and dependency needs, conflicts, and attitudes (Fahmy 2012; Kleiger and Khadivi 2012; Ritzler and Sciara 2012; Rose, Kaser-Boyd and Maloney 2001; Wood, Nezworski and Garb 2003).

Ganellen (1994) administered the MMPI and the Rorschach to a group of commercial pilots who were in the process of applying for medical reinstatement following their discharge from an inpatient program for treatment of substance abuse. The research found that although the pilots had an incentive to create a favorable impression, problems in psychological adjustment were evident on the Rorschach that were consistent with the pilot's clinical histories.

When administering the Rorschach, the examiner should rely upon standardized instructions, coding of responses, and interpretation (Viglione and Rivera 2003). Reliance upon research-based computer scoring programs such as the R-PAS (Myer and Eblin 2012), Rorschach Interpretation Assistance Program: Version 5 (RIAP; Exner, Weiner and PAR Staff 2003) or the ROR-SCAN 7 (Caracena 2008) will substantially enhance the scoring and the interpretation process.

This listing of mental health assessment measures is not comprehensive and does not include measures of effort/malingering, neurocognitive functioning, substance abuse/dependence, academic functioning, or health issues. Most mental health assessments will include administration of tests that also assess areas of selected neurocognitive functioning that may be impacted by emotional and/or personality factors (see Chapter 10, this volume for information on the neurocognitive assessment of aviators). Information on measures which are focused on neurocognitive factors, health issues, academic functioning, and/or specific areas of psychopathology is available in numerous publications on psychological testing (for example, Lezak et al. 2012; Spies, Carlson and Geisinger 2010).

Conclusions/Comments

Mental health issues and psychological disorders are common within the general population and may significantly impact an individual's ability to function within community, family, and social settings. Commercial as well as private pilots are not immune to the same stresses and struggles that the general population faces on a day-to-day basis. Commercial pilots are a generally healthy group of individuals and are monitored on a regular basis not only by their aviation employer but also by flight training departments, flight operations, their AME, and other crewmembers. Military aviators are monitored by their flight surgeon, other health providers, squadron commander, operations officer, and also by other crewmembers.

Psychiatric disorders and mental health problems are a potential threat to pilot performance and aviation safety. Every few years incidents involving air crew (pilot or flight attendant) behavior and/or mental health issues are reported by the media. Many of these incidents resurrect calls for mandating periodic mental health assessment of flight crew. In cases where mental health problems have been identified the problems may disqualify a pilot from being medically certified. The airlines, pilot unions, FAA, military and/or airline agencies depend on aviation mental health experts, including aviation psychologists and psychiatrists, to conduct mental health and fitness-for-duty assessments to determine the extent and severity of any behavior or affective characteristics which could compromise flight and passenger safety.

When a mental health evaluation of an airman is requested, due to safety implications and/or regulatory requirements, the quality of the mental health report must be of the highest caliber and meet standards that would be defensible in a court of law. The report will play a pivotal role in whether or not an individual is able to maintain an aviation career. Therefore, it is not surprising that there are frequent challenges to adverse recommendations made by aviation psychologists and psychiatrists. The report that is generated must be comprehensive, based upon valid and reliable assessment methods, and appropriate for use with pilots. All sources of information including job/performance records, information from collateral resources, clinical impressions/observations, and test scores and profiles must be considered in the conclusions reached and recommendations rendered. Findings from multiple sources will provide the most solid foundation for conclusions and interpretations. When data are inconsistent caution must be exercised when generating interpretations. Diagnostic inferences or statements made on the basis of insufficient information and/or data should be avoided. Recommendations must be based on the information documented in the report. The recommendations should describe realistic and practical intervention goals and treatment strategies.

References

American Psychiatric Association. 2000. *Diagnostic and Statistical Manual of Mental Disorders, 4th Edition, Text-Revision (DSM-IV-TR)*. Washington, DC: Author.

American Psychiatric Association. 2012. *DSM-5 Development: Personality Disorder*. Washington, DC: Author.

American Psychological Association. 1999. *The Standards for Educational and Psychological Testing*. Washington, DC: Author.American Psychological Association. 2002. *Ethical Principles of Psychologists and Code of Conduct*. Washington, DC: Author.

American Society of Aerospace Medicine Specialists. 2008. Mood disorders. *Clinical Practice Guidelines*. Available at: www.asams.org/guidelines/COMPLETED/New Mood Disorders update.htm [accessed 1 April 2012].

Beck, A.T., Steer, R.A. and Brown, G.K. 1996. *Beck Depression Inventory-II Manual*. San Antonio, TX: Pearson Education Inc.

Bennett, B.E., Brinklin, P.M., Harris, E., Knapp, S., VandeCreek, L. and Younggren, J.N. 2006. *Assessing and Managing Risk in Psychological Practice*. Rockville, MD: The Trust.

Berry, M.A. 2010. New SSRI certification guidelines in place. *Federal Air Surgeon's Medical Bulletin*, 48(2), 1, 4.

Bills, C.B., Grabowski, J.G. and Li, G. 2005. Suicide by aircraft: a comparative analysis. *Aviation, Space, and Environmental Medicine*, 76(8), 715–719.

Blaney, B. 2012 (July 10). Doctor: lack of sleep prompted pilot's breakdown. Available at: http://www.msnbc.msn.com/id/48145997/ns/business/t/doctor-lack-sleep-prompted-pilots-breakdown/[accessed 23 July 2012].

Boeing. 2009. Statistical summary of commercial jet airplane accidents (Technical Report). Seattle, WA: Author.

Boone, K., Lu, P. and Herzberg, D. 2002. *The Dot Counting Test Manual (DCT)*. Torrance, CA: Western Psychological Services.

Bor, R. and Hubbard, T. 2006. Aviation mental health: an introduction (chapter 1), in *Aviation Mental Health: Psychological Implications for Air Transportation*, edited by R. Bor and T. Hubbard. Burlington, VT: Ashgate Publishing Company. 1-9.

Briere, J. 2010. *Trauma Symptom Inventory*. 2nd Edition (TSI-2). Lutz, FL: Psychological Assessment Resources.

Brienza, V. 2011. The Ten Most Stressful Jobs of 2011, *Careercast* [Online 30 March 2012] Available at: http://careercast.com [accessed 20 March 20 2012].

Bush, S.S. 2010. Determining whether or when to adopt new versions of psychological and neuropsychological tests: Ethical and professional considerations. *The Clinical Neuropsychologist*, 24(1), 7–16.

Bush, S.S., Ruff, R.M., Troster, A.I., Barth, J.T., Koffler, S.P., Pliskin, N.H., Reynolds, C.R. and Silver, C.H. 2005. NAN position paper. Symptom validity assessment: Practice issues and medical necessity. NAN policy and planning committee. *Archives of Clinical Neuropsychology*, 20(4), 419–426.

Butcher, J.N. 2002. Assessing pilots with 'the wrong stuff', a call for research on emotional health factors in commercial aviators. *International Journal of Selection and Assessment*, 10(1-2), 168–184.

Butcher, J.N. and Han, K. 1995. Developmental of an MMPI-2 scale to assess the presentation of self in a superlative manner: the S scale, in *Advances on Personality Assessment, Volume 10*, edited by J.N. Butcher and C.D. Spielberger. Hillsdale, NJ: LEA Press. 25-50.

Campbell, J.S., Moore, J., Poythress, N.G. and Kennedy, C.H. 2009. Personality traits in clinically referred aviators: Two clusters related to occupational suitability. *Aviation, Space, and Environmental Medicine*, 80(12),1049–1054.

Caracena, P.F. 2008. *ROR-SCAN*. Edmond, OK: Rorschach Interpretative System.

Centers for Disease Control and Prevention. 2011. *International Classification of Diseases, Clinical Modification (ICD-9-CM)*. Atlanta, GA: Author.

Conners, C.K. 2008. *Conners' Continuous Performance Test II, Version Five User's Manual*. San Antonio, TX: Pearson Education Inc.

Costa, P. and McCrae, R. 2005. *NEO Inventories: NEO Personality Inventory-3 (NEO-PI-3)*. Lutz, FL: Psychological Assessment Resources.

Davis, A.D. 2001. Overview, in *Essentials of Rorschach Assessment*, edited by T. Rose, N. Kaser-Boyd, and M.P. Maloney. New York: John Wiley and Sons, Inc. 1-16.

DeFilippis, N.A. and PAR Staff. 1993. *Category Test: Computer Version, Research Edition*. Lutz, FL: Psychological Assessment Resources.

Delis, D.C., Kramer, J.H., Kaplan, E. and Ober, B.A. 2000. *California Verbal Learning Test, 2nd Edition, Adult Version*. San Antonio, TX: Pearson Education Inc.

Department of Transportation. 2013. *User's Guide for the Federal Aviation Administration MedXPress System*. Washington, DC: Author.

Derrogatis, L. and Unger, R. 2010. Symptom checklist-90-revised, in *Corsini Encyclopedia of Psychology*, 4th Edition, edited by I. Weiner and W. Craighead. Hoboken, NJ: John Wiley and Sons, Inc. 4, 1743-1744.

Dewa, C., Lin, E., Kooehoorn, M. and Goldner, E. 2007. Association of chronic work stress, psychiatric disorders, and chronic physical conditions with disability among workers. *Psychiatric Services*, 58(5), 652–658.

Edwards, D.C. 1990. *Pilot Mental and Physical Performance*. Ames, IA: Iowa State University Press.

Elliott, R.W. 2007. Psychiatric and psychological evaluations. Presentation at Air Line Pilots Association HIMS Seminar, Denver, CO, September 10.

Elliott, R.W. 2011. *Aviation Assessments*, Workshop presented at National Academy of Neuropsychology 31st Annual Conference, San Marco Island, FL, October 16–19.

Ellis, S., Moore, J. and Dolgin, D. 2001. Aviation personality assessment: part I-aeronautical adaptability. *Aviation, Space, and Environmental Medicine*, 72(3), 254.

Evans, S. and Radcliffe, S. 2012. The incapacitation rate of pilots. *Aviation, Space, and Environmental Medicine*, 83(1), 42–49.

Exner, J.E., Weiner, I.B. and PAR Staff. 2003. *Rorschach Interpretation Assistance Program: Version 5 (RIAP 5)*. Lutz, FL: Psychological Assessment Resources, Inc.

Fahmy, A. 2012. Rorschach technique. *Encyclopedia of Mental Disorders*. Flossmoor, IL: Advameg, Inc.

Faust, D. 2012. *Coping with Psychiatric Testimony*. 6th Edition. New York: Oxford University Press.

Federal Aviation Administration. 2013a. *Guide for Aviation Medical Examiners*. Washington, DC: Department of Transportation. Available at: http//: www.faa.gov/about/office_org/headquarters_offices/avs/office/aam/ame/guide [accessed 3 April 2013].

Federal Aviation Administration. 2013b. *Aeronautical Information Manual* (AIM). Washington, DC: Department of Transportation. Available at: http//:www.faa.gov/air_traffic/publications/atpubs/aim [accessed 3 April 2012].

Federal Aviation Administration. 2008. *Application for Airman Medical Certificate or Airman Medical and Pilot Certificate* (Form 8500-8). Washington, DC: Department of Transportation.

Federal Judicial Center, 2011. *Reference Manual on Scientific Evidence*. 3rd Edition. Washington, DC: National Academic Press.

Fitzgibbons, A., Davis, D. and Schutte, P.D. 2004. Pilot personality profile using the NEO-PI-R. *National Aeronautics and Space Administration, Langley Research Center* [NASA/TM-2004-23237]. Hampton, VR: NASA.

Flight Safety Foundation. 2012. *Aviation Safety Network Accident Description EgyptAir Flight 990*. Alexandria, VR: Author.

Front, C.M. 2013. *Personality assessment of pilots and air traffic control specialists*. Workshop presented at the FAA Aviation Psychology Seminar, Oklahoma City, Oklahoma, March 9, 2013.

Ganellen, R.J. 1994. Attempting to conceal psychological disturbance: MMPI defensive response set and the Rorschach. *Journal of Personality Assessment*, 63(3), 423–437.

Grant, I. and Adams, K.M. 2009. *Neuropsychological Assessment of Neuropsychiatric and Neuromedical Disorders*. 3rd Edition. New York: Oxford University Press.

Greenberg, L., Leark, R.A., Dupuy, T.R., Corman, C.L. and Kindschi, C.L. 2012. *Test of Variables of Attention, Version 8 (TOVA 8) Professional Manual*. Lutz, FL: Psychological Assessment Resources.

Greene, R.L. 2011. *The MMPI-2/MMPI-2-RF: An Interpretative Manual.* 3rd Edition. Boston, MA: Pearson Education, Inc.

Greiffenstein, M.F. 2010. Noncredible neuropsychological presentation in service members and veterans, in *Military Neuropsychology*, edited by C.H. Kennedy and E.A. Zillmer. New York, NY: Springer. 81-100.

Hess, D.W., Kennedy, C.H., Hardin, R.A. and Kupke, T. 2010. Attention Deficit/Hyperactivity Disorder and learning disorders, in *Military Neuropsychology*, edited by C.H. Kennedy and E.A. Zillmer. New York, NY: Springer. 199-226.

Jamison, K. 2000. Suicide and bipolar disorder. *Journal of Clinical Psychiatry*, 61 (Supplement 9), 47–51.

Jansen, B. 2012. *USA Today*. Pilot outburst raises safety questions, March 28, 1-A.

Jensen, R.S, 1991. Editorial. *The International Journal of Aviation Psychology,* 1(1), 1–3.

Johnson, E.W. and Lovell, M.R. 2011. Neuropsychological assessment (chapter 8), in *Textbook of Traumatic Brain Injury*. 2nd Edition, edited by J.M. Silver, T.W. McAllister and S.C. Yudofsky. Washington, DC: American Psychiatric Association. 127-141.

Jones, D.R. 2008. Aerospace Psychiatry, in *Fundamentals of Aerospace Medicine*. 4th Edition, edited by J. Davis, R. Johnson, J. Stepanack, and J. Fogarty. Philadelphia: Lippincott Williams and Williams. 406-424.

Jones, D.R. and Marsh, R.W. 2001. Psychiatric considerations in military aerospace medicine. *Aviation, Space, and Environmental Medicine*, 72(2), 129–135.

Jones, D.R. and Patterson, J.C. 1989. Medical or administrative? Personality disorders and maladaptive personality traits in aerospace medical practice. *Aviation Medicine Quarterly*, 2, 83–91.

Kaplan, R.M. and Saccuzzo, D.P. 2009. *Psychological Testing: Principles, Applications, and Issues*. 7th Edition. Belmont, CA: Wadsworth, Cengage Learning.

Kay, G.G. 2002. Guidelines for the psychological evaluation of air crew personnel. *Occupational Medicine*, 17(2), 227–245.

Kay, G.G. 1999. *CogScreen: Professional Manual*. Washington, DC: CogScreen LLC.

Kivimäki, M., Hotopf, M. and Henderson, M. 2010. Do stressful working conditions cause psychiatric disorders? *Occupational Medicine*, 60(2), 86–87.

Kleiger, J. and Khadivi, A. 2012. Assessing psychiatric conditions with the Rorschach and clinical interview. Workshop presented at Society for Personality Assessment Annual Meeting. Chicago, March 14, 2012.

Lezak, M.D. Howieson, D.B., Bigler, E.D., and Tranel, D. 2012. *Neuropsychological Assessment*. 5th Edition. New York: Oxford University Press.

Lollis, B.D., Marsh, R.W., Sowan, T.W. and Thompson, W.T. 2009. Major depressive disorder in military aviators: a retrospective study of prevalence. *Aviation, Space, and Environmental Medicine*, 80(8), 734–737.Mazure, C. (Editor). 1995. *Does Stress cause Psychiatric Illness?* Washington, DC: American Psychiatric Press, Inc.

McCaffrey, R.J., Palav, A.A., O'Bryant, S.E. and Labarge, A.S. (Editors). 2003. *Petitioner's Guide to Symptom Base Rates in Clinical Neuropsychology*. New York: Kluwer Academic/Plenum Publishers.

McClellan, J.M. 2010. Left Seat: The Psychology of Safety, *Flying*. 11 June. Available at: http://www.flyingmag.com/safety/left-seat-psychology-safety [accessed 30 March 2012].

McClellan, S.F. and Bernstein, S. 2006. The role of aeronautical adaptability in the disqualification of a military flyer. *Aviation, Space, and Environmental Medicine*, 77(11), 1188–1192.

Meyer, G. J. and Eblin, J. J. 2012. An overview of the Rorschach Performance Assessment System (R-PAS). *Psychological Injury and Law, 5,* 107-121.

Miller, H. 2001. *Miller Forensic Assessment of Symptoms Test (M-FAST) Professional Manual*. Lutz, FL: Psychological Assessment Resources.

Mitrushina, M., Boone, K.B., Razani, J. and D'Elia, L. 2005. *Handbook of Normative Data for Neuropsychological Assessment*. 2nd Edition. New York: Oxford University Press.

Morey, L. 2003. *Essentials of PAI Assessment*. New York: John Wiley and Sons, Inc.

Morrison, J. 1995. *DSM-IV Made Easy: The Clinician's Guide to Diagnosis*. New York: Guilford Press.

Morse, J.S. and Bor, R. 2006. Psychiatric disorders and syndromes among pilots, in *Aviation Mental Health*, edited by R. Bor and T. Hubbard. Burlington, VT: Ashgate Publishing Company. 107-126.

Murphy, L.L., Conoley, J.C. and Impara, J.C. (Editors). 1994. *Tests in Print IV, Volume I*. Lincoln, NE: The University of Nebraska Press.

Musa, A. and Mayer, J. 2012. Experts cite flimsy FAA mental testing, *Amarillo Globe News*. March 28. Available at: http//:www.Amarillo.com/news/local-news/2012-3-28/experts-cite-flimsy-FAA-mental-testing. [accessed March 28, 2012]

Musson, D.M. 2006. Psychological aspects of astronaut selection, in *Aviation Mental Health: Psychological Implications for Air Transportation*, edited by R. Bor and T. Hubbard. Burlington, VT: Ashgate Publishing Company. 243-254.

National Institute of Mental Health. 2010. *Suicide in the US: Statistics and Prevention*. Baltimore: NIMH. Available at: http//: http://www.nimh.nih.gov/health/publications/suicide-in-the-us-statistics-and-prevention/index.shtml [accessed 23 July 2012].

National Institute of Mental Health. 2012. Any anxiety disorder among adults. *Health Topic Statistics Prevalence*. April 17. Available at: http://www.nimh.nih.gov/statistics/index.shtml. [accessed 18 April 2012].

Nichols, D.S. and Greene, R.L. 1997. Dimensions of deception in personality assessment: The example of the MMPI-2. *Journal of Personality Assessment*, 68(2), 251-266.

Pagura, J., Katz, L.Y., Mojtabai, R. Druss, B.G., Cox, B. and Sareen, J. 2011. Antidepressant use in the absence of common mental disorders in the general population. *Journal of Clinical Psychiatry*, 72(4), 494–501.

Paulhus, D.L. 1984. Two-component models of socially desirable responding. *Journal of Personality and Social Psychology*, 46(3):598-609.

Paulhus, D.L. 1998. *Paulhus Deception Scales*. Toronto: Multi-Health Systems.

Pella, R.D., Hill, B.D., Singh, A.N., Hayes, J. and Gouvier, W.D. 2012. Noncredible performance in mild traumatic brain injury, in *Detection of Malingering during Head Injury Litigation*. 2nd Edition, edited by C.R. Reynolds and A.M. Horton. New York: Springer Science and Business Media. 121-150.

Piedmont, R., Sherman, M. and Barrickman, L. 2000. Brief psychosocial assessment of a clinical sample: an evaluation of the Personal Problem Checklist for Adults. *Assessment*, 7(2), 177–187.

Pratt, L.A., Brody, D.J. and Gu, Q. 2011. Antidepressant use in persons aged 12 and over: United States, 2005–2008. *NCHS Data Brief*, 76, 4–5.

Richardson v NTSB [2010] Docket No. SM-4893, Westlaw 2010 WL 4262641 (N.T.S.B.). Anchorage, AK.

Ritzler, B. and Sciara, A. 2012. The Rorschach comprehensive system: advance coding and administration. Workshop presented at Society for Personality Assessment Annual Meeting. Chicago. March 15, 2012.

Roscoe, S.N. 1980. Concepts and definitions. In *Aviation Psychology*. Edited by S. Roscoe and A. Williams. Ames, IA: Iowa State University Press. 3-11.

Rose, T., Kaser-Boyd, N. and Maloney, M.P. 2001. *Essentials of Rorschach Assessment*. New York: John Wiley and Sons, Inc.

Shappell, S.A., Detwiler, C., Holcomb, K., Hackworth, C., Boquet, A. and Wiegmann, D.A. 2007. Human errors and commercial aviation accidents: an analysis using the human factors analysis and classification system. *Human Factors*, 49(2), 227–242.

Shappell, S.A. and Wiegmann, D.A. 2001. Applying reasons: The human factors analysis and classification systems (HFACS). *Human Factors and Aerospace Safety*, 1(1), 59–86.

Shipper, L.J., Sollman, M.J. and Berry, D.T. 2010. NEO personality inventory (NEO-PI-R), in *The Corsini Encyclopedia of Psychology*. 4th Edition. Volume 3, edited by I.B. Weiner and W.E. Craighead. New York: Oxford University Press. 1068-1070.

Sladen, F.J. (Editor). 1943. *Psychiatry and the War*. Springfield, IL: Thomas.

Slick, D., Hopp, G., Strauss, E. and Thompson, G. 1995. *Victoria Symptom Validity Test (VSVT)*. Lutz, FL: Psychological Assessment Resources.

Slick, D., Tan, S. Strauss, E. and Hultsch, D. 2004. Detecting malingering: a survey of experts' practices. *Archives of Clinical Neuropsychology*, 19 (4), 465–473.

Society for Personality Assessment (SPA). 2005. The status of the Rorschach in clinical and forensic practice: An official statement by the board of trustees of the society for personality assessment. *Journal of Personality Assessment*, 85(2), 219–237.

Spies, R.A., Carlson, J.F. and Geisinger, K.F. (Editors). 2010. *The Eighteen Mental Measurements Yearbook*. Lincoln, NE: University of Nebraska Press.

Strack, S. 2008. *Essentials of Millon Inventories*. 3rd Edition. New York: John Wiley and Sons, Inc.
Strauss, E., Sherman, E. and Spreen, O. 2006. *A Compendium of Neuropsychological Test*. 3rd Edition. New York: Oxford University Press.
US Department of Health and Human Services. 1999. Adults and Mental Health: Anxiety disorders. In *Mental Health: A Report of the Surgeon General*. Pittsburgh, PA: NIMH. Author. 4. 233-243.
Viglione, D.J. and Rivera, B. 2003. Assessing personality and psychopathology with projective methods, in *Handbook of Psychology, Volume 10, Assessment Psychology*, J.R. Graham and J.A. Naglieri (Volume Editors) and I. Weiner (Editor-in-chief). Hoboken, NJ: John Wiley and Sons, Inc. 531-549.
Vinokur, A. and Selzer, M.L. 1975. Desirable versus undesirable life events: Their relationship to stress and mental distress. *Journal of Personality and Social Psychology*, 32(2), 329–337.
Weiner, I.B. and Craighead, W.E. (Editors). 2010. *The Corsini Encyclopedia of Psychology*. 4th Edition. Volume 3. New York: John Wiley and Sons, Inc.
Wood, J.M., Nezworski, M.T. and Garb, H.N. 2003. Focus on empirically supported methods: what's right with the Rorschach? *The Scientific Review of Mental Health Practice*, 2(2), 1–7.
World Health Organization. 2005. *Promoting Mental Health: Concepts, Emerging Evidence, Practice: A Report of the World Health Organization Department of Mental Health and Substance Abuse in Collaboration with the Victorian Health Promotion Foundation and the University of Melbourne*. H. Herrman, S. Saxena, and R. Moodie (Editors). Geneva, WHO.

Appendix A

The following psychological report is fictitious and included for illustration purposes only.

SAMPLE REPORT
AIRMAN PSYCHOLOGICAL EVALUATION REPORT

AIRMAN NAME: P.M. **EVALUATION DATES**: May 5–6, 2012
BIRTH DATE: August 2, 1964 **CHRONOLOGICAL AGE**: 47
EMPLOYER: Major Airline
REFERRAL SOURCE: Airline EAP Representative
AIRCREW POSITION: Capt. B757-300

REASON FOR REFERRAL

Capt. P.M. was referred to this examiner for an Airman Psychological Evaluation to determine his current level of cognitive resources, to identify emotional and/

or personality factors that appear to be interfering with his job performance as a commercial airman, and to render an opinion whether the airman is suitable for active flight status.

TESTS ADMINISTERED AND SCORES

Minnesota Multiphasic Personality Inventory-2 (MMPI-2)
Welsh Code Profile: Significant, 46'2+370-158/9: F/KL:
Millon Clinical Multiaxial Inventory-III (MCMI-III)
Significant psychopathology (Profile: Dysthymic Disorder)
Trail Making Test
Nonsignificant for neurological indices (A: 29 secs/0 error; B: 90 secs/1 error)
Dot Counting Test
Normal Effort (E-Score: 11)
Green's Word Memory Test (WMT)
Good Effort (Profile Scores: IR & DR 90%; LDFR 60%)

CogScreen-Aeromedical Edition (Windows Version)

Test Performances at/below 5th percentile for age/education

	Raw	T-scores	%iles
LRPV	0.9999*	------	------
Speed	7	<30	<2.5
Accuracy	3	34	5
Thruput	7	<33	<5
Process	5	<30	<2.5

Taylor Aviation Factor T-Scores

Attribute Identification	36.54
Motor Coordination	24.45
Visual Association memory	32.78
Speed/Working Memory	38.90
Tracking	51.37

SOURCES OF INFORMATION

In addition to the results of the Airman Psychological Evaluation conducted on May 5–6, 2012, a number of records, provided by Major Airline, were reviewed for completion of this report.

Records reviewed included:
Major Airline Medical Department Medical Leave report (4/12/12).
Notes from Major Airline Medical Department for the period 3/12/12 through 5/18/12.
Memo from psychiatrist outlining treatment efforts (3/15/12).
FAA First Class Airman Medical Certificate (7/1/11).

Telephone conferences were held with the airman's spouse regarding her husband's home and community-related behavior and with the company chief pilot and base chief pilot, regarding the airman's work performance.

CURRENT STATUS

Capt. P.M. is a Major Airline B757-300 pilot who is based in Los Angeles. He is currently on sick leave because of a disabling orthopedic condition and a mental health disorder which is being treated with psychotropic medications which are not authorized for active duty pilots. Capt. P.M. last flew as a cockpit crewmember during October 2011. He is currently living in a suburb of Los Angeles, California with his spouse and two young children.

INITIATING EVENTS (Airman Perspective)

The following information was provided by Capt. P.M. during our clinical interview. Capt. P.M. stated that he has been struggling with depression-related symptoms since his childhood years. He did not participate in a counseling program until he was in his young adult years.

Capt. P.M. participated in outpatient psychiatric treatment program with Dr. R.S., psychiatrist, during 2008 for treatment of depression and alcohol-related issues. During 2008 Capt. P.M. was prescribed Wellbutrin, which he took until 2011. When he started taking Wellbutrin, he stopped drinking. He developed back problems, which grounded him in October 2011. During November 2011 after Capt. P.M. was informed that Wellbutrin was a disqualifying psychotropic medication he stopped taking the medication but his depression symptoms reemerged. After physically relocating his family in January 2012 Capt. P.M. changed psychiatrists (Dr. T.U.) and was placed on Lithium, 300 mg tid during March 2012. After further discussion with the company physician, the airman resumed taking Lithium.

REVIEW OF RECORDS

Records indicate that on July 1, 2011, P.M. underwent an FAA First Class Airman Medical Examination. His airline medical examiner (AME) issued a medical certificate with a "must wear corrective lenses" limitation. There is no disclosure of a mental health condition, meetings with a psychiatrist, or use of any psychotropic medication.

Dr. R.S., psychiatrist, in a March 30, 2012 memo, stated that he met with P.M. six times for diagnosis and treatment of a mood disorder which included symptoms of depression and mania. The psychiatrist noted that use of psychotropic medication might jeopardize Capt. P.M.'s career as a pilot. Dr. T.U. prescribed Lithium and added that the prognosis was uncertain and that treatment would be a long-term process.

On April 5, 2012, P.M. drafted a memo to the company medical department requesting long-term disability status. The medical department drafted request for leave memo stating that Capt. P.M. had limitations and was not eligible to fly. Capt. P.M. was placed on medical leave.

SYMPTOMS

P.M. has been experiencing numerous symptoms that are associated with or a result of his mental health condition. He reports the following symptom cluster:

Psychiatric. P.M. reports that he has been experiencing symptoms of depression and mood swings since his childhood years. Although he had not engaged in counseling or taken psychotropic medication until 2002, his symptoms were apparent to himself, in retrospect, much earlier.
Depression. Off medication Capt. P.M. describes his depression as significant and experienced on a daily basis.
Suicidal Thoughts. During his college years P.M. made a suicidal gesture by ingesting unknown pills. There was no medical treatment rendered and no loss of consciousness or medical consequences associated with that attempt. He harbored suicidal-related thoughts on a daily basis more recently but has never formulated a suicide plan. The primary deterrent has been concern about the potential impact on his children.
Social. P.M. stated he has never been a good "small talker." When he was drinking heavily he was able to easily engage in casual conversation but with "sobriety" small talk has become much more difficult. He has a few casual friends but no close friends.
Crying. Experienced on a daily basis but crying is also cathartic according to P.M.
Anxiety. Anxiety is experienced daily when he is around his wife. The couple has been sleeping in separate quarters for the last three years.
Manic episodes. P.M. acknowledged that he has experienced frequent manic episodes historically but none since he started taking his medications two months ago. Some of his manic episodes/behaviors have included making plans to sell their home, to take extended trips globally, remove his children from the public school and home teach them himself, and purchase an airplane he would fly around the world. Other manic behaviors historically have included becoming involved in starting up a contracting business, inventing/developing/selling a special soap, writing a book on some of his life experiences, and getting up at 0300 a.m. to conduct research on the plot involving 9/11.
Nightmares. Experienced nightly.
Interests. P.M. has no history of significant interests or hobbies.
Feelings of guilt/worthlessness. Experienced on a daily basis.
Death thoughts. Experienced on a daily basis.
Psychosis: P.M. denied any history of auditory or visual hallucinations or delusions.

Physical. P.M. acknowledges that his physical condition has deteriorated since he developed back problems. He used to be able to run for exercise, which kept him fit and alleviated many of his depression-related symptoms. He is no longer able to run.
Weight gain. Forty pounds gained during the last year.
Back problems. Developed last year and will require further surgery.
Sleep. Sleeps 4 to 6 hours a night. Feels rested most days but his sleep is disrupted and includes nightmares and occasional symptoms of insomnia.
Fatigue. Tired all the time and getting worse with lack of exercise.

P.M. denies any symptoms of restlessness or agitation.

Cognitive. P.M. stated that he has had a history of difficulties focusing and concentrating. He has always been "slightly weak" in multitasking capabilities.
Concentration. Problems historically, but improved with Lithium.
Memory. Intact.
Multitasking. Has never been good at multitasking. Is more efficient with sequential problem-solving approaches.
Distractibility. Frequently experiences distractions but not disruptive to flight responsibilities.

Flight Performance Perception. P.M. stated that he has never felt that his mental health condition or his symptoms has adversely impacted his flight performance. He had a couple of disputes with gate agents and a flight attendant historically, but none of these disputes resulted in disciplinary action. P.M. has had no difficulty transitioning aircraft or passing his proficiency checkrides. During his last checkride he acknowledged that he experienced a problem with the check airman but he eventually passed the checkride. P.M. acknowledged that his anger may have played a role in the flight difficulties he experienced during his last checkride. P.M. rated himself as an "above average" pilot. He has logged 15,000 accident-free flight hours.

BACKGROUND

Early Development. P.M. was born in Memphis. As far as he knows he was healthy as an infant and had only the normal illnesses and injuries.

Family History. P.M. was raised in Memphis by his mother and father with a younger brother and older sister. His father was a college graduate who worked as a marketing supervisor. The father was described as an alcoholic who hated everybody. The mother was a college graduate who worked as a school counselor. She was described as a quiet individual who developed serious mental health problems requiring multiple hospitalizations.

P.M. moved away from home at 18 years of age to attend college. His parents divorced in 1986 after his mother tried to commit suicide and his father was getting drunk on a nightly basis. Neither parent remarried. His father is now 76 years of age and living in poverty in the south.

P.M.'s 76-year-old mother is in good health and living in Ohio. Capt. P.M. maintains contact with his mother. His mother is stable on Lithium and other medications. P.M.'s 51-year-old brother is in good health. He works in law enforcement in the Memphis area and is stable on medication for his mood disorder. He has been hospitalized on three occasions for treatment of his mood disorder.

P.M.'s 43-year-old sister is living in Memphis where she works as a nurse. At one time P.M. was close to his sister but they have drifted apart during the last few years. His sister has no history of a mental health condition.

Education. P.M. described himself as an average student academically. He was active in high school sports and had a number of friends. During his high school years he began consuming alcohol in social settings. He graduated from high school in 1982.

P.M. attended University where he majored in communications. In college he began taking flying lessons. After he joined a social fraternity he became a heavy drinker. By the time he graduated from college he had earned his private pilot's license and CFI, multi-engine, and instrument ratings. He graduated in 1987 with a 3.2 GPA.

Career Development. After graduating from college, P.M. taught flying at an FBO for two years. He worked on a contract basis for a number of regional airlines and corporate flight departments until 1995 when he was hired by Major Airline as a B727 Second Officer. He was promoted to first officer in 1995 on the B727 and progressed into the B747-200 in 2001. He was promoted to captain on the B727 in 2003, progressed to the B747-400 and then the B757-300. His last assignment was as a captain on the B757-300. He has had no other flying assignments with Major Airline.

Marriages/Children. P.M. has been married twice. He married his first spouse in 1991 but from the beginning of the relationship the marriage was unstable. The couple had no children. The couple participated in marriage counseling for one year but this was unsuccessful. The couple divorced in 1995.

P.M. met his second wife in 1998 and married her in 2001. She was a high school graduate who had been working in retail sales for a number of years. The beginning of the relationship was rocky because of his drinking, her mistrust of his faithfulness, and her recognition of his mood swings. She had a family history of parental alcoholism and an emotionally disturbed sister.

The couple has two children, seven and ten years of age. The younger daughter is happy and appears to be doing well in school. The older son struggles in school, although he has made the honor roll and is a good athlete.

P.M. stated that his wife has a history of alcohol-related problems. She has had two DUIs and has been involved with AA historically. She continues to struggle with her drinking on a daily basis. She is participating in psychiatric treatment because of depression, panic attacks, and sleep problems.

Alcohol/Drug Use. P.M. reported that he had a long history of abuse of alcohol which began during his college years and continued until 2008 when he began taking Wellbutrin. He was also getting tired of the withdrawal symptoms associated with heavy drinking so he stopped drinking. He denied any history of drug use.

General Medical History. P.M. denied any history of hospitalizations. He has undergone back surgery and more surgeries are planned for the future. He smokes two cigarettes and consumes three cups of coffee a day. His last FAA flight physical examination was during July 2011. No problems were noted.

COLLATERAL INTERVIEWS

Spouse. A telephone conference was held with his wife regarding her perceptions and her concerns. She noted that when P.M. took his medication he was stable emotionally. Off the medication his mental health problems are quite evident. He has trouble communicating fully and he finds it difficult to open up emotionally. She stated that P.M.'s manic behaviors create instability in the family relationship and cause uncertainty on the part of the children. She stated that depression is evident in P.M.'s behavior, time management, and thinking. He is scattered in his thinking and has difficulty focusing. His frustration level becomes easily elevated and he lacks motivation. On medication he seems calmer and not as sporadic in his behavior. She believes that his primary problems are depression and anger management.

Flight Management. A brief interview was conducted with the company chief pilot and the base chief pilot regarding P.M.'s work performance. No work or training-related problems are noted in Capt. P.M.'s personnel folder. Neither manager had personal knowledge of Capt. P.M.'s work performance.

OBSERVATIONS DURING TESTING

The assessment process proceeded over a period of two days. P.M. is a 6', 210 lb., Euro-American male adult who is right-hand dominant. There was no evidence of delusional thought processes, he was fully oriented, and his gait and posture were unremarkable. During the clinical interview on both Day 1 and 2 he would frequently cry and needed periodic breaks in order to recover.

P.M. was advised that the test results would be shared with the Major Airline EAP and the company physician. He signed an Authorization for Exchange/ Release of Information with the company physician, EAP, flight management, his spouse, and his treating psychiatrist. He was on time for the evaluation sessions and was cooperative in providing detailed information about his background and difficulties in his life situation.
Capt. P.M. stated that he had noticed that his ability to focus, calculate math problems, and generally perform cognitively challenging tasks was compromised on the medication he had been taking for the last couple of weeks. He was not surprised that his performance on *CogScreen* and other neurocognitive measures was compromised. There was some discussion about the possibility that his medications may impact his ability to test performance.

On Day 2 additional testing was scheduled and planned. Capt. P.M. reported on time to the office but complained that he was fatigued and had gotten only one hour sleep the night before. He further stated that he had "suspicions about my intentions" and that "I was not after his best interests." He further stated that the entire interview process and testing conducted the previous day made him feel anxious and despondent. Throughout the Day 2 narrative he was tearful. We discussed his concerns about distribution of the report and "my intentions." After one hour he requested that no further testing be conducted.

ASSESSMENT/TEST RESULTS
Validity Indicators
In an attempt to establish the degree of effort put forth on testing, the *Dot Counting Test*, *Green's Word Memory Test*, and validity measures on the *MMPI-2* were administered. All measures indicated a normal effort. The results of this validity testing indicate that Capt. P.M.'s responses, particularly on personality measures, accurately reflect his perceptions and opinions of his own personality attributes and structure.

Cognitive Functioning
CogScreen: Aeromedical Edition results indicate that compared to commercial pilots, 45 to 49 years of age who fly for major US carriers, Capt. P.M.'s performance was in the impaired range (LRPV 0.9999). Capt. P.M. had seven scores at or below the fifth percentile on measures of Speed (2.5 percentile) and ten scores at or below the fifteenth percentile (less than 2.5 percentile). This level of performance on speed measures is below average for healthy aviators. Specific weaknesses in speed were evident on measures of motor coordination, concentration, computational math skills, visuoperceptual speed, visual scanning, and mental flexibility.

Emotional and Personality Functioning
P.M. was administered the *MMPI-2* and the *MCMI-III* to assess his current emotional development and personality features. His responses were determined to be valid and reflect his current emotional status.

Moods. P.M. is experiencing mild to moderate emotional distress that is characterized by brooding, dysphoria, and anhedonia. He broods and worries constantly over what is happening to him and he feels that he is "no good at all." His dysphoria is permeated with anger, stubbornness, and oppositional behavior. He wishes that he could be as happy as others seem to be. He feels unable to "get going" and to get things done.

Because he feels misunderstood and unappreciated he is likely to become touchy and irritable when he is experiencing even mild levels of stress. This reaction may set into motion further withdrawn behavior and depressive moodiness, beginning the vicious cycle anew. A pattern of dysthymia is an integral part of his characterologic structure. He exhibits a cluster of chronic, general traits in which feelings of uselessness, dejection, pessimism, and discouragement are intrinsic components.

Interpersonal Relations. Interpersonal relationships are apt to be a major problem area. He tends to withdraw and remain aloof in his interpersonal relationships. He believes that no one understands him. His disappointments may lead him to want to withdraw even from supportive personal relationships. He is suspicious of the motives of others and senses that he knows who is responsible for most of his trouble.

Other Problem Areas. P.M. is concerned about his physical health and lack of exercise. He experiences occasional suicidal ideation/thoughts which need to be closely monitored by his treating psychiatrist. He broods over problems that will increase the probability of acting out, either towards himself or others if they provoke him.

TEST INTERPRETATION
Neurocognitive. P.M. is demonstrating impairment in his cognitive skills. He shows significant weaknesses in motor coordination speed, speed and accuracy in his math skills, slow processing on tasks that require mental flexibility, slow visual scanning, weak visual learning and recall skills, weak visual memory, and weak spatial orientation accuracy measures.

Affective/Personality. Affective and personality assessment results identified dysthymic disorder characteristics. The *MMPI-2* 462 profile suggests that hostility and depression are apt to form a cyclical pattern. His expressions of hostility are apt to lead to guilt which sets into motion a recurring cycle.

On the *MMPI-2* it is likely that the normal range Mania scale (Ma) results are a product of the therapeutic/beneficial effects of P.M.'s current medication regime. The Depression scale (D) results are elevated. Capt. P.M.'s clinical history and his family history support a diagnosis of Bipolar Disorder as evidenced by meeting the following criteria:

1. Depressed mood
2. Weight gain
3. Insomnia
4. Fatigue
5. Feelings of guilt
6. Recurrent thoughts of death
7. Grandiosity
8. Decreased need for sleep
9. Flight of ideas
10. Distractibility
11. Foolish business involvements

IMPRESSIONS

P.M. is a 47-year-old, Major Airline B757-300 captain, who initiated a formal treatment program for symptoms of depression and Alcohol Abuse beginning in 2008. He stopped drinking and was prescribed Wellbutrin for treatment of depression symptoms. He continued to take Wellbutrin into 2011 when he developed orthopedic problems which grounded him. As his life began to unravel he developed severe depression symptoms. He began meeting with a psychiatrist who prescribed Lithium and Wellbutrin.

P.M. reported a history of depression symptoms dating back to his childhood and young adult years which remained untreated until 2008. Intertwined in his early emotional and behavioral difficulties was heavy drinking, which continued until 2008.

P.M. is struggling with depression and manic symptoms at the present time. His current medication regime, including Lithium Carbonate and Wellbutrin, has had a significant and beneficial impact on relieving his manic symptoms and has lowered his level of depression, although depressive symptoms remain evident. He continues to struggle with a number of issues in his life which impact his condition.

Psychological testing indicated numerous cognitive weaknesses, primarily due to problems in concentration. His affective and characterologic features indicate depression and avoidant and schizoid traits. Manic behaviors are evident in his clinical history. He is unhappy with himself and others, manifests irritable moodiness, and he is having difficulty with anger control, which has not been manifested behaviorally. Dysthymia is an integral part of his characterologic structure, which includes feelings of dejection, pessimism, and discouragement.

DIAGNOSTIC IMPRESSIONS

AXIS I 1) 296.52 Bipolar I Disorder, most recent episode depressed, moderate severity
2) V62.2 Occupational Problem (use of disqualifying psychotropic medication)

AXIS II V71.02 No diagnosis Axis II

AXIS III Orthopedic problems involving the back

AXIS IV Psychosocial and Environmental Problems: Loss of work assignment, medical condition, financial stresses, marital issues, and relationship with father

AXIS V Current GAF: 60

RECOMMENDATIONS

P.M. is manifesting evidence of a Bipolar Disorder, which includes depression and manic symptoms. Associated with this condition are significant weaknesses in selected areas of cognitive functioning. Based on a diagnosis of Bipolar Disorder, use of disqualifying psychotropic medications (Lithium and Wellbutrin), and significant cognitive weaknesses, P.M. is not fit for duty as a pilot at this time.

I recommend a cognitive and psychological re-evaluation in 12 months in order to document changes in cognitive, affective, and characterologic functioning.

Robert W. Elliott, Ph.D.
Consulting/Examining Psychologist

Chapter 5
Substance Abuse in Aviation: Clinical and Practical Implications

Carlos R. Porges

A captain at a large regional airline was senior enough to consistently bid the same three-day trip, on the same days of the week, to the same cities, for a number of years. On this occasion, a particular mix of irregular operations due to weather and maintenance problems led to his being rescheduled beyond his usual three days and to different layover cities. The first officer and the flight attendant noted and shared their concern about the captain's increasingly erratic behavior as the third day of flying progressed. When deplaning at the end of the day, at a new city, he asked the flight attendant for a few airline-sized whisky bottles for the overnight. The flight attendant replied she could not comply because she had already sealed the liquor cart, as per standard procedure. The captain became upset, belligerent and verbally abusive when she refused, and then proceeded to break into the liquor cart. She reported his behavior to his superiors. Subsequent investigation revealed the captain, who always arranged to stay in the same room at the same hotel, had pre-positioned caches of whisky secreted above the ceiling panels in his room. The opinion of his treating doctors at the rehabilitation facility was that he was beginning to experience withdrawal symptoms as he faced a third day away from his stealthy, and until then reliably available, alcohol supply.

Pilots are not impervious to substance abuse and dependence, and like other people dependent on substances, may go to great lengths to conceal their struggles with alcohol or drugs. But pilots are not like other people, in that they are entrusted with people's lives every time they fly. Piloting stands out from other human endeavors in that it comprehensively engages and taxes all neurocognitive domains and is accompanied by a variety of stressors inherent to aviation. A pilot, regardless of the aircraft or the context in which it is flown, operates a complex machine that freely moves, with no "pause" button, in an infinitely variable 3-D environment colored by weather, mechanical, geographical, procedural, traffic, time, and human factors. Once an aircraft starts to taxi, the pilot's crucial task is to never let the airplane go to a place his/her mind has not been at least ten minutes before. The pilot, simply put, must think faster than the aircraft moves. The pilot who cannot keep mentally apace with events will find him/herself increasingly overwhelmed by a logjam of information—an overload—that will saturate him or her to the point of ineffectiveness.

It follows, then, that substance abuse represents a serious threat to aviation safety because of the negative affects on cognitive, emotional, personal, and interpersonal function. The assessment and treatment of individuals who are abusing or dependent on substances are therefore critical elements of aerospace psychological practice.

Global Substance Use

In 2005, worldwide per capita alcohol consumption equaled 6.13 liters (about 1.62 gallons) consumed by every person aged 15 years old and over (World Health Organization (WHO) 2011). Many millions, however, are abstemious: about half of all men and two-thirds of all women worldwide have not consumed alcohol in the previous year. Abstention rates are highest in countries with high Muslim populations. Though the world's wealthier and more developed nations tend to have the highest consumption levels, they do not necessarily have the highest alcohol-related problems and high-risk drinking rates. These include nations in the Northern Hemisphere, as well as Australia and New Zealand. In the US, more than 60 percent of adults reported past-year use of alcohol and about 6 percent of these were dependent. Though it is less addictive than some other drugs, the popularity of alcohol is quite high: there are about eight million dependent users per year, about five times the number of people dependent on all illicit drugs combined (Grant, Dawson and Moss 2011)

WHO (2011) estimates that harmful use of alcohol leads to the net loss of 2.25 million lives a year and that alcohol consumption is the third largest risk factor for death and disability. Heavy drinkers have a greater risk for hypertension, gastrointestinal bleeding, sleep disturbance, mental health disorders, cerebrovascular accidents, hepatic cirrhosis, and cancer (Rehm et al. 2003). Alcohol is a causal factor in 60 types of disease and injuries and a component cause in 200 others. Worldwide, more deaths are attributed to alcohol than to HIV/AIDS, violence, or tuberculosis.

Patterns of alcohol consumption vary among different countries and regions, as well as by gender and age. Underage drinking was reported to be on the increase in 71 percent of a sample of 73 countries. The same pattern holds true for young adults (18–25 year olds). Among a sample of 80 countries, 80 percent showed an increase in drinking (World Health Organization 2011). Worldwide, four times as many men engage in weekly episodes of binge drinking than women. The same report further notes a gender differential, with more male than female deaths attributable to alcohol (6.2 versus 1.1 percent). In the US, according to a Substance Abuse and Mental Health Services Administration (SAMHSA) 2010 survey, close to a quarter of persons aged 12 and older (about 58 million people) engaged in binge drinking. Binge drinking is defined as having five or more drinks on the same occasion on at least one day in the prior 30. The same survey found that 6.7 percent of that population (16.9 million people) engaged in heavy drinking—

defined as binge drinking—on at least five of the previous 30 days. Alcohol is the leading risk factor for death in males ages 15–59, due in large part to injuries, violence, and cardiovascular disease.

One of every five deaths in the former Soviet Union (that is, the Commonwealth of Independent States; CIS) can be attributed to drinking. Countries with the highest risky drinking patterns include Kazakhstan, the Russian Federation, and the Ukraine, as well as non-CIS countries including South Africa and Mexico. An intermediate ranking is held by most South American nations (excepting Argentina), as well as many African and South-East Asian nations (World Health Organization 2011). Worryingly, Alcohol Dependence often goes unnoticed. In US primary care settings, McGlynn, Asch and Adams (2003) report that only 10 percent of alcohol-dependent patients receive the recommended level of care.

Substance Abuse in Aviation

There is a widely held misconception that based on their presumed socioeconomic status and/or other superficial appearances of high functioning the incidence/prevalence of substance use problems is reduced for aviators. However, the reality is that the prevalence of substance use disorders in aviation is potentially no different than the general population, about 10–12 percent. Among US Navy and Marine Corps pilots, Alcohol Abuse and Dependence are the fifth most commonly occurring disqualifying diagnoses (Bailey, Gilleran and Merchant 1995). In Mexico, Salicrup (personal communication, 25 July 2012) reports an increase in the alcohol referrals of pilots coincident with recently implemented work rule changes allowing longer duty days, with fewer and shorter time off periods.

In the US, recorded incidents of flying while intoxicated or of performing the duties of an air traffic controller (ATC) while intoxicated are rare. Between 1995 and 2002 random alcohol tests of commercial pilots and ATCs revealed 329 out of 511,745 examinations (0.06 percent) to have BAC levels above the legal limit of 0.04 %. The rate was highest in non-federally employed air traffic controllers (0.19 percent) which was considerably more than the 0.03 percent rate found in flight crew (Li et al. 2007). By contrast, post-accident testing covering the same time period showed that 0.13 percent of aviation accidents were attributable to alcohol (Li et al. 2007). Additional prevalence studies using random testing data for the period 1995–2005 revealed an overall .09 percent prevalence rate for positive alcohol tests and a .06 percent rate for detection of illicit drugs (Li et al. 2010). Those pilots found to test positive for drugs were determined to be three times more likely to be involved in a mishap (Li et al. 2011). It was reported that 11 percent of fatal general aviation crashes were associated with alcohol. These pilots were more likely to crash at night and when flying from visual flight rules into instrument meteorological conditions (Li et al. 2005a). Canfield, Dubowski, Chaturvedi and Whinnery (2012) analyzed specimens from 1,353 pilots who died in US aviation accidents between 2004 and 2008. A total of 7 percent of these pilots were found to have alcohol at or above the 0.04% BAC limit. None of these were flying Part 121 (that is, scheduled airline) operations at the time of the crash. Of 208 pilot fatalities involving a pilot with an air transport pilot rating, 6 percent had ethanol in excess of FAA regulation. Of 397 pilots with commercial ratings, 6 percent exceeded ethanol levels. Finally, 8 percent of private pilot fatalities exceeded ethanol levels. It has been reported that for general aviation pilots, with a history of a Driving While Intoxicated (DWI) charges, the risk of crashing an aircraft is 43 percent higher than for a pilot without such a history (Li et al. 2005b).

In addition to the acute cognitive and emotional effects of substances, about 50 percent of patients who meet criteria for Alcohol Dependence evidence neurocognitive impairments, even when abstinent for about a month (Rourke and Grant 2009). Please see below for a comprehensive discussion of the effects of chronic substance abuse on the cognitive abilities required for aviation.

Diagnostic Definitions

There is no single set of definitions or criteria used to describe substance use disorders. Different approaches are offered by scientific, regulatory, and clinical organizations. Definitions and criteria have evolved over time, to create a useful clinical and regulatory framework to describe, organize, and categorize the signs and symptoms of these disorders. The following are definitions of both addiction and substance abuse and substance dependence.

Addiction

The key elements underlying any definition of addiction include an acknowledgment that addiction is a primary, chronic disease, with genetic, psychosocial, and environmental factors that shape its development and manifestations. Behavioral markers include impaired control, over use, compulsive use, craving, and/or continued use despite harm. Recently, the American Society of Addiction Medicine (2011) published a new definition of addiction: addiction is a primary, chronic disease of brain reward, motivation, memory, and related circuitry. Dysfunction in these circuits leads to characteristic biological, psychological, social, and spiritual manifestations. This is characterized by an inability to consistently abstain from use, impairment in behavioral control, craving, diminished recognition of significant problems with one's behaviors and interpersonal relationships and a dysfunctional emotional response. Addiction, like most chronic diseases, commonly involves cyclic relapse and remission. Lack of treatment, or of long-term engagement in recovery activities, will highlight its progressive, disabling, and even lethal nature.

Substance abuse and substance dependence

The Diagnostic and Statistical Manual of Mental Disorders Fourth Edition, Technical Revision (DSM IV-TR; American Psychiatric Association 2000), distinguishes between substance abuse and substance dependence. Abuse is characterized by recurrent problems with work, school or home obligations, dangerous behavior (for example, drunk driving), legal involvement, or social/interpersonal problems caused by continued use of the substance in a 12-month period. Dependence is characterized by physiological dependence, as demonstrated by tolerance or withdrawal, lack of control, inability to quit or cut down, spending significant periods of time using/recovering from use of the substance, and interference with important social or occupational activities due to the substance.

A somewhat different definition of substance abuse and substance dependence is used in aeromedical certification in the US (see Chapter 6, this volume for information related to standards and substance diagnoses within the US military). Regulatory guidance can be found in the Federal Aviation Regulations (FARs), specifically 14 CFR, Part 67, "Medical Standards and Certification." According to these regulations, dependence is manifested by tolerance, withdrawal, lack of control of use, or continued use of the substance despite damage to health, social relationships, personal functioning, or occupation. The regulations indicate that abuse is characterized by two (or more) instances of substance use in a physically hazardous situation or a positive alcohol or drug test result or refusal to submit to a test.

Decisions regarding medical certification in the U.S. are based on Part 67 criteria—not the *DSM.* The aviation psychologist needs to refer to FAA criteria not DSM criteria in providing their diagnoses and recommendations. Under Part 67, Substance Abuse is a disqualifying diagnosis, generally leaving the pilot with two choices:

1. Show no substance abuse, and have no certificate, for two years before applying for a regular medical certificate. This course of action requires the airman to show no further substance abuse during that two-year period and requires re-evaluation by a substance abuse professional when reapplying for the medical certificate; or.
2. Enter a Human Intervention Motivation Study (HIMS)-type abstinence-based treatment program including aftercare and monitoring, leading, if successful, to a Special Issuance medical certificate no earlier than six months later that would allow the airman to return to flying. This is, generally, the recommended approach.

For a pilot meeting FAA criteria for Substance Dependence the diagnosis is medically disqualifying. To become re-certified the airman is required to participate in a HIMS-type program of abstinence-based treatment, aftercare, and monitoring that leads, if successful, to a Special Issuance medical certificate that will include formal monitoring for several years as well as career lifetime abstinence once released from formal monitoring as a condition of continued medical certification.

International Policies

Alcohol policies vary by country. In some countries, funds derived from alcohol sales can be used in unexpected ways; public education systems in some US states and Colombia, for example, receive funding from alcohol sale taxes. In countries such as South Africa the issue of personal alcohol use may be imbued with constitutional issues. The Aeromedical Committee's balance of public safety with individual rights frequently leads to cases undergoing successful legal challenge. The end result in South Africa is that there is considerably more leeway in how far pilots push regulatory limits than in many other parts of the world (Reynolds 2012). At the other end of the spectrum, some countries choose to monitor all pilots very thoroughly and frequently. In Kazakhstan, for example, airline pilots must submit to a medical exam with an airport-based physician prior to the start of every duty period.

Most countries have drug and alcohol regulations similar to the US FARs described above. There are some variations, for example, in New Zealand, where there is no legal eight-hour "bottle to throttle" rule, similar rules are implemented by individual airlines. The only requirement is that pilots not fly under the influence of alcohol. Additionally, there is no threshold BAC and no state-mandated alcohol and drug-testing program.

Alcohol/Drug Regulations and the Pilot

In the US, federal law mandates, among other things, random and post-accident alcohol and drug testing by certain categories of air carriers of individuals involved in safety-sensitive positions. Regulations issued by the FAA contain specific provisions pertaining to pilot use of alcohol and drugs. The alcohol prohibitions are in relation to the performance of safety-sensitive duties. It should be noted that not all duty for which pilots are paid involves safety-sensitive functions. Such prohibitions include (Kalfus 2003):

> *Requirements to report alcohol-related arrests:* Pilots are directed by the FARs to report any motor vehicle action to the FAA's Security Division within 60 days of the date of the action; pilots are also required to report such actions on their next medical application. The definition of a motor vehicle action includes a conviction for operating a vehicle while intoxicated, impaired, or under the influence of alcohol or a drug. The definition is more expansive for what must be reported on the medical application. Very importantly, both definitions include any action in which the pilot's

driver's license was suspended, revoked, or denied for something related to operating a vehicle while intoxicated, impaired, or under the influence. This is relevant because in many states refusing to submit to testing results in automatic suspension of driving privileges. This means that a pilot has to report an incident to the Security Division even if there was no conviction, but where driving privileges were suspended or revoked. The FAA routinely searches the National Driver's Registry (NDR) to monitor pilots' compliance with their reporting obligations. This regulation supports the integration of law enforcement data with aeromedical certification data. This is critical to the identification and treatment of the alcohol dependent pilot before the disease progresses and becomes evident in the flying environment. Many countries, for example, New Zealand, South Africa, Mexico, Kazakhstan and Panama do not currently have a system by which information regarding a pilot's drunk driving arrests become available to the aeromedical certification authority.

Reporting for duty within 8 hours after consuming alcohol. Most pilots understand the eight-hour "bottle to throttle" rule, though most airlines have stricter rules (and some militaries use a 12-hour rule). Many pilots are unaware, however, that merely observing this rule may be insufficient because a pilot who stops drinking eight hours before duty can still be unfit for duty as a result of the amount of alcohol remaining in their blood after they stop drinking. The reason for this is the extremely low level of acceptable blood alcohol content. Pilots are prohibited from performing safety-sensitive duties with a breath alcohol level as low as .02. A reading between .02 and .039 will require the pilot to wait eight hours or retest at .02 or lower before flying.

Reporting for duty or remaining on duty with a breath alcohol concentration of .04 or greater. Violation of this rule will likely lead to the emergency revocation of the pilot's certificate by the FAA.

Using alcohol while on duty or while on call for duty. A pilot on duty is prohibited from consuming alcohol from any source, including medications. A pilot on reserve (that is, on call) is subject to this rule. A deadheading pilot, one traveling as a passenger on aircraft in the passenger cabin on company business, may or may not be considered to be on safety-sensitive duty. Pilots using their "jumpseat" privileges can be considered to be an additional crewmember, even if out of uniform, sitting in the passenger cabin, traveling on an airline different from their own, and on an aircraft in which they are not qualified.

Reporting for duty under the influence of alcohol. This regulation specifically prohibits acting or attempting to act as a crewmember. It is a different regulation from that prohibiting an alcohol level of .04 or greater. This regulation is pertinent because it does not require an alcohol test per se to demonstrate that the pilot is under the influence of alcohol. Evidence of prior consumption (video or credit card records, for example) or other evidence of impairment can be used. Take for example a 48-year-old airline pilot who was going through airport security with the intention of operating his flight, which was scheduled for departure at 0600. An airport security officer, concerned by what he suspected was a faint odor of alcohol, informed airport police and airline operations. Contact was made with the pilot while he was performing his preflight checks. Further investigation revealed credit card transactions and security camera video indicating that the pilot had purchased and consumed drinks at the hotel bar until about 0100 the prior evening.

Refusing to submit to an employer-requested alcohol test under the FAA testing regulations or refusing a lawful request to test directed by a law enforcement officer. The Department of Transportation (DOT) and the FAA mandate airlines and other commercial operators (for example, Part 135 Air Taxi operators) to maintain a program in which pilots and other safety-sensitive personnel are subjected to random, post-accident, reasonable suspicion, return to duty, and follow-up testing. Refusal to submit to the test is a rule violation and will likely lead to FAA certificate action. Similarly, it is a violation for a pilot to refuse a lawful request by a law enforcement officer who is either enforcing a law prohibiting flying or a law prohibiting driving under the influence of alcohol.

Finally, it is worth noting that, consistent with the FAA's definition of Substance Abuse (see above), a pilot with a second instance of using alcohol or drugs in a situation that is physically hazardous, such as operating a vehicle under the influence, is at risk of losing their medical certificate. The event may be considered a second instance by the FAA no matter how much time has elapsed between the incidents. For example, a 52-year-old pilot who has a Driving Under the Influence (DUI) arrest may lose medical certification if the pilot had a prior DUI, even if the prior DUI had occurred 20 years earlier. Also, in the US, the FAA can take emergency revocation action against a pilot's certificates if results on a mandatory DOT/FAA test are reported to be above the prescribed limits or if other alcohol prohibitions have been violated (Kalfus 2003). A pilot whose license has been revoked cannot reapply for a year, and then has to earn all his (flight) certificates and ratings again, though his/her accumulated flight time remains intact. The revocation permanently remains in the pilot's FAA file. This has extremely serious repercussions on a pilot's career and livelihood. It is intended not only to heighten aviation safety, but also to deter substance abuse in the aviation population.

Neuropathology

Cerebral atrophy, to a mild or moderate degree, has been consistently identified in alcoholic patients. As far back as the mid-1950s, Courville (1955) concluded that alcoholism was probably the most common cause of cortical atrophy in patients under the age of 50. Studies over the past several years, bolstered by continuously improving neuroimaging methods, have consistently described cortical atrophy or shrinkage as well as involvement of the dorsolateral aspect of the frontal lobes, with enlargement of the frontal horns of the lateral ventricles. This is pertinent because frontal network functions are exquisitely germane to the ability to conduct safe flight operations. Some of these difficulties can be so clinically subtle as to be undetected by standard psychological measures though they are critical to flight safety. Changes have been noted not only in brain structure but also at cellular and metabolic levels. Altered cerebral blood flow has been demonstrated by neuroimaging. Scans show decreased flow to areas involved in thinking, learning, and memory. Studies have also demonstrated cellular damage and decreased brain function due to reduced brain glucose. A little over half of chronic alcoholics evidence either cortical shrinkage or ventricular dilation or both. Frontal changes can be observed even in fairly young alcoholics, appearing well within the first decade of their history of alcohol abuse. These changes become more marked in older age groups. For older alcoholics the changes are more pronounced, particularly for those with longer drinking histories (Bergman 1980). The above-described neuropathological aspects of alcoholism correlate well with the cognitive deficits found on neuropsychological assessment of alcoholics.

Alcohol-related Neurocognitive Deficits

The impact of acute alcohol consumption on aviation-related abilities has been demonstrated by flight simulator studies. Even at very low BAC levels (0.01–0.03) flight tasks which are impaired are "terrain separation, aircraft descent and angular acceleration." At 0.03-0.05 impaired abilities include "tracking radio signals, managing heavy workload conditions, vectoring airport traffic control, observing and avoiding air traffic, and performing linear acceleration." Higher BACs result in incrementally worsening problems with half of all pilots losing control of the aircraft at a BAC of .12 (Li et al. 2007, 510; see Billings, Wick, Gerke and Chase 1973). So-called hangover effects are also of concern. Positional alcohol nystagmus can be seen up to 34 hours post-consumption (Ryback and Dowd 1970).

Chronic use of alcohol also has significant consequences on neurocognitive functioning, often long after cessation of drinking. Careful consideration of cognitive function in pilots has long been considered important, even in the early stages of identification and treatment. As far back as the mid-1970s, Gregson and Taylor (1977), for example, concluded that an alcoholic's cognitive status on admission to the hospital was the best single predictor of response to treatment. Cognitive areas affected include frontal network/executive functions, speed of information processing, working memory, visuospatial abilities, psychomotor speed, and memory and learning. Generally speaking, language and language-based skills tend to be better preserved than visual/non-language-based skills. For example, word knowledge, word finding, confrontational naming, fluency, prosody, grammar, syntax, and abstract verbal thinking tend to be resilient to the effects of Alcohol Abuse. Similarly, language-based memory and learning tends to be better preserved than visual/non-language-based memory and learning. This is particularly pertinent because a detailed mental status examination and interview, and even some standard psychological tests that rely on overlearned verbal skills may fail to elicit precisely those deficits that are most pertinent to aviation safety.

Aviation-pertinent deficits are most detectable on measures of speed of information processing, visuospatial and psychomotor skills and, most importantly, on frontal network functions. The types of tasks involve problem recognition, abstract reasoning, hypothesis generation, implementing problem-solving strategies, and measures of mental flexibility and perseveration. Tests of frontal network functions known to be sensitive to alcohol dependence include the Halstead Category Test, Wisconsin Card Sorting Test, and Iowa Gambling Test.

Substance Evaluation

The aviation substance use evaluation differs from what most traditional substance use evaluations are likely to include. In broad terms, the key differences are that the evaluator must have expertise in both aviation and the FARs and must produce a more detailed and extensive report than is the norm in the non-aeromedical milieu. In addition the aviation psychologist must avoid taking an advocacy position.

More specifically, the evaluation needs to encompass the components of a comprehensive aviation mental health examination (see Chapter 4, this volume). In addition, the assessment will always include a neuropsychological evaluation, even for young aviators. The instruments included in the FAA's core battery include measures of working memory, speed of information processing, reasoning, hypothesis generation, memory and mental flexibility (see Chapter 10, this volume for more information on aeromedical neuropsychology and a sample substance evaluation). A fundamental, overriding element has to be an assessment of how the airman has integrated—when appropriate—his/her alcoholism into their sense of self. Complete abstinence is necessary but not sufficient—life change is required.

Of course clinical judgment must be used in the assessment. Take for example the case of a 22-year-old newly winged military pilot who went out drinking with his classmates upon completion of two years of intensive training during which almost no alcohol was consumed. That night, after consuming six beers, he drove up to the gate of the Air Force Base and asked (seriously) the military policeman on duty how much the toll was. He was charged with a DUI (.11 BAC), was grounded, and referred for a substance abuse evaluation to determine if he should be allowed to maintain flight status. In this case, a substance abuse evaluation, taking into consideration social features of the military aviation community, his specific circumstances and history, substance-related screening instruments validated for use in military populations (for example, Gates et al. 2007), and the policies related to his particular branch of service, needed to be conducted. Given his age and his lack of a significant drinking history the aviator was not required to undergo neuropsychological assessment. In the end he was found to not meet criteria for any substance use disorder and was sent to attend an alcohol education program. He remained grounded until this was completed.

In contrast, take for example the case of a 32-year-old pilot who was in a motor vehicle crash. A passing motorist, driving at 0300 on a lonely road, called 911 after he observed what appeared to be an unconscious man lying next to his crashed motorcycle. When police arrived the pilot had regained consciousness. He reported to the police that he had gone for a ride around midnight and crashed as he swerved to avoid a raccoon. It was not known how long he had been unconscious at the scene. At the hospital, he was treated for a hairline skull fracture. Later, when reviewing the airman's medical chart the regional FAA flight surgeon noticed that the airman's BAC the morning following the crash had been 0.30. When asked about his drinking the airman reported that he had consumed two beers that evening.

The aviation psychology evaluation to be conducted in such a case will differ considerably from the evaluation conducted in the prior example. Factors to be evaluated will include; the elevated BAC level, the grossly inaccurate self-report of alcohol consumed, the history of the motorcycle crash, the pilot's age, and the possibility of residual brain dysfunction from a traumatic brain injury. In this case, the pilot was diagnosed with Alcohol Dependence and entered treatment. Any BAC level that exceeds 0.20 percent is considered by the FAA to be indicative of tolerance and therefore dependence. The pilot was referred to a 28-day inpatient treatment program, followed by aftercare, attendance at 90 AA meetings in 90 days, participation in biweekly group meetings, weekly individual therapy meetings, monthly meetings with the chief pilot and HIMS Aviation Medical Examiner (AME), as well as weekly drug and alcohol testing. He was then referred for psychiatric and psychological evaluations (the P&P examination; see below). Both evaluators independently assessed the pilot as per detailed FAA protocols, reviewed treatment records, discussed the case with the pilot's HIMS AME and sponsor, and generated (separate) reports for the FAA. In this case both clinicians were of the opinion that the airman's recovery was robust, with good cognition and adjustment and supported a Special Issuance contingent upon continued abstinence, daily AA meetings, weekly therapy, and frequent random testing for alcohol and/or drug use. The airman was granted a Special Issuance medical certificate.

Intervention and Treatment

A 42-year-old 757 first officer called in sick for a trip, stating she fell getting out of her hot tub and sprained her ankle. Three months earlier she reported tripping in her living room and breaking a wrist. A number of sick calls in the previous two years had completely depleted her bank of sick days. Her ex-husband, a pilot at the same airline, had reported to the chief pilot that he had concerns about her drinking and her ability to care for their daughter. An intervention was planned and executed at work, after which she agreed to enter treatment.

As can be seen by this example, and the others interspersed throughout this chapter, there are varied circumstances in which a pilot or other flight crew's substance abuse issues may come to the attention of aeromedical personnel. The most frequent scenarios that uncover substance abuse problems are positive drug/alcohol screens, contact with law enforcement in the form of driving infractions, excessive sick calls and unexplained injuries, and family and co-worker concerns.

Within the aviation population, early intervention is critical. In the absence of a sophisticated network of monitoring, treatment, and support, the first indication of an active substance abuse problem may come (too) late in the disease process. This outcome is most undesirable because it usually underlies an advanced disease process.

Early intervention requires early information. One rich source of contextual early warning information comes from law enforcement records such as the NDR. Any history of an arrest (not just a conviction) for infractions such as drunk and disorderly conduct, domestic violence and, especially, DWI or DUI merits a close examination of the airman's substance abuse status. Another source of information is found in the pilot's general medical records. The provision of complete medical records describing, for example, a seemingly unrelated medical condition or procedure may reveal evidence of an ongoing substance abuse issue. To achieve the goal of detecting people early, three elements are necessary. These include close cooperation between disparate government agencies, careful management of privacy and informed consent issues, and a cooperative relationship between the aeronautical authority, the pilot's unions and interest groups (Aircraft Owners and Pilots Association, for example), and airline management. This cooperation includes the commitment to help the pilot. In other words, if pilots trust that compliance and successful rehabilitation will provide a path back to flight status they will be more likely to ask for help, or encourage a fellow pilot to obtain treatment. Indeed, one way to gauge commitment to sobriety is to discuss the concept that in order to fly he/she cannot drink again and ask their perceptions of this issue. Pilots motivated to fly and attain sobriety can discuss this confidently, sincerely, and in clear terms regarding sobriety. Pilots, who have not committed to sobriety, are not convinced they have a substance problem, or who are not motivated to fly, often answer in an unconfident, ambiguous way, signaling potential problems with compliance and poor prognosis.

While not a formal treatment modality, Birds of a Feather can offer significant assistance to substance-dependent flight personnel. This group, formed in the 1970s by pilots and airmen, is based on Alcoholics Anonymous (AA) principles and is restricted to flight personnel. The group meets in a variety of countries in North and South America, Europe, and the Middle East.

The Human Intervention and Motivation Study (HIMS) began in the US in 1974 as a rehabilitation program for pilots. The model has been applied in other countries, with well-established programs in New Zealand and The Netherlands and a new program in Mexico. This program embodies a triad of cooperation between the FAA, airline management, and the pilot's union. The goal is to provide each airman the best opportunity for a successful recovery experience. This has been an extraordinarily successful program, with about 4,000 pilots treated and relapse rates as low as 10 percent (Steenbilk 2007). The true HIMS success, however, is philosophical: instead of focusing on catching and punishing a few pilots, HIMS proactively identifies and treats many more. Coming to treatment without fear of retribution or job loss makes it less likely the disease will progress undetected to the point of showing up for duty under the influence of alcohol. The keys to success are commitment and trust, the provision of a structured program run by aviation-sophisticated experts and, above all, extensive monitoring.

If a substance abuse problem is suspected, the pilot is referred for an evaluation with a specialized clinician or at an alcohol treatment facility. A confirmed diagnosis will result in the pilot being grounded along with an offer of participation in the HIMS program, which typically begins with inpatient treatment. The inpatient treatment program involves at least a 28-day rehabilitation program, the successful completion of which leads to aftercare. This, in turn, involves at least 90 AA meetings in 90 days, group and individual therapy, developing and maintaining a constructive relationship with an AA sponsor and continuing the process of working the steps outlined by AA. Additionally, the pilot has to attend meetings with the designated chief pilot from his airline, who monitors his recovery. This can also include, separately or in combination, meeting with the airline's Employee Assistance Program (EAP), as well as union representatives. Complete, permanent and verifiable abstinence is a necessary requirement for the Special Issuance of pilots diagnosed with Alcohol Dependence. No earlier than five or six months from the beginning of rehabilitation, and in accord with the HIMS AME working the case, the pilot is referred for the P&P examination. These are two independent evaluations, conducted separately by a board certified psychiatrist and a board certified (or board eligible) clinical neuropsychologist, both of whom have completed HIMS training in substance abuse and who are familiar with the cognitive and personality demands of aviation, and with FAA regulations pertaining to substance abuse and dependence. The report of the P&P evaluators are submitted to the pilot's HIMS Aviation Medical Examiner (AME) who submits the package to the FAA for review and possible authorization of a Special Issuance medical certificate. The particular limitations and provisions of the Special Issuance will vary depending on the particulars of the case but invariably is conditioned upon permanent and complete abstinence. The pilot will be followed up frequently by the HIMS AME and will usually follow up with the psychiatrist on an annual basis. Follow-up neuropsychological services may also be considered.

Threats to Human Intervention and Motivation Study success

While HIMS has been successful, largely due to the evaluation and monitoring by aviation and addiction professionals, there are challenges inherent to the program. The personality traits described above will often result in pilots focusing on the behavioral aspects of recovery (that is, not drinking) without a corresponding focus on the psychological and spiritual aspects of recovery. Key indicators a clinician should consider when deciding whether a pilot needs continued rehabilitation are a sense of anger and victimization, as well as minimization of drinking or of the consequences of drinking. In brief: resentment is the kryptonite of recovery.

A second threat to success is the lack of a structured program at the pilot's airline. Some airlines do not have a HIMS program, leaving pilots to provide their own structure. These situations require considerable support from the HIMS AME and the counselors working with the patient. Other airlines may have a HIMS program in place but not the corporate culture or funding required.

A third threat to success involves the use of rehabilitation institutions or clinicians that cannot provide necessary care or produce detailed, individualized clinical records. The P&P clinicians and the HIMS AME, as well as the reviewing FAA physicians need individualized, personalized, and pertinent records outlining in detail the pilot's progress through rehabilitation.

The successful experience of ASPA (Asociación Sindical de Pilotos Aviadores), the Mexican pilots' union, in developing a HIMS program in that country highlights many of the challenges faced by nascent HIMS programs. Under the aegis of Dr. Carlos Salicrup Díaz de León, an aerospace physician and airline pilot, the program became active in 2010. According to Dr. Salicrup (personal communication, 25 July 2012) challenges include a resistance to the disease model of alcoholism and to shedding the culture of blame, punishment, and shame. As of 2012, Aeroméxico is the only Mexican carrier accepting the medical model of alcoholism and offering confidential treatment to pilots. The company's policy is to support pilots who self-disclose but to have a zero tolerance approach to pilots who are found to violate alcohol or drug regulations. The pilot group was described as quite distrustful of the process in its initial stages.

Another challenge to implementing a HIMS-type program is to secure cooperation from governmental agencies and the airlines. ASPA was successful in having the IMSS (government medical service) formally recognize addiction as a medical disease, allowing pilots in treatment to be considered disabled and thus receive their full salary for four months. This is followed by vacation credit and when this is exhausted pilots are placed on long-term unpaid leave. Their jobs are preserved provided they comply with treatment, follow-up, and monitoring. ASPA has successfully negotiated an agreement with both the aeronautical authority and the airlines to conduct random alcohol and drug screens only before flights. In this manner, a positive result is treated as a work-related issue and not a federal crime. This does not eliminate the aeronautical agency or law enforcement's right to conduct a screen at any time if there is reasonable suspicion.

ASPA has worked to develop a treatment and support system that is both competent and available. They have followed the HIMS model in that pilots are referred for up to five weeks of inpatient rehabilitation, followed by aftercare in the form of daily AA meetings, weekly individual therapy, and monthly medical follow-ups. ASPA has successfully started a Birds of a Feather chapter, as well as a peer pilot support network. The Mexican HIMS program experience is very new and does not have enough data regarding success and relapse rates. Nonetheless, it does highlight the universal nature of the challenges and threats to the development of a new HIMS program.

Conclusion

The modern pilot must not just properly manipulate controls but has to judiciously manage a complex system comprised of the aircraft itself, crewmembers, air traffic control, and a host of other elements. An airplane moves fast; its pilot must think faster. Information from multiple sources must be processed efficiently in order to generate and implement a problem-solving strategy that has to be at once rigid and appropriately flexible. Personality and affective variables provide the framework within which information-processing and problem-solving take place: the pilot has to have integrity, conscientiousness, emotional stability, and social skills. Both cognitive and personality variables are critical to Crew Resource Management (CRM), the cornerstone of safety in airline operations.

Alcohol and drug abuse represent a threat to aviation safety due to the detrimental effect on cognitive, affective and interpersonal functioning. In terms of cognition, frontal network functions and non-verbal/spatial skills tend to be at risk. A decline in aviation-pertinent cognitive skills may not be initially detected in line flying or recurrent simulator checks, because of their highly regimented, overlearned, repetitive nature. Additionally, relatively spared language-based skills may allow the pilot to correctly verbalize complex procedures, make seemingly insightful statements, and give a favorable impression that may lead an observer to conclude that no deficits exist. However, in reality, with respect to affect and personality function, the addicted pilot has lost control over his/her compulsive use of a substance, which persists despite negative consequences. Affective stability and interpersonal relationships are fundamentally affected. To assess these consequences of a substance use disorder a specialized comprehensive aviation–neuropsychological evaluation is required.

Pilots, as a group, are uniquely successful in their substance recoveries. The extraordinary track record of the HIMS program is a testament to this. The particular neuropsychological traits pilots possess are the very vehicle by which they are so uniquely successful at recovery. They are intelligent, proud, competitive, and hard-working individuals with a mission-accomplishment mentality, who derive great personal pride and satisfaction from a job well done.

References

American Psychiatric Association. 2000. *Diagnostic and Statistical Manual of Mental Disorders*. 4th Edition, Text Revision (DSM-IV-TR). Washington DC: American Psychiatric Association.

American Society of Addiction Medicine. 2011. Public Policy Statement: Definition of Addiction. Available at: http://www.asam.org/docs/publicy-policy-statements/1definition_of_addiction_long_4-11.pdf?sfvrsn=2 [accessed: 25 August 2012].

Bailey, D.A., Gelleran, L.G. and Merchant, P.G. 1995. Waivers for disqualifying medical conditions in U.S. naval aviation personnel. *Aviation, Space, and Environmental Medicine*, 66 (5), 401-407.

Bergman, H., Borg, S., Hindmarsh, T., Ideström, C-M and Mützell, S. 1980. Compound Tomography of the brain and neuropsychological assessment of male alcoholic patients and a random sample from the general male population. *Acta Psychiatrica Scandinavica*, 62, Suppl. 286, 77-88.

Billings, C.E., Wick, R.I., Gerke, R.J. and Chase, R.C. 1973. Effects of ethyl alcohol on pilot performance. *Aerospace Medicine*, 44, 379–382.

Canfield, D.V., Dubowski, K.M., Chaturvidi, A.K. and Whinnery, J.E. 2012. Drugs and alcohol found in civil aviation accident pilot fatalities from 2004–2008. *Aviation, Space and Environmental Medicine*, 83(8), 764–770.

Courville, C.B. 1955. *The Effects of Alcohol on the Nervous System of Man*. Los Angeles, CA: San Lucas Press.

Gates, T., Duffy, K., Moore, J., Howell, W. and McDonald, W. 2007. Alcohol screening instruments and psychiatric evaluation outcomes in military aviation personnel. *Aviation, Space, and Environmental Medicine*, 78(1), 48–51.

Gregson, R.A.M. and Taylor, G.M. 1977. Prediction of relapse in men alcoholics. *Journal of Studies on Alcohol*, 38, 1749-1760.

Grant, B.F., Dawson, D.A. and Moss, H.B. 2011. Disaggregating the burden of substance dependence in the United States. *Alcoholism: Clinical and Experimental Research*, 35(3), 387–388.

Kalfus, S. 2003. Alcohol prohibitions for pilots and flight instructors, *Air Line Pilot Magazine*, April, 72(4) 26-29.

Li, G., Baker, S.P., Lamb, M.W., Qiang, Y. and McCarthy, M.L. 2005a. Characteristics of alcohol-related fatal general aviation crashes. *Accident Analysis and Prevention*, 37, 143–148.

Li, G., Baker, S.P., Qiang, Y., Grabowski, J.G. and McCarthy, M.L. 2005b. Driving-while-intoxicated history as a risk marker for general aviation pilots. *Accident Analysis and Prevention*, 37, 179–184.

Li, G., Baker, S.P., Qiang, Y., Rebok, G.W. and McCarthy, M.L. 2007. Alcohol violations and aviation accidents: Findings from the US mandatory alcohol testing program. *Aviation, Space, and Environmental Medicine*, 78(5), 510–513.

Li, G., Baker, S.P., Zhao, Q., Brady, J.E., Lang, B.H., Rebok, G.W. and DiMaggio, C. 2011. Drug violations and aviation accidents: Findings from the US mandatory drug testing programs. *Addiction*, 106, 1287–1292.

Li., G., Brady, J.E., DiMaggio, C., Baker, S.P. and Rebok, G.W. 2010. Validity of suspected alcohol and drug violations in aviation employees. *Addiction*, 105, 1771–1775.

McGlynn, E., Asch S., Adams J, et al. 2003. The quality of health care delivered to adults in the United States. *N Engl J Med* 348(26): 2635-45.

Rehm, J., Room, R, Graham, K., Monteiro, M., Gmel, G. & Sempos, C.T. 2003. The Relationship of Average Volume of Alcohol Consumption and Patterns of Drinking to Burden of Disease: An Overview. *Addiction*, 98(10), 1209–28

Reynolds, T. 2012. When Office-Based Assessment and Simulators Collide: A Case Study. Presentation to the Biennial Conference of the South African Aerospace Medicine Association, Cape Town, South Africa, 27–28 July 2012.

Rourke, S.B. and Grant, I. 2009. The neurobehavioral correlates of alcoholism, in *Neuropsychological Assessment of Neuropsychiatric and Neuromedical Disorders*. 3rd Edition, edited by I. Grant and K. Adams. New York: Oxford University Press, 398–454.

Ryback, R.S. and Dowd, P.J. 1970. Aftereffects of various alcoholic beverages on positional nystagmus and coriolis acceleration. *Aerospace Medicine*, 41, 429–435.

Steenbilk, J.W. 2007. HIMS: The quiet success story. *Air Line Pilot*, November–December, 25–29.

Substance Abuse and Mental Health Services Administration. 2011. *Results from the 2010 National Survey on Drug Use and Health: Summary of National Findings*, NSDUH Series H-41, HHS Publication No. (SMA) 11-4658. Rockville, MD: Substance Abuse and Mental Health Services Administration.

World Health Organization. 2011. *Global Status Report on Alcohol and Health.* Available at: http://www.who.int/substance_abuse/publications/global_alcohol_report/en/. [accessed 25 August 2012].

Chapter 6
US Military Standards and Aeromedical Waivers for Psychiatric Conditions and Treatments

Arlene R. Saitzyk, Christopher A. Alfonzo, Timothy P. Greydanus, John R. Reaume and Brian B. Parsa

Aeromedical physical standards are developed to enhance safety of flight and successful mission completion, and to ensure the most qualified personnel are accepted and retained for military aviation duties. There is a higher standard for military aviation, that is, while some individuals may be Physically Qualified (PQ) to enlist or commission in the military for general duty, they may not be qualified for some special duties or assignments, such as aviation. In order to meet physical standards for general duty, prospective service members must be physically capable of performing duties without unnecessary risk of injury or harm to themselves or others, without assignment limitations or modifications to existing equipment or systems, and not likely to incur physical disabilities as a result of military service (Bureau of Medicine and Surgery (BUMED) 2005). For some conditions, the simple presence of a problem is cause for disqualification (for example, hearing loss), whereas for other conditions (for example, headaches) the impact on the health or functionality of the service member is more critical in evaluating and determining whether the service member meets the standard and is PQ. Mental health is an important component of physical qualification.

With respect to aviation psychiatric standards, service members are evaluated to determine whether they are both PQ and aeronautically/aeromedically adaptable to engage in duties involving flight. These two terms will be elaborated later in this chapter, but briefly, they generally correspond with the *Diagnostic and Statistical Manual of Mental Disorders, Fourth Edition–Text Revision (DSM-IV-TR)* Axis I and Axis II disorders (American Psychiatric Association 2000). If no disorder/maladaptive personality characteristics are found, then service members are considered to have met the standard.

When service members do have a mental health history or current diagnosis, the military may consider waiving certain conditions. The waiver process was developed to ensure the consistent and proper management of disqualified individuals, with the ultimate goal of having an effective and ready fighting force. A waiver does not make individuals PQ, but rather provides an opportunity to enlist, commission, or maintain a special duty despite the fact that a disqualifying condition existed/exists.

With regard to aviation personnel in particular, service members who have earned their wings or who perform other duties involving flying are high-value assets with millions of dollars invested by taxpayers. Thus, the waiver system allows the military to continue to utilize the unique abilities of trained aviation personnel who may not meet established physical standards but who are able to safely complete their demanding missions. Initial aviation applicants may also be considered for waiver, though typically more rigorous standards are applied to applicants. This optimizes the chances that the Department of Defense (DoD) is investing in candidates who are capable of completing training, and remaining combat-ready in the physically demanding environment of aviation. Losing trained aviation personnel through medical attrition adversely affects manning requirements, wastes fiscal resources, and impacts personal and career development for service members.

Our goal in this chapter is to outline the general requirements for mental health standards for military aviation, highlight the specific guidelines for the most common mental health disorders, and discuss the processes for submitting waivers in accord with each of the military service's guidelines. It should be noted up front that the focus of this chapter in essence makes it a living document as each service's waiver guides are ever-evolving with the advent and incorporation of new data and practices in both the medical and aviation communities as well as advances in military aviation technology. We conclude with points of contact for each of the services for the most current guidance.

General Standards

As mentioned above, there are general physical standards for enlistment and commissioning in the military. These are published and maintained in the US Air Force (USAF) Instruction 48-123 (2009), Manual of the Medical Department (MANMED)—Chapter 15 (BUMED 2005) for the US Navy, Marine Corps, and Coast Guard (which fall under Navy regulations for flight status), and US Army (USA) Regulation 40-501 (2007), and the reader is directed to these references. In addition, all aviation applicants must meet aviation standards. If an applicant does not meet these standards, the applicant is considered to have a disqualifying condition and a waiver of standards, if appropriate, is required for flight status. Designated aviation personnel must remain fit for full general duty and continue to meet the aviation standards published in the above noted regulations for their respective service.

Any medical condition, disqualifying diagnosis, or chronic medication use requires a waiver of standards. Waivers may be granted based upon the needs of the service, consistent with training, experience, performance, and proven safety of the aviation personnel. Once waivers are granted, they will often have unique stipulations that must be maintained in order to monitor the service member and ensure that the waiver remains appropriate for the condition. Due to the military's challenging operational environments and possible limited access to treatment resources, exceptions to certain waiver requirements may be considered by the appropriate waiver process authorities in each service, on a case-by-case basis. Air Force waivers are often time limited and will frequently require specific evaluations be performed before expiration. Navy/Marine Corps/Coast Guard and Army waivers often contain stipulations for a continuance to be issued by the Naval Aerospace Medical Institute (NAMI) Aeromedical Physical Qualifications Department or the US Army Aeromedical Activity (USAAMA), respectively. Each service maintains a waiver guide. For the Air Force it is the Air Force Waiver Guide (US Air Force 2012), for the Navy/Marine Corps/Coast Guard it is the Aeromedical Reference and Waiver Guide (ARWG; Naval Aerospace Medical Institute 2010), and in the Army, the US Army Aeromedical Policy Letters (US Army 2008) provide guidance for the appropriate disposition of soldiers applying for waivers.

Just as stricter standards and waiver criteria may be applied to applicants versus designated aviation personnel, requirements vary by branch of military service, and by Flying Class (FC). Table 6.1 overleaf provides a comparison of the different FCs between different career fields, levels of training, and respective services.

Standards, Requirements, and Nomenclature

Diagnoses on both Axis I and Axis II from the *DSM-IV-TR* have implications for flight status. For psychiatric issues in the Air Force, aviation personnel are considered either medically qualified or not medically qualified with respect to flying duties when diagnosed with Axis I disorders. Axis II diagnoses, however, are administratively managed, and the service member can be found unsatisfactory under the Adaptability Rating for Military Aviation (ARMA). In the Navy/Marine Corps, aviation applicants are considered either PQ or Not Physically Qualified (NPQ) for any Axis I diagnosis. For Axis II considerations, applicants are either Aeronautically Adaptable (AA) or Not Aeronautically Adaptable (NAA). Applicants are considered AA when they demonstrate the potential to adapt to the rigors of aviation by possessing stable mental health and behavior to allow full attention to flight, adaptive coping skills, leadership qualities, and absence of impulsivity. Designated aviation personnel are either Aeronautically Adapted (AA) or Not Aeronautically Adapted (NAA), based on demonstrated performance, ability to tolerate operational stress, and long-term use of adaptive coping (BUMED 2005). In the Army, psychiatric conditions are dispositioned as either qualified or disqualified for aviation duties. A history of maladaptive personality traits or current personality disorder may be cause for an unsatisfactory Aeromedical Adaptability (AA) rating.

Table 6.1 Flying Classes, Levels of Training and Military Services

Service Member Class	US Air Force	US Navy, Marines, and Coast Guard^	US Army
Pilot Trainee	Flying Class I (FC-I)	Student Naval Aviator (SNA)	Class 1A; Class 1W##
Navigator/CSO* trainee	Flying Class IA (FC-IA)	Student Naval Flying Officer (SNFO)	N/A
Trained Pilot, Flight Surgeon (Air Force, Army)	Flying Class II (FC-II)	Class I Service Group (SG) I, II, III **	Class 2
Trained Navigator/ CSO, Aeromedical Officer (US Navy)	FC-II	Class 2	N/A
Officer/enlisted aircrew career fields not in direct control of aircraft (for example, loadmaster, flight nurse)	Flying Class III (FC-III)	Class 2	Class 3
Unmanned Aircraft Systems Operator (UAS)***	FC II-U	Class 3	Class 3
Ground Based Controller (GBC)#	GBC	Class 3	Class 4
Space and Missile Operations Duty (SMOD)	SMOD	N/A	N/A

*CSO—Combat System Operator
** Service Groups (SG) SG-I—Unrestricted pilot
SG-II—Pilot restricted from carrier take-offs and landings
SG-III—Pilot restricted to dual control aircraft
*** Large (Group 4 or 5) aircraft only IAW Joint Concept of Operations for UAS
# Air Traffic controllers, weapons controllers/directors, combat controllers, Aerospace Control and Warning Systems, Tactical Air Control Party, Air Liaison officer, and UAS Sensor Operator
US Army Class 1A applies to officer pilots. Class 1W—Warrant Officer pilots. Terms now being consolidated into single standard
^ The Marines and Coast Guard are included with the US Navy because both utilize the Navy nomenclature

All services will consider waivers if the following general criteria are met: the disorder does not pose a risk of sudden incapacitation; there is minimal potential for subtle performance decrement, particularly with regard to the higher senses; the disorder is resolved, or stable, and is expected to remain so under the stresses of the aviation environment; if the possibility of progression or recurrence exists, the first symptoms or signs are easily detectable and do not pose a risk to the individual or the safety of others; treatment does not require unconventional, non-standard, or not easily accessible tests, non-routine medications, regular invasive procedures, or frequent absences to preclude completion of training or military service to monitor for stability or progression; finally, the condition must be compatible with the performance of sustained flying operations, and not put mission completion at risk (US Air Force 2012; BUMED 2005; US Army 2008).

Specific Guidance for Psychiatric Conditions and Treatments

This section will cover the most commonly encountered mental health disorders seen in the military aviation community, the associated impact upon aviation duties, and the processes for requesting a waiver as applicable for each service, when appropriate. It is important to note that severe and chronic mental health conditions (for example, Schizophrenia, Bipolar Disorder) are never considered for waivers by any branch of service. In general, all waiver requests require an Aeromedical Summary (AMS) which is completed and submitted by the flight surgeon. An AMS is a detailed summary of the service member's condition and describes how it relates to his/her current flying duty. The AMS includes a detailed history of present illness (HPI), directed physical exam, and results of all pertinent ancillary studies. An aeromedically tailored psychological evaluation is often the key resource for the flight surgeon in constructing a good AMS as well as in the ultimate determination of flight status. The reader is directed to Chapter 4, this volume regarding conducting the aviation mental health evaluation. The AMS should provide enough detail so that the reviewer can make an appropriate aeromedical decision based on this document.

It is also worth noting here that any licensed provider may ground a service member from flight status, though it is the flight surgeon, through consultation with pertinent specialists and review authority, who has the sole mandate to put someone "up." The flight surgeon should consult with a mental health provider early on in the process in accordance with DoD Directive 6490.1 (1997a) Mental Health Evaluations of Members of the Armed Forces and DoD Instruction 6490.4 (1997b) Requirements for Mental Health Evaluations of Members of the Armed Forces. These guidelines establish procedures for service members referred for mental health evaluations, and help protect the rights of service members referred.

Adjustment disorders

Adjustment disorders involve a response to an identifiable stressor and result in significant emotional and behavioral symptoms, such as depressed mood, anxiety, fatigue, changes in social relationships, problems with concentration, attention, decision-making, and at least temporary functional impairment. While these disorders are among the most common mental health conditions for aviators (often referred to as the common cold of mental health), symptoms are incompatible with aviation duties. Therefore, if diagnostic criteria for an adjustment disorder are met, service members are immediately removed from duties involving flying until the disturbance is resolved. Most do experience full recovery, and once resolved may return to aviation duties. If the symptoms do not resolve, or if the condition worsens, then the service member may be considered for longer-term or permanent disqualification.

All services allow for temporary disqualification during the symptomatic period, followed by return to flying duties without need for a waiver when problems resolve relatively quickly. In the Air Force, if the disorder resolves within 60 days, the service member is placed back on flying status and no waiver is required. However, if the disorder persists beyond 60 days they are considered disqualified and a waiver is required upon resolution. With the exception of a one-year period after resolution for FC I/IA applicants (that is, pilot/navigator trainees) and for other untrained applicants for all FCs, there is no mandated recovery period before waiver application. The period of remission for trained personnel should be of such length that the flight surgeon and aviation mental health consultant believe they will not suffer a significant recurrence. Evaluation by a mental health professional is necessary prior to waiver consideration.

In the Army, fitness for flight status is determined by the severity of the adjustment disorder and the required treatment. A mild adjustment disorder with complete recovery within 90 days can be considered "information only," and no waiver is required. However, if the disorder persists beyond 90 days, the service member is disqualified and a waiver is required upon resolution. Complete recovery without chronicity or medication supports waiver consideration. Mental health evaluation is required prior to waiver consideration. In addition, psychotherapy cannot occur while a service member is in an "up" status if the therapy is for symptom treatment. However, follow-up or "booster session" psychotherapy may occur while the service member is on flight status when it is for mild symptoms or for stress management after the service member no longer meets criteria for the diagnosis.

An AMS is required in a waiver package for the Air Force and Army, and should only be submitted after all appropriate treatments have been implemented. The AMS should include all clinical diagnoses requiring a waiver, symptom frequency and duration, treatment, precipitating factors, action taken to prevent recurrence, social, occupational, administrative, or legal problems, copies of psychiatric evaluation and treatment summary, current functioning, and letters from the service member's squadron commander or operations officer and treating mental health providers supporting or refuting a return to flying status. The Air Force also requires an AMS for any necessary waiver renewals, which should include interval history, any changes in the service member's condition and copies of any applicable evaluations. It should be noted that certain psychiatric disorders, including adjustment disorders with unsatisfactory duty performance, may render an individual unsuitable for duty, and are subject to administrative separation.

Under Navy regulations, a service member with an adjustment disorder is temporarily considered NPQ for aviation until symptoms are resolved, and similar to the Army, psychotherapy during the symptomatic period is not compatible with aviation duties. Adjustment disorders diagnosed by mental health personnel are not considered resolved until a mental health provider makes that statement in the service member's health record. Once fully resolved, the individual is PQ, the condition is Not Considered Disqualifying (NCD), and no waiver is required. Upon return to an "up" status, the flight surgeon must submit a brief summary of pertinent symptoms and treatment, and any mental health records or Medical Evaluation Board (MEB) reports (if applicable). Follow-up is at the discretion of the mental health provider.

Alcohol use disorders

Alcohol use causes acute and chronic effects that are deleterious to cognitive and physical performance. Acute alcohol intoxication can compromise G-tolerance by 0.1-0.4 G (Ernsting and King 1988), visual acuity, and coordination, as well as cause oculovestibular dysfunction or positional alcohol nystagmus (Modell and Mountz 1990). Even low doses of alcohol ingestion which may not cause overt intoxication symptoms can still negatively affect reaction time, attentiveness, and the ability to rapidly assimilate multiple sensory organ input. Alcohol consumption can cause impairments that persist well after the blood alcohol level has returned to zero, including the risk of cardiac dysrrhythmias up to one week following last use. Alcohol use is correlated with a higher accident rate in both ground and flight operations, and alcohol intoxication has been implicated in roughly 16 percent of fatal aviation mishaps (Modell and Mountz 1990). For more information on substance disorders and aviation see Chapter 5, this volume.

The waiver requirements for all DoD military services are largely similar, but there are some distinct service-specific requirements and nomenclature that must be known and understood. A diagnosis of either Alcohol Abuse or Dependence is considered permanently disqualifying for all aeronautically designated personnel, students or applicants in the Air Force, Army, and Navy. Please note that no military service considers waivers for illicit drug problems, consistent with DoD's zero tolerance policy. For service members to be considered for waivers, they must meet several conditions: successful completion of the appropriate level of treatment; abstinence (which must continue for the remainder of their flying career) without the need for amethystic agents; maintained a positive attitude and an unqualified acknowledgment of the alcohol disorder; and, complied with all aftercare requirements. Service members must wait at least 90 days after completion of treatment before an initial waiver request is submitted.

The Army requires the service member complete "the appropriate treatment program" for their particular alcohol disorder, while the Air Force specifically states its requirement for completion of the indicated level of treatment per the American Society of Addiction Medicine (ASAM) patient placement criteria. Navy regulations also require treatment completion in accordance with ASAM patient placement criteria but clearly stipulate a minimum of Level 1 (Outpatient) treatment for a diagnosis of Alcohol Abuse, and Level 2 (Intensive Outpatient) treatment for Alcohol Dependence.

For all military services, the member must have documented participation in an organized alcohol recovery program. The Navy urges the program be Alcoholics Anonymous (AA), but the Army and Air Force allow for participation in other lesser-established programs such as Rational Recovery (RR) and Secular Organizations for Sobriety (SOS) as approved or allowed by their respective substance abuse prevention and treatment programs. The Air Force and Navy require recovery program attendance at least three times a week for the first year of aftercare, and once weekly for the second and third years of aftercare. The two services do not require, but strongly recommend, continued participation following the third year of aftercare. The Army requires participation at least three to five times a week for the first 90 days of recovery and then one to three times per week thereafter for a total of five years in aftercare. Participation must be in person vice other resources (for example, computer online AA meetings).

The service member must also meet with designated professionals per established timelines during their period of aftercare. The Air Force and Navy have similar aftercare requirements in that the service member must meet with the flight surgeon monthly for the first year of aftercare, quarterly for the second and third years, and at least annually thereafter. The Army requires essentially the same timeline but requires the service member meet with the flight surgeon at least every six months for the third through fifth aftercare years, and then annually thereafter. The flight surgeon meetings should be documented, include appropriate physical examination and laboratory studies if indicated (for example, blood alcohol level, cell blood count, liver function test, carbohydrate deficient transferrin, and so on). The flight surgeon should also document pertinent comments that address the service member's abstinence, level of functioning, performance, stressors, attitude toward recovery, aftercare participation, and a mental status examination.

The Air Force and Navy require the service member meet with certain substance abuse prevention and treatment teams. The Air Force program is known as Alcohol and Drug Abuse Prevention and Treatment (ADAPT). The Navy utilizes Drug and Alcohol Prevention Advisors (DAPAs), and the Marine Corps has Substance Abuse Control Officers (SACOs). The Coast Guard uses Collateral Duty Alcohol Representatives (CDARs) to initiate alcohol screening, and to provide contact while the service member awaits treatment. As appropriate to their respective branch of service or depending on inter-service operations, service members must have documented meetings with a DAPA, SACO, or ADAPT staff monthly for the first three years of aftercare. The Army requires that service members meet with the local Army Substance Abuse Program (ASAP) clinical director monthly for the first year, on a quarterly basis for the second year, and annually for the third through fifth years of aftercare. Staff assist service members in meeting aftercare requirements, maintaining abstinence, and collecting and maintaining alcohol recovery program attendance documentation. The Air Force and Navy further require that the service member meet with a privileged licensed independent practitioner (for example, clinical psychologist, psychiatrist, or licensed clinical social worker) annually for the first three years of aftercare. The purpose of these annual mental health evaluations is to evaluate for continued remission of the service member's alcohol disorder and potential for sustained recovery, as well as discern any other comorbid psychopathology or underlying concerns.

If the service member meets the requirements for a waiver request and it has been at least 90 days since the completion of treatment, he/she may apply for an aeromedical waiver of the alcohol disorder. The waiver request package must include the following information and documents: a complete flight physical, including a mental status examination; an AMS by the flight surgeon; a copy of the alcohol treatment program summary; a statement from the ADAPT/ASAP/DAPA/SACO documenting aftercare and alcohol recovery program attendance; copy of the mental health evaluation, preferably performed close to the 90-day mark after completion of the treatment program (this is not required by the Army); a copy of an internal medicine evaluation (if indicated); a letter of recommendation from the service member's commanding officer endorsing support for a waiver; and a statement letter by the service member for request of a waiver that provides evidence of their knowledge of aftercare requirements and compliance of such. Navy regulations specifically direct the following statement to be included in the member's letter, per BUMED instruction series 5300.8 (1992):

> I have read and received a copy of BUMEDINST 5300.8 series. I understand that I must remain abstinent. I must meet with my Flight Surgeon monthly for the first year, then quarterly for the next two years of aftercare. I must meet with the DAPA monthly and receive an annual mental health evaluation for the first three years of aftercare. And I must document required attendance at alcoholics anonymous (AA).

For annual waiver continuation, the Air Force requires that the service member's flight surgeon submit the following: an AMS; copies of consultation notes from any evaluating providers; and a copy of a signed abstinence letter by the service member. For all five years of the Army's aftercare program requirements, the Army requires annual submission of the flight surgeon's recommendations, ASAP counselor's recommendations, documentation of alcohol recovery program attendance, and a letter of support from the service member's commanding officer.

For the first three years of aftercare, the Navy requires annual submission of the following: a complete long-form flight physical examination; flight surgeon's statement addressing the service member's safety of flight, performance of duties, potential for sustained recovery, any symptoms of comorbid diseases, and compliance with aftercare requirements; DAPA statement documenting AA attendance; and a copy of the annual mental health evaluation. After three years of aftercare, the Navy requires annual submission of only a short-form flight physical examination and a flight surgeon's statement addressing the pertinent information noted above.

After being granted an initial waiver for an alcohol disorder, service members can be declared an aftercare failure if they fail to abstain, do not fully comply with all aftercare requirements, or are in denial of the disorder. In these instances the service member is grounded immediately, and re-evaluated by the flight surgeon and alcohol treatment facility to determine potential for further treatment. Under Navy regulations, the service member's command must also formally recommend revocation of the current waiver in accordance with the BUMEDINST 5300.8 series. If further treatment is appropriate, the processes for an initial waiver and aftercare are followed. However, the earliest the service member may apply for another waiver is 12 months following completion of treatment in the Air Force, and six to 12 months in the Navy. Waiver requests following a treatment failure are considered only on a case-by-case basis. Typically, the service member is permanently grounded from all flight-related duties if there is another treatment failure occurrence after getting a second waiver.

An isolated alcohol-related incident is NCD for flight status in the military. However, there are specific guidelines for each service directing what must be done regarding administrative action and substance use evaluation, in such a circumstance. Multiple alcohol-related incidents are considered indicative factors of an underlying alcohol disorder in the Air Force and Navy and merit formal evaluation for such. If an Army member has multiple incidents of alcohol misuse and an ASAP evaluation does not reveal an underlying diagnosis of Alcohol Abuse or Dependence, a waiver will still be required and evaluated on a case-by-case basis.

Anxiety disorders

Anxiety disorders manifest symptoms that can produce sudden and dangerous distractions in flight operations. Panic attack episodes are especially hazardous due to the risk of symptoms that can appear unexpectedly and lead to incapacitation.

Most anxiety disorders, with the exception of Specific and Social Phobias, are considered permanently disqualifying for all aeronautically designated personnel, students or applicants in the Air Force, Army, and Navy. While it is theoretically possible for service members to obtain aeromedical waivers for any of the disqualifying anxiety disorders in the Air Force and Navy, the Army will not allow waivers for Panic Disorder and Obsessive–Compulsive Disorder.

In addition to waiving the disorder itself, in some instances the Army will also grant a waiver for pharmacotherapy. Specifically, the Army will consider granting a waiver for Post Traumatic Stress Disorder (PTSD), Acute Stress Disorder, Generalized Anxiety Disorder and Anxiety Disorder Not Otherwise Specified (NOS) as part of the Army's "Selective Monoamine Reuptake Inhibitor Surveillance Program" (US Army 2008). Please see Psychotropic Medications section later in this chapter for additional information as well as Chapter 12, this volume. For Army members to be waiver eligible for these conditions under this program, they must meet the following criteria for at least four months: remain free of aeromedically significant symptoms and experience no medication side-effects on a stable dosage of a medication approved for the program. If the service member has any further recurrences of symptoms, they will be disqualified with permanent termination of flight duties.

The Air Force and Navy will allow service members to request a waiver for an anxiety disorder once they have completed treatment successfully and remained asymptomatic without treatment for six months (Air Force) or 12 months (Navy). Like the Air Force however, the Navy only requires six months of being asymptomatic for Acute Stress Disorder. The Army does allow for members to request a waiver for Acute Stress Disorder without partaking in the Selective Monoamine Reuptake Inhibitor Surveillance Program if they are asymptomatic without active treatment for at least three months.

As noted above, Specific and Social Phobias are not inherently considered disqualifying unless these conditions have an impact on flight performance or safety. If symptoms do adversely affect flight-related duties and safety (for example, water-related phobias, claustrophobia), these conditions are then considered disqualifying for flight; waiver request packages may be submitted to the appropriate reviewing authorities after successful treatment for consideration on a case-by-case basis.

For waiver consideration, the initial request package should contain a thorough AMS, discussing the particular disorder's symptoms, frequency, duration, treatment, precipitating factors, actions taken to mitigate recurrence, and any other psychosocial concerns pertaining to personal and occupational functioning. The package should also contain copies of all relevant psychiatric and medical records, along with a copy of a recent mental health evaluation ideally completed within the three months prior to waiver submission. The mental health records should clearly reflect when the mental health provider declared the condition fully resolved. If the member required any medical evaluation and workup to rule-out any organic etiology for anxiety symptoms (for example, laboratory studies, internal medicine consult), copies of those records should be submitted for review as well. The Air Force also requires submission of a letter from the service member's commanding officer supporting return to flight-related duties. For untrained personnel, the Air Force will only consider a waiver for cases where the service member's condition has well-defined precipitating factors that have been determined unlikely to recur. The Navy does not make any distinctions between trained and untrained personnel.

Once a waiver has been granted, the flight surgeon will submit an annual AMS for Air Force members, or a detailed flight surgeon statement on the flight physical examination for those falling under Navy regulations; these reports should discuss the service member's interval history and any changes in condition. Copies of any relevant medical records should also be included in these waiver continuance submissions each year.

Attention Deficit/Hyperactivity Disorder (ADHD)

ADHD, a persistent pattern since early childhood, characterized by an inability to marshal and sustain attention, modulate activity level, and/or moderate impulsive actions, is incompatible with aviation. However, because the diagnosis of ADHD may be inappropriately assigned and medication prescribed to children and adolescents who do not satisfy diagnostic criteria, waivers for a history of ADHD are entertained by all services. Nevertheless, since many children diagnosed with ADHD continue to exhibit symptoms into adulthood, the waiver process is rather involved for all services.

In the Air Force, applicants for initial training with a possible past history of ADHD do not require a waiver if they successfully meet the following accession criteria: have not required an Individualized Education Program (IEP) or work accommodations since the age of 14; no history of comorbid mental disorders; never taken more than a single daily dosage of medication or been prescribed medication for this condition for more than 12 cumulative months after the age of 14; during periods off medication after the age of 14 have been able to maintain at least a 2.0 GPA without accommodations; and, can provide documentation from the prescribing provider that continued medication is not required for acceptable occupational performance. The applicant is also required to pass service-specific training periods with no prescribed medication for ADHD (US Air Force 2012). If record review and clinical interview suggest normal attention or inappropriate diagnosis of ADHD in childhood or adolescence, this may be noted as "information only" and no waiver is required. For designated aircrew, a waiver may be considered if the service member is symptom-free, has not manifested degradation in performance of aircrew duties, and has the ability to function effectively and safely without need for medication for at least one year. Mental health and neuropsychological evaluations, along with a review of pertinent medical, neurological, or mental health records, and an AMS detailing any social, occupational, administrative, or legal problems, including an analysis of the aeromedical implications of this particular case history is also required. For FC I/IA (that is, student pilots/navigators), a detailed history of academic achievement and use of any educational accommodations is needed, and for FC II (that is, trained pilots, navigators, or flight surgeons) and FC III (that is, aircrew not in direct control of the aircraft), a letter from the aviation supervisor or commander supporting a return to flying status is required. For the Army and Navy, a history of ADHD for aviation applicants and current ADHD diagnosis for designated aircrew are considered disqualifying; waivers are sometimes granted after thorough review. In the Army, a history of ADHD is disqualifying for initial applicants for Class 1A/1W (that is, pilot trainee), though Exceptions to Policy (ETP) are considered on a case-by-case basis. A history of ADHD for initial applicants for Classes 2, 3, and 4 (that is, all Army FC categories with the exception of pilot trainees) is also considered disqualifying, but a waiver is possible. For rated aviation personnel for all classes, a new diagnosis of ADHD may be considered for a waiver on a case-by-case basis.

In order for a waiver to be considered by the Army, evaluation by a clinical psychologist or psychiatrist must be completed, to include developmental, academic, employment, psychiatric, social, drug/alcohol, criminal, driving infractions, and medication history, and evaluation of any other psychiatric conditions that may contribute to symptoms along with a review of treatment records. If interview and record review suggests positive findings for ADHD, a detailed neuropsychological assessment is required. It should be noted that while the Air Force and Navy consider treatment with stimulant and non-stimulant medications for ADHD incompatible with flying and not waiverable, in the Army, treatment of ADHD is considered on a case-by-case basis. Specifically, waivers for the use of the Selective Monoamine Reuptake Inhibitors, bupropion (Wellbutrin®) and atomoxetine (Strattera®) may be considered if supported by assessment (indication of mild to moderate ADHD with demonstrably improved performance on objective testing) and the service member has remained free of aeromedically significant symptoms and medication side-effects on a stable dosage for a minimum of three months. Thus, if treatment includes the use of medication, this assessment should be conducted both on and off medication. Continuous Performance Testing, along with an in-flight performance evaluation in either actual aircraft or simulator is recommended to add ecological validity. Those diagnosed with ADHD require annual follow-up with a treating psychologist or psychiatrist. The diagnosis of ADHD in childhood that has not required treatment since adolescence, and does not prove to be currently impacting the individual based on testing, does not require annual follow-up.

For the Navy, applicants with ADHD who have not taken medication for 12 months and who demonstrate no symptoms may be considered for a waiver. All waiver packages require an AMS documenting prior symptoms, absence of persistent features, course of the disorder, medication use, and current level of functioning. Level of evaluation (that is, record review, mental health evaluation, neuropsychological evaluation) and supporting documentation required varies by education and history of accommodations. Briefly, for non-college graduates and graduates of non-traditional colleges (for example, online education), mental health evaluation by a clinical psychologist or psychiatrist is required, and should include interview with a parent or primary caregiver, review of report cards, high school transcripts, college transcripts (if applicable), IEPs (if applicable), and childhood medical records relevant to ADHD, as well as include all standard elements of a mental health evaluation (mental health/medical/social history, legal issues, substance use, mental status examination). For traditional college graduates (campus-based, full-time education) who did not require or use either medication or academic accommodations for the entire college experience, no mental health evaluation is required. However, a letter from the college/university academic learning center (or equivalent) stating the applicant was never evaluated for or provided academic accommodations during their entire college experience is required. If the applicant transferred to another college, letters from both academic centers are required. As well, applicants must provide a statement attesting to the fact they did not require or use medications for ADHD throughout college. Given that there are no available academic accommodations at the US Naval Academy, midshipmen in their senior year must provide a statement attesting to the fact he/she did not require or use medications for ADHD throughout college.

For college graduates who required or used ADHD medication and/or academic accommodations at any time while in college, a neuropsychological evaluation is required. This evaluation should include administration of the full current edition of the Wechsler Adult Intelligence Scale with all index scores, verbal and visual memory testing, vigilance testing, tests of executive function (four tests are required), ADHD self-report measures, alcohol screening, depression screening, personality testing, and a comprehensive mental health evaluation. Evaluation must be performed by a credentialed neuropsychologist, with the individual off of all ADHD medications. Any indication upon neuropsychological testing of continuance of the disorder is disqualifying. For more information on aerospace neuropsychology, please see Chapter 10, this volume.

Learning disorders

Learning disorders, which involve cognitive deficits that interfere with learning and academic achievement in one or more areas such as reading, spelling, writing, and arithmetic, in the context of average or above average intelligence, can compromise flight safety and mission performance in military aviation. A confirmed diagnosis of learning disorder is disqualifying, as these individuals are likely to have great difficulty keeping up with the rigors of training and operational flying without accommodations, and thus incompatible with aviation duties. All services allow for exception to policy and waivers if evaluation simply indicates history of learning disorder rather than current difficulties.

In the Air Force, for FC I/IA applicants to receive a waiver, their academic record must have been achieved without any accommodations and there must be no evidence of current problems. Waiver may be considered for aircrew with a history of learning disorder, providing they are symptom-free and have not manifested a degradation of their performance of aircrew duties.

For the Army, exception to policy and waiver can be considered if medication (that is, stimulants) is not needed to maintain adequate performance and if behavioral characteristics do not hinder flight performance or flight safety. Waiver packages must include an AMS detailing any social, occupational, administrative, or legal problems, including an analysis of the aeromedical implications of the case, mental health evaluation summary, specifically including psychological and neuropsychological evaluation reports, any pertinent past medical, mental health records, or current neurological or other medical consultation reports. FC I/IA also requires detailed history of academic achievement and use of any accommodations. Trained FC II or III individuals require a letter from the aviation supervisor or commander supporting a return to flying status. When submitting a waiver for the Army, the AMS must include psychiatric, psychological, and educational evaluations as indicated.

The Navy waiver process requires that the AMS document all prior symptoms, absence of persistent features, course of the disorder, and current level of functioning. Childhood medical and school records documenting the diagnosis and any academic interventions, grade school report cards, high school and college transcripts (if applicable) are also required. If the absence of academic/functional impairment cannot be determined from available records, or if there are residual problems or history of a persistent learning disorder, then a neuropsychological evaluation, conducted by a credentialed neuropsychologist is required. Any indication upon neuropsychological testing of continuance of the disorder is disqualifying.

Mood disorders

Mood disorders are associated with a wide variety of symptom manifestations that may significantly affect cognitive processes, behaviors, emotions, and somatic functions. Mood disorders include a spectrum of psychopathology ranging from depression to mania, and in severe cases, even symptoms of psychosis. Depression adversely affects the service member's ability to focus and sustain attention and concentration, and blunts psychomotor coordination, memory, and reaction time. Bipolar spectrum disorders are especially concerning due to loss of insight, impaired judgment and reality-testing, often coupled with compromised judgment and treatment compliance. Mood disorders are often treated with psychotropic medications; all psychotropic medications have side-effects that pose safety concerns and risks for flight-related duties (Ireland 2002; Rayman, Hastings and Kruyer 2006).

All mood disorders are considered permanently disqualifying for all designated personnel, students, or applicants in the Air Force, Army, and Navy, though it is possible for service members to obtain aeromedical waivers for certain conditions in all of the services. These conditions include Major Depressive Disorder (MDD), Dysthymia, and Depressive Disorder Not Otherwise Specified (NOS). For these particular disorders, the Army will consider waivers for aviation duties as part of the Army's "Selective Monoamine Reuptake Inhibitors Surveillance Program." The indicated treatment is psychotherapy or psychotropic medications, either alone or in combination. Service members may obtain a waiver after completing either intervention. More specifically, a waiver under this surveillance program requires that service members remain free of any aeromedically significant symptoms and medication side-effects, while on a stable dosage of an approved medication for a period of at least four months before a waiver request may be submitted for consideration. The Air Force and Navy share policies in which service members may only request waivers once completely asymptomatic in full-duty status for a minimum of six months after completion of all treatment, to include psychotherapy and pharmacotherapy. For depressive disorders with a history of recurrent episodes, the Air Force and Navy consider these conditions permanently disqualifying and not waiverable.

For waiver consideration pertaining to use of an approved medication to treat depression under the Army's "Selective Monoamine Reuptake Inhibitors Surveillance Program," the waiver request package must include a detailed clinical interview by an aeromedically trained clinical psychologist or psychiatrist, review of psychiatric treatment records, neuropsychological evaluation, and in-flight performance assessment. If the Army grants a waiver for medication use, the service member must have a consultation with an aeromedically trained psychiatrist at least every six months, and before returning to aviation duties following any changes in the dosing regimen or discontinuation of the medicine.

For an initial waiver request of a depressive disorder, the Air Force and Navy require submission of a detailed AMS, with comments describing symptoms, frequency, duration, treatment, precipitating factors, and psychosocial consequences. The Air Force and Navy require a current mental health evaluation and treatment summary; the evaluation should be done close to waiver submission and clearly state when the condition was fully resolved, as well as affirm sustained remission of symptoms. If the service member required treatment under the auspices of a Limited Duty MEB due to significant debilitation by the condition, a copy of the MEB narrative should be submitted with the waiver package. The Air Force also requires a letter from the service member's squadron commander or operations officer supporting or refuting any return to flight status. For waiver continuation, Navy regulations require annual submission of detailed comments by the service member's flight surgeon discussing interval history and confirming sustained remission of symptoms. The Air Force requires a similar statement by the service member's provider but also mandates that the waiver continuation package include a mental health evaluation done within three months of the package submission.

Bipolar spectrum disorders are considered permanently disqualifying and are not waiverable for all services. The services all recommend that service members with such disorders be referred to a central Physical Evaluation Board (PEB) to determine fitness for general military duty and retention. The Air Force further declares that a family history of Bipolar Disorder in both parents is disqualifying for FC I/IA (pilots/navigators); a waiver may be considered pending the results of a mental health evaluation.

Personality disorders

Personality disorders involve an enduring pattern of inner experiences, behaviors, and interpersonal interactions that deviate markedly from an individual's culture, are pervasive and inflexible, stable over time, and lead to distress or social and/or occupational impairment (American Psychiatric Association 2000). In aviation, this can lead to multiple difficulties, including working closely and productively with others and responding appropriately to authority under stressful conditions, and can ultimately result in problems with flight safety, aircrew coordination, and mission completion (Rayman, Hastings and Kruyer 2006). Treatment typically requires long-term intensive psychotherapy, which is incompatible with aviation duties.

For all military services, maladaptive personality traits and personality disorders that negatively impact performance may affect a service member's "suitability" for general military service (distinct from "fitness" for duty), and administrative separation due to psychological unsuitability for military service may be pursued. With respect to aviation duties, maladaptive personality traits or personality disorder impact a service member's adaptability rating, and as explained in a prior section, although each service has a unique label for this, all center on the effect on performance in this highly demanding specialty. Clinicians must be careful to avoid quick diagnosis of personality disorders in service members with idiosyncratic personality traits because these diagnoses are generally considered disqualifying for all flying classes.

All services are wary about granting waivers for a history of maladaptive personality traits or personality disorders, as personality issues are typically long-standing. In the Air Force, service members must be psychologically stable, with manifestations no longer interfering with duty, and review/evaluation must occur at the Aeromedical Consultation Service (ACS). Waiver is not recommended for any initial FC for service members with a history of personality disorder, and there are no waivers for FC I/IA, though ACS will review if requested by Air Education & Training Command (AETC) who has waiver authority. For FC II/III, the Major Command (MAJCOM) has waiver authority, but there are no indefinite waivers. Information required for an initial waiver package for the Air Force includes history of symptoms and impact on work and home, time line of events, treatment, mental health consultation reports, psychological testing results, substance use history if applicable, letters of support from the squadron commander, and MEB results if applicable. Waiver renewals require history since last waiver submission, including any legal or job-related problems, current treatment for the condition (if any), and mental health consultations.

For the Army, personality disorders are disqualifying for flying duties with no waiver recommended, and further, exception to policy is generally not recommended. As well, reversal of personality disqualification at a later date is difficult. Service members may be evaluated by any aeromedically trained psychiatrist or psychologist, but may also be referred to NAMI Psychiatry in Pensacola, Florida, or ACS at Wright Patterson Air Force Base in Dayton, Ohio for further evaluation. Requirements for Army waiver package include mental health evaluation, results of psychological testing, and a letter from the service member's commander regarding work performance and social adjustment. Follow-up evaluations are at the discretion of the treating mental health provider.

For the Navy, once a service member is found NAA, it is unlikely he/she will be found AA at a later date. However, if the service member demonstrates over a period of two to three years substantial improvement in terms of ability to sustain stressors of the aviation environment, work in harmony with others, and stability in his or her personal life, then with strong support from the chain of command and flight surgeon, they may be considered for re-evaluation by an aeromedically trained psychiatrist or psychologist. For the Navy, consultation with NAMI Psychiatry must be done. Questions regarding the aeronautical adaptation of designated aviation personnel should be referred to NAMI Psychiatry for consultation or in-person evaluation. Information required for the waiver package includes mental health evaluation, which must also clarify suitability for general and special duty.

Other Psychiatric Conditions of Concern

This section covers psychopathology less commonly encountered in the aeromedical setting.

Attempted suicide

A suicide attempt by itself is a behavior and not a psychiatric diagnosis. Suicidal behaviors constitute a very serious aeromedical concern. In the aviation community, it is possible that the availability of aircraft may serve as a means for self-destruction, and thus the safety of others in the air and on the ground may be placed in jeopardy as well.

As suicidal behaviors in and of themselves do not constitute a psychiatric disorder but rather a manifestation of such, aeromedical waivers are based upon the actual condition that evoked the suicidal behavior. The Air Force, Army, and Navy all have similar aeromedical policies regarding suicide attempts. If the suicide attempt was secondary to psychopathology from a diagnosed MDD, the service member would be permanently disqualified from all aviation duties; a waiver may be possible depending upon the particular nature of that MDD (that is, single episode versus recurrent) and the waiver requirements posed by the service member's military branch. If the suicidal behaviors were secondary to a severe adjustment disorder or some other temporarily disqualifying or non-disqualifying condition, the service member must be symptom-free and all treatment completed for a period of at least six months before a waiver may be requested.

To request a waiver, the package must include a complete AMS discussing the events leading up to the suicide attempt and all subsequent care provided to the service member. The AMS should include clear evidence that the underlying stressors that precipitated the attempt have all been thoroughly addressed and resolved. The waiver package must also include all relevant mental health and medical records, and preferably a recent mental health evaluation discussing the service member's current condition and prognosis.

It must be noted that recurrent episodes of suicidal ideation, actions, or attempts are considered grounds for permanent disqualification from all aviation duty—regardless of the underlying diagnosis.

Eating disorders

Eating disorders involve a significant disturbance in eating behavior and self-image, and can cause potentially life-threatening metabolic alkalosis, hypochloremia, and hypokalemia, which can have drastic implications for aviation safety (US Air Force 2012, US Army 2008). As well, a significant aeromedical concern is the comorbidity of emotional difficulties that lower stamina for managing the high stress of military flying. Anxiety and depression are highly associated with eating disorders, and there is an increased risk of suicide. Treatment is very difficult and involves intensive long-term therapy and possibly pharmacotherapy, which are incompatible with aviation duty. The course and outcome of these disorders is highly variable and marked by relapse. As a result, the disorder is considered disqualifying for aviation duty.

Nevertheless, all services will consider waivers on a case-by-case basis. For the Air Force, eating disorders are specifically disqualifying for FC I/IA, II/IIU, and III. For Space and Missile Operations Duties (SMOD) if the flight surgeon believes the condition or history will interfere with the performance of duty, a waiver is necessary. For Air Traffic Control/Ground Based Control (ATC/GBC) duties, a waiver is required for any personality disorder, or mental condition that may render the individual unable to safely perform controller duties. Initial waivers should be granted for only one year due to the high rate of relapse, and an indefinite waiver is not recommended. Waiver review and authority varies by FC, and candidate versus designated status, as does wait period when symptom-free prior to consideration (one to two years). The AMS for the initial waiver should include history (to address pertinent symptoms of amenorrhea, constipation, abdominal pain, cold intolerance, lethargy, excess energy), any social, occupational, administrative, or legal problems, physical exam (height and weight, stability of weight, blood pressure, skin, cardiovascular, abdominal and neurologic systems), lab work (cell blood count, chemistry panels, urinalysis, electrocardiogram), psychiatric evaluation and treatment summary by a doctoral level provider that includes psychological testing of the service member's emotional and cognitive disposition (for example, Minnesota Multiphasic Personality Inventory-2 (MMPI-2), most recent edition of the Wechsler Adult Intelligence Scales), dental evaluation for eating disorders with purging, MEB reports if applicable, and input from the service member's commander/supervisor regarding their current status. The AMS for the waiver renewal should include assessment for recurrences, including stability of the service member's weight, physical exam (height and weight, blood pressure, skin, cardiovascular, abdominal, neurologic), and a psychiatric evaluation for the first renewal, and if clinically indicated on subsequent renewals.

For the Army, a waiver will be considered if the individual meets minimum aviation weight standards, is symptom-free, off medication, and fully functional in an alternate duty assignment for one year. The AMS should include psychiatric and/or psychological evaluation, copy of MEB if applicable, and the flight surgeon's narrative outlining any social, occupational, administrative or legal problems.

For the Navy, service members must also meet minimum weight standards, be symptom-free, off medication, and out of active treatment for one year. On-site evaluation by NAMI Psychiatry is required before waiver consideration. The AMS should outline any social, occupational, administrative, or legal problems, and include the aviation mental health evaluation and a copy of the MEB if applicable.

Impulse control disorders

These disorders involve an inability to resist acting on an impulse that may be dangerous to the individual or to others. While these disorders occur very infrequently among military aviators, when present, such stereotyped or impulsive behaviors include features that are incompatible with mission readiness and flight safety, and may also cause concerns amongst aircrew about leadership, reliability, and trustworthiness. For example, behaviors such as compulsive gambling may disrupt sleep, consume time and mental energy, and cause anxiety or stress-related distractions, and any of these factors can affect primary flying duties (US Army 2008). Psychotropic medications used with these disorders are incompatible with aviation duty.

In general, impulse control disorders are considered disqualifying, and though waivers are handled on a case-by-case basis, they are generally not recommended. In the Army, impulse control disorders (including Intermittent Explosive Disorder, Kleptomania, Pathological Gambling, Pyromania, Trichotillomania) are considered permanently disqualifying with no waiver recommended. These individuals are also considered unsatisfactory AA. In the Navy, impulse control disorders are considered for waivers on a case-by-case basis. Information required for the waiver package includes mental health evaluation and AMS with flight surgeon's narrative outlining any social, occupational, administrative, or legal problems. For those falling under Navy regulations, contact NAMI Psychiatry for telephone consultation or referral for formal evaluation. Though there is no specific guidance on impulse control disorders in the Air Force waiver guide, this disorder would disqualify the service member if it were causing functional impairment; waivers for designated aviation personnel would be considered on a case-by-case basis (US Air Force 2009).

Psychotic disorders

The majority of psychotic disorders are incompatible for general military duty and will require a PEB to determine fitness for retention in the armed forces. The spectrum of possible symptoms with psychosis especially present aeromedical concerns. Examples of such symptoms include: impaired reality testing; disorganized speech or behaviors; social withdrawal; and, elevated risk of suicide. Furthermore, the psychotherapy and psychotropic pharmacotherapy typically utilized to treat these conditions are incompatible with aviation duty.

The Air Force, Army, and Navy all have similar aeromedical policies regarding psychotic disorders. Psychotic disorders are considered disqualifying for all aviation personnel with the exception of three specific psychotic conditions. These exceptions include: Brief Psychotic Disorder With Marked Stressors (Brief Reactive Psychosis); Substance-Induced Psychotic Disorder; and, Psychotic Disorder Due To General Medical Condition. All other psychotic disorders are considered permanently disqualifying for flight and should be referred to a central PEB for determination of fitness for general duty/retention.

Brief Psychotic Disorder With Marked Stressors (Brief Reactive Psychosis) is disqualifying for all aviation duties, but may be considered for an aeromedical waiver if the condition completely resolves and the service member remains asymptomatic with no further needed therapy (that is, psychotherapy, pharmacotherapy) for a year in a full general duty status. These cases are considered for a waiver on a case-by-case basis depending on prognostic factors. For waiver consideration, an AMS should thoroughly discuss the history of the illness, all treatments administered, current status of occupational and psychosocial functioning, and comment upon potential aeromedical implications associated with the particular case. A recent mental health evaluation should be included in the waiver package, commenting on the continued remission of symptoms over the past year, overall prognosis, and fitness for general duty.

A Substance-Induced Psychotic Disorder that has clear evidence from medical and psychiatric workup (that is, history, physical examination, laboratory, or radiologic studies) that demonstrates the disorder is etiologically related to substance use, is considered disqualifying for aviation duties throughout the duration of the active disorder. The disorder is no longer considered disqualifying once it has fully resolved and as long as the inciting substance was neither alcohol nor illicit drugs.

Psychotic Disorder Due To General Medical Condition is also disqualifying for aviation duties throughout the duration of the active disorder. The condition is no longer considered disqualifying once the general medical condition has resolved and if the organic etiologic factors are identified and determined unlikely to recur.

For Substance-Induced Psychotic Disorder and Psychotic Disorder Due To General Medical Condition, an AMS discussing pertinent details and all relevant medical and psychiatric records should be submitted to the appropriate aeromedical authority (for example, NAMI) for review. The service member will also require appropriate evaluation and waiver request for any underlying disorders leading to the psychotic symptoms.

Sexual disorders

Sexual dysfunctions, such as Sexual Desire, Aversion, Arousal, Pain, or Orgasm Disorders, typically do not impact a service member's aviation performance, and standard treatment usually involves behavioral techniques that should not preclude duties involving flight, though it should be noted that use of medication for these disorders is incompatible with aviation duty. Paraphilias, however, may impact aviation performance as such service members often demonstrate questionable judgment, poor impulse control, and compulsive behavior. Legal ramifications may cause the service member to be inattentive to detail and thus become a safety risk. Treatment is usually less successful with this group. All services may consider sexual disorders disqualifying depending on their impact on performance. Notably, sexual dysfunctions may be NCD if they do not affect aviation duties, whereas paraphilias are generally considered disqualifying.

The Air Force does not consider sexual dysfunctions and other unspecified sexual disorders necessarily disqualifying unless in association with another Axis I disorder. They do not consider sexual paraphilias medically disqualifying, though service members meeting diagnostic criteria for these disorders are dealt with administratively through their chain of command.

In the Army, to request a waiver, mental health evaluations with statements from the flight surgeon and the commander regarding the service member's aviation performance are required. In the case where a service member becomes unfit for general duty due to a sexual disorder, they are referred to a PEB or MEB for review. Many such cases are handled by administrative disposition due to legal implications and impact on good order and discipline.

In the Navy, waiver requests are handled on a case-by-case basis after the service member has completed treatment and been asymptomatic for one year. Information required includes psychiatric evaluation and treatment summary, and AMS documenting any social, occupational, administrative, or legal problems of the service member. Factors that are considered in waiver requests for paraphilias include the type of paraphilia, duration and frequency, type of treatment required, and the adequacy of follow-up care. As with the other branches of service, many of these cases are handled by administrative disposition due to legal implications.

Sleep disorders

Sleep disorders, including Somnambulism, Obstructive Sleep Apnea, Primary Insomnia, Idiopathic Hypersomnia, Narcolepsy, Periodic Limb Movement Disorder, Restless Extremity Syndrome, and Circadian Rhythm Disorders, are generally covered in the neurology sections of waiver guides. Because there is often an association between disorders of sleep architecture and timing, and underlying psychiatric disorder, cognitive disturbance, or other pathology, these conditions are often addressed by aeromedical mental health professionals.

Sleep disorders frequently result in demonstrable deficits in cognitive and psychomotor performance. Fatigue, sleepiness, and circadian rhythm disturbances can have a critical effect on aviation safety. Cognitive function and neuromuscular coordination may be affected by both the sleep disorder and/or the treatment modalities used. Service members with sleep disorders may have more than the usual difficulty in adjusting to the circadian rhythm disruption that occurs with travel across time zones. This presents an additional hazard to a service member who may deploy several time zones away and is expected to perform flying duties. For more information on fatigue and aviation, please see Chapter 9, this volume.

For the Air Force and Army, waivers are considered after full recovery for transient cases related to life crises or medical conditions that may be treated by short-term surgical or medical means (not to include medication). Waivers are unlikely for those who need any significant follow-up other than routine annual exam. Service members with Restless Extremity Syndrome may be considered for waiver if the cause has been defined and permanently cured, and the secondary sleep disorder has resolved.

For the Air Force, service members with a documented sleep disorder require an ACS evaluation prior to returning to flying status. This includes a polysomnography at Wilford Hall Medical Center Sleep Disorders Laboratory and neuropsychological testing to evaluate cognitive functioning. Service members with mild to moderate Sleep Apnea may be considered for waiver for FC II duties if results of neuropsychological testing are normal. Sleep Apnea must be controlled by either an oral appliance or surgery before waiver will be considered. Medications are not waiverable. Nasal continuous positive airway pressure (CPAP) is incompatible with worldwide qualification because the device needs a continuous power supply. To submit a waiver, the disorder must be documented at the ACS, and while initial diagnostic workup need not be performed at Wilford Hall Medical Center, this is encouraged, particularly when there is suspicion of Sleep Apnea. If narcolepsy is diagnosed by an outside sleep laboratory, the service member should be referred to the ACS for confirmation of the diagnosis. Although this diagnosis, if confirmed, will result in permanent disqualification, there are cases of service members who have been improperly diagnosed.

For the Army, the waiver package must include a complete AMS including sleep disorder workup with polysomnography as needed. For the Navy, a history of sleep disorders is generally considered permanently disqualifying, due to the persistent nature and impact on psychomotor and cognitive performance. However, waivers may be considered in cases with successful treatment. Information required for the waiver package includes neurology/sleep specialist consultation with polysomnography, vigilance testing, and mental health evaluation (as indicated).

Somatoform and Factitious disorders

The common feature of the somatoform disorders is the presence of physical symptoms that suggest a general medical condition, but are not fully explained by any medical condition, direct effects of a substance, or another psychiatric disorder (American Psychiatric Association 2000). Aviation performance is affected, as evaluation and treatment for these disorders take time away from aviation duty, may limit effectiveness on duty, and it is expected that the service member's inherent preoccupation with physical symptoms reduces time devoted to the mission.

Identifying the somatoform nature of the problem allows the physician to avoid unnecessary, expensive, or invasive diagnostic procedures. Equally important is to evaluate for recent stressors surrounding flying, as somatoform symptoms or disorders can obscure an unspoken fear of flying. The symptoms presented are sometimes preceded by the words, "I'd like to fly, but …"—a striking contrast to the attitude of most service members who insist on flying *in spite of* their symptoms. The service member may also describe symptoms in terms of their effect on flying, with no particular anxiety about being significantly ill, and have little interest in specific treatment. If asked, "Will you go back to flying when you are well?" they may equivocate or signal reluctance (US Air Force 2012). For more information on fear of flying and motivation to fly, please see Chapter 7, this volume.

Distinctly, Factitious Disorders are characterized by the *intentional* production or feigning of physical or psychological symptoms. The service member may seriously injure themselves, which presents an extreme risk in the aviation environment. Treatment offers little hope of return to flight status. These service members are rarely motivated for psychotherapy, and frequently change physicians when confronted.

All services consider somatoform disorders disqualifying. In the Air Force, consideration for a waiver will only be entertained if the service member is successfully treated and remains off all psychotropic medication for 12 months, and procedures vary by FC. Factitious Disorder, when it significantly interferes with performance of duty, is considered an unsuiting condition, that is, an administrative condition, and not a disability subject to an MEB. Initial training (for any FC) and FC I/IA are handled by the AETC and generally not considered for waiver. The AMS for initial waiver should include the history of the disorder and all treatments administered, the current status of any social, occupational, administrative, or legal problems associated with the case, psychiatric conditions, mental health consultation reports (need all treatment notes from treating mental health professionals as well as a summary of the mental health record, and all psychological testing, if performed), an analysis of the aeromedical implications of the case, and a letter of support from the service member's supervisor. The AMS for waiver renewal should include an interim history since last waiver, current treatment for the condition if any, and mental health consultation reports if accomplished since last waiver request.

In the Army, somatoform and Factitious Disorders are considered disqualifying, and waivers are rarely recommended. The service member should be referred to a PEB or MEB to review for retention. A waiver may be considered for those who are successfully treated, provided they remain asymptomatic and off medications for one year in a full-duty status. An unconscious fear of flying is a disqualifying condition but may be waiverable with successful treatment and if the service member remains asymptomatic for one year.

In the Navy, waivers may also be considered for those rare cases that are successfully treated on a Limited Duty Board and remain asymptomatic and off medications for one year in a full-duty status. The Army and Navy also require mental health evaluation, a copy of the MEB if applicable, and the flight surgeon's narrative outlining any social, occupational, administrative, or legal problems.

V Code diagnoses:

V Codes are conditions that come to the attention of a clinician though they may not necessarily constitute clinical disorders. In other words, they may prompt a patient visit, result from another mental disorder, or potentially cause or worsen a mental disorder. With respect to aviation, depending upon the circumstances, V Codes may interfere with safe or effective flying. Most relevant to aeromedical decision-making is the response of the service member rather than the severity of the stressor. Typically, V Code problems resolve satisfactorily and have no permanent impact on flight status; however, chronicity or need for medication could lead to permanent disqualification. While they are not necessarily disqualifying, attention should be paid to potential underlying psychopathology as well as maladaptive personality traits. Assessment or intervention must consider whether the service member should continue to fly, as numerous small stressors can produce fatigue, irritability, early task saturation, distraction, inability to compartmentalize (that is, keep one's mind in the flight box), and cognitive inefficiency as much as a single major stressor. If another diagnosis seems warranted, it should be established and documented, so that the service member receives proper treatment in a timely manner. Avoiding proper diagnosis solely in order to avoid a "down" chit will only delay treatment and return to flight status.

For both the Air Force and Army, if a service member is placed in a down status due solely to a V Code, a waiver will be considered once the symptoms have subsided, the service member has completed use of any medications, and the service member has returned to full functioning, such that return to flight-related duties is possible. "Talk therapy" may continue for check-in or "booster" sessions, or possibly for marital counseling, so long as symptoms remain sufficiently relieved. If the V code is an additional diagnostic code listed for completeness during the treatment of a disqualifying mental disorder, waiver action should be taken primarily in accord with the requirements for the primary disqualifying diagnosis.

Waiver submissions for the Air Force and Army should include an AMS with pertinent social, occupational, legal, or financial information, history of the particular stressor, and rationale as to why the service member should be safe to return to flying status especially if the situational stressor is not completely resolved or if it could reasonably be expected to recur. A current mental health evaluation, including all treatment notes, summary of the mental health record, and any psychological testing or evaluation reports should also be included. The Air Force also requires a letter from the service member's supervisor rendering an opinion about their readiness to return to flying status. The Navy does not require waivers for return to flight status. A service member may instead be considered NPQ during the symptomatic period, but upon resolution, the V Code is NCD.

Psychotropic medications

Use of psychotropic medications is considered incompatible with aviation duties in the Air Force and Navy. Psychotropic medications prescribed by a competent medical authority are considered sufficient cause for grounding from flight status for the duration of medication use and its drug effects. However, the Army's "Selective Monoamine Reuptake Inhibitor Surveillance Program" permits aviation personnel to be waivered for flight-related duties while being treated with certain psychotropic medicines. The Army's policy regarding the use of selective monoamine reuptake inhibitor (SMRI) medicines, for the treatment of mood disorder, was first drafted and presented for review by the Aeromedical Consultants Advisory Panel (ACAP) in June 2005 and has since been updated following feedback from flight surgeons and mental health personnel. The rationale behind the development of the policy was to allow a way for aviation personnel to be treated for mood disorders while maintaining flight status. It was felt that a significant number of aviators may be suffering in silence, self-medicating, using unproven herbal remedies, or using medications obtained from unapproved sources. With an approximate 6 percent prevalence of depression in the general population, the assumption was that there are a significant number of aviation personnel with a mood disorder who were not seeking treatment in order to avoid being removed from flying duties. Adherence to the policy may allow a return to flight status in as little as four to five months after presenting for care.

The term SMRI was chosen to include the use of selective norepinephrine reuptake inhibitors that have similar efficacy and side-effect profiles as selective serotonin reuptake inhibitors (SSRIs). Except on rare occasions, treatment must be limited to a single agent. Medications must be approved by the US Food and Drug Administration (FDA) for the treatment of depression or anxiety with a minimum three years of favorable post-market surveillance. The psychiatric conditions that fall under this policy include but are not limited to: MDD, Dysthymia, Depressive Disorder NOS, medication use in support of Adjustment Disorder with affected mood, PTSD, Acute Stress Disorder, Generalized Anxiety Disorder, Anxiety Disorder NOS, and Premenstrual Dysphoric Disorder. The inclusion of the less severe anxiety disorders allows them to be included in this policy which would avoid patients with an anxiety disorder being diagnosed with depression solely in order to benefit from the program.

Side-effect profiles are similar in this class of medications, but may vary somewhat in relative frequency. These tend to occur early on in treatment and diminish as patients become accustomed to the medication. The most common side-effect with prolonged treatment is sexual dysfunction, which may not be aeromedically significant but may lead some patients to alter or discontinue therapy. If the dose of a medication is changed or discontinued, then temporary grounding and close monitoring for aeromedically significant symptoms is initiated.

The use of a SMRI and the diagnoses that the SMRI is being utilized for are disqualifying for all classes of aviation. Initial aviation applicants will be considered for an exception to policy on a case-by-case basis. Experienced rated aircrew may apply for a waiver which is considered on a case-by-case basis. There is a significant amount of information required for a SMRI waiver request to be considered, including a detailed clinical interview by a psychiatrist or psychologist, preferably aeromedically trained (note that if the clinical interview is performed by a non-aeromedically trained professional, the Army's aviation psychiatry consultant or designee will also review the case); narrative summary of treatment records from the treating psychiatrist or prescribing non-psychiatrist provider, with documented evidence of uncomplicated illness, and no evidence of psychosis or suicidal behavior; evidence that medication has been at a stable dose for a minimum of four months without aeromedically significant side effects; neuropsychological assessment, including cognitive domains and motor skills testing to demonstrate functional ability; and, an operational assessment demonstrating aeronautical ability after three months of maintenance therapy. Rated aviation personnel are required to perform an in-flight evaluation or job specific equivalent assessment, in either an aircraft (preferred) or a simulator. After having successfully met these requirements and with command and local flight surgeon endorsement, temporary aeromedical clearance for up to 90 days may be granted while the waiver is being processed (US Army 2008).

If a waiver is recommended and granted, the service member will be required to follow-up with their treating psychiatrist as recommended in accordance with their clinical condition. They will also be required to undergo evaluation by a psychologist or psychiatrist (aeromedically-trained preferred) every six months for the duration of treatment, plus six months after medication cessation to insure stability. After that period, the flight surgeon shall note mental health status on an annual basis to continue the waiver. Results should be annotated on the annual flight physical. Any change in mental health status, including relapse or exacerbation of symptoms, or return to medication after cessation is disqualifying and a new AMS will need to be submitted. The US Army Aeromedical Research

Laboratory (USAARL) is developing a study in order to evaluate if these medications are associated with any decrements to flight performance. For more on psychopharmacology and aviation, please see Chapter 12, this volume.

Waiver Request Processes and Granting Authorities

In the Air Force and Army, waivers for disqualifying conditions are usually requested by service members and their flight surgeons. Line commanders can also request waivers via the local Flight Medicine office. Waivers for Flying and Special Duty personnel are granted by the medical community, in contrast to the Navy where waiver authority is held by the Line, that is, the Bureau of Naval Personnel (BUPERS).

In the Air Force, waiver authority is held by the Surgeon General, who delegates waiver authority to the Air Force Consultant for Aerospace Medicine (AF/SG3PF). For most conditions, waiver authority is further delegated to the MAJCOM Surgeons and Physical Standards components (SGPS). HQ Air Force (AF/SG3PF) retains the waiver authority for certain circumstances, such as conditions requiring an initial categorically restricted waiver (that is, pilot restricted to a multi-place aircraft), or conditions that fail to meet retention standards. MAJCOM Chief Flight Surgeons (SGPs) can further delegate waiver authority for certain conditions and waiver renewals to local SGPs, particularly if they are graduates of the Air Force Residency of Aerospace Medicine. In exceptional circumstances, the Air Force has an ETP process that is requested by line Air Force commands and adjudicated by the Air Force Chief of Staff. In cases like this, aeromedical authorities have already denied a waiver for the condition in question, but the ACS may be asked to review the case in detail and provide a qualified aeromedical risk assessment to assist the Line in determination of flight status disposition.

For the Army, waivers are granted by the United States Army Personnel Command (PERSCOM); Chief, National Guard Bureau; and by local commanding officers, depending upon the status of the aircrew member. USAAMA makes recommendations to the waiver granting authorities with input from the local flight surgeons.

As noted above, in the Navy waivers are granted by BUPERS, the Commandant of the Marine Corps (CMC), or other appropriate waiver granting authority. NAMI Aeromedical Physical Qualifications Department must review all waiver requests and forward their recommendations to BUPERS or CMC as appropriate. It is important to note that the BUMED endorsement notification recommending a disposition on an aircrew member is not the final action and requires BUPERS or CMC endorsement. In other words, a waiver is not truly granted until BUPERS or CMC acts. Until that time, the waiver is still in a "recommended" status.

Routing of Waiver Requests

For the Air Force, all waiver requests will be initiated in the Aeromedical Information Waiver Tracking System (AIMWTS) and the Physical Exam

Processing Program (PEPP) by the local flight surgeon. After being reviewed and digitally signed by the senior flight surgeon of the Flight Medicine Clinic, the waiver will be forwarded to the appropriate MAJCOM. From there, if required, the waiver request will be forwarded to the ACS for evaluation and recommendation, or to AF/SG3PF, if HQ Air Force retains waiver authority. It is important to note that the ACS has no waiver authority, but only provides expert ACS upon request of the waiver authorities.

In the Navy, waivers may be requested in a number of ways. Most commonly the service member requests a waiver via the local Aviation Medicine Clinic and flight surgeon. Also, the service member's commanding officer may request a waiver for one of his/her fliers. A medical officer can initiate a waiver request, but obviously, with the service member's concurrence and support. Finally, BUMED, the Naval Reserve Center, CMC, or Navy Personnel Command (NAVPERSCOM) can take the initiative to request a waiver for a service member or candidate. It is important to note, however, that this action as well as any waiver request must be undertaken with the knowledge and endorsement of the service member's commanding officer. After review by the commanding officer, waiver requests are submitted to NAMI Aeromedical Physical Qualifications Department for review in the web-based program AERO. NAMI Aeromedical Physical Qualifications Department reviews all waiver requests and forwards a recommendation to the appropriate waiver granting authority (BUPERS or CMC) via AERO. The waiver authority then acts on the waiver recommendation provided by NAMI Aeromedical Physical Qualifications Department and BUMED. Local flight surgeons can electronically access the waiver recommendation letter in AERO.

For the Army, the waiver process is initiated at the local flight surgeon's office at the time of the discovery of a disqualifying medical condition. Local evaluations and consultations must be obtained which support or fail to support a waiver recommendation. Once this packet is forwarded to USAAMA, it can take several different routes depending on the nature of the disqualification. Most waiver requests are considered routine waivers (those that have clear policy established) and require little more than review and endorsement, and then are forwarded with recommendations for appropriate follow-up or restrictions to the waiver authority.

Waiver Submission Requirements

For the Air Force, waiver requests for all initial flying and special duty examinations will be submitted using an AMS in AIMWTS and a current physical examination in PEPP. Supporting documents must be uploaded as attachments into these applications and forwarded to the reviewing/certification authority. However, a physical in PEPP is not required for waiver submission for trained aviators unless specifically requested by the waiver authority. The AMS format in AIMWTS is utilized when requesting waivers for trained aircrew or for aircrew in training, and required documents are uploaded as attachments. Forms such as the SF 88, DD Form 2808, or Preventive Health Assessment (PHA) do not need to be completed solely for the purpose of a waiver submission unless the flight surgeon deems it necessary, or as directed by other authority. The requirements for these ongoing, routine evaluations are determined elsewhere, and while important, do not necessarily impact submitting a waiver request. Waiver submissions in the Navy require a current flight physical and AMS be submitted in AERO for both applicants and designated fliers. The applicable portion of the Navy's ARWG should be consulted frequently and thoroughly for the inclusion of specific requirements for each disqualifying diagnosis. Applicant submissions should include a detailed history, a review of systems, and a complete applicant physical examination recorded on the appropriate forms (DoD 2807, DoD 2808, and DoD 507 forms within AERO). Designated waiver submissions will include the most recent flight physical (either long or short) and all continuation requirements for previously-granted waivers. Both applicants and designated personnel should have an AMS completed, and all required supporting documentation in accordance with the ARWG uploaded into AERO. The local flight surgeon's recommended disposition and the commanding officer's endorsement should be included as well. The AERO website should be reviewed prior to submission to ensure that the member is current with all prior waivers and physical exams.

For the Army, an AMS is required for any action that requires a waiver, permanent medical disqualification (permanent termination from flying), termination of permanent termination from flying (that is, reinstatement), or request for aeromedical consultation. The information needed to process a waiver may vary but usually needs to include any available specialty consultations, reports of all surgeries, tissue examinations, pathology and laboratory reports, diagnostic studies, hospital summaries, past medical documents, and any letters of recommendation.

Waiver Continuation

For the Air Force, the waiver authority establishes the term of validity for each waiver, and the appropriate expiration date is placed on a waiver for conditions that may progress or require periodic re-evaluation. Before the waiver expires, the local flight surgeon is required to submit an AMS in AIMWTS for waiver renewal. If the condition worsens, the waiver is invalidated, and a new waiver must be requested. If the condition resolves and the service member meets medical standards, the waiver can be retired in AIMWTS with concurrence of the waiver authority.

In the Navy, waiver continuation requests must be submitted to NAMI Aeromedical Physical Qualifications Department, and include all required information set forth in the BUMED/CMC letter, as well as any pertinent information in the ARWG. Waiver continuance requests must also include the current long (every five years) or short physical exam (annual).

In the Army, once a waiver has been granted, a request for waiver continuation must be submitted on each subsequent flight physical complete with documentation verifying compliance with waiver requirements. Continuation requirements vary depending on the specific condition for which the waiver is required.

Aeromedical Clearance

In the Air Force, once the waiver has undergone the appropriate disposition by the waiver authority, the entry in AIMWTS is certified with the waiver information and term of validity. This document is then downloaded by the local Flight Medicine office, entered into the service member's record, and an AF Form 1042 (up/down chit) with the appropriate information is completed and provided to the service member, entered into the medical record, and sent to the Host Aviation Resource Management (HARM) Office. The Air Force has no mechanism for temporary authorizations to fly while waiver disposition is pending.

In the Navy, only BUPERS and the CMC can grant a waiver. Recommendation letters from NAMI authorize a temporary "upchit" for 90 days pending BUPERS and CMC action. On occasion, a Local Board of Flight Surgeons (LBFS) can meet and issue a temporary 90-day upchit pending NAMI and BUPERS/CMC action. An LBFS can only be utilized for certain conditions and must be authorized by the squadron's commanding officer. Temporary "upchits" cannot be issued if the member has been grounded by BUPERS or CMC; only these agencies can reverse their own grounding action; therefore, an upchit cannot be issued until their letter arrives.

In the Army, there are four possible dispositions for waiver requests: (1) Qualified, no waiver is required; (2) Qualified, information only. These conditions are tracked in a database and reported on the annual summary sheet. No waiver is required; (3) Disqualified, waiver recommended. Waiver recommendation will be forwarded to PERSCOM for final approval; and (4) Disqualified, waiver not recommended. Waiver/exception to policy is not recommended. If this final disposition is approved by PERSCOM, it will result in termination of aviation service.

Clearance for flying duty is obtained after submission of a waiver request to USAAMA and by positive acknowledgement by USAAMA that the request for waiver has been approved. Prior to obtaining an approved waiver, local flight surgeons may allow temporary clearance to fly, depending on the condition in question. Returning disqualified service members to full flight duty prior to receipt of waiver raises several possible concerns especially if done without coordination with USAAMA. First, waiver requests are not always granted. Second, waivers may be granted with certain flight restrictions, and finally, waiver policy is frequently changing to keep pace with current medical knowledge. Minor disqualifications (for example, service member who had an adjustment disorder for less than 90 days, is temporarily grounded locally, recovered without incident, and was able to resume flying duties), however, may be granted local clearance when following established policy, thereby expediting the return to full duty for many service members.

Processes for Appeals or Special Case Considerations

Depending on the particular military branch, there are certain boards and processes available for evaluating special medical cases and appeals of aeromedical disposition. Under Article 15-81 of the MANMED, a Special Board of Flight Surgeons may be convened to evaluate special medical cases which, "due to their complexity or uniqueness, warrant a comprehensive aeromedical evaluation." It is important to note that a Special Board of Flight Surgeons may not be requested "to challenge a physical standard or disqualification without evidence of special circumstances" (BUMED 2005). Requests for aviation applicants are typically not granted. Under MANMED Article 15-82, a Senior Board of Flight Surgeons (SBFS) may be convened at BUMED to serve as the final appeal board in review of aeromedical dispositions requested by NAVPERSCOM, the Chief of Naval Operations (CNO), or CMC. The Air Force has an ETP process, which is an avenue of recourse for a flier who has been disqualified. The ETP process is a means of appeal initiated by line commands and forwarded to the Air Force Chief of Staff to overrule a medical recommendation, typically when a waiver has not been granted. The ACS is often asked to provide risk assessment appropriate to the situation. ETPs usually number less than five per year, and are mostly for applicants to flying programs. The Army utilizes the ACAP to review special or complex cases. The Army also utilizes USAAMA to review the majority of aeromedical cases—this is managed via the computerized online AERO system. In any circumstance, it is best advised to contact the designated waiver authority for a member's branch of service to receive further guidance regarding appeals of aeromedical disposition or review of special cases.

Process for Updating Waiver Standards

Each military service has a particular process for updating their waiver guides and aeromedical standards. Updates are vetted by aerospace medicine specialty consultants based on extensive research of the best aeromedical and general medical data to make recommended changes. These updates are subsequently forwarded to the appropriate higher authorities for review and approval. Once approved, the updates are then incorporated into the particular governing standards and/or waiver guides.

Conclusions

This chapter provides an introduction to general military physical standards, as well as the additional standards required for aviation duty. We illustrate how physical and mental health conditions and symptoms are considered in the aviation environment, particularly with respect to their impact on crew coordination, safety of flight, and mission completion. Waiver and appeals procedures for individual cases in accord with each of the military services are discussed, as well as larger-scale process changes (for example, the Army's "Selective Monoamine Reuptake Inhibitor Surveillance Program"). We wish to re-emphasize that this chapter should be considered a "living document" as each service's waiver guides are continually updated. We advise the reader to seek further assistance from points of contact for each of the services listed below.

Waiver Guides and Points of Contact

US Air Force: Air Force Waiver Guide, 28 Mar 2012; http://airforcemedicine.afms.mil/waiverguide

The Air Force does not have a centralized point of contact at the time of this publication. Contact the local Air Force Base flight surgeon office for assistance.

US Army: Army waiver guide: Army Aeromedical Policy Letters, 31 Mar 2008;http://usasam.amedd.army.mil/dl/Flight%20Provider%20Refresher/References/APLs%20Guide%20Mar08_v4.pdf

Army waiver requests are submitted to USAAMA, headquartered at Fort Rucker, AL. All physicals and AMSs should be submitted through the web-based program AERO: https://vfso.rucker.amedd.army.mil/. For assistance, you may also contact Director, Aeromedical Activity, ATTN: MCXY-AER, Fort Rucker, AL, 36362-5333; (334) 255-7575.

US Navy: Navy waiver guide: Aeromedical Reference and Waiver Guide (ARWG, 2010); http://www.med.navy.mil/sites/nmotc/nami/arwg/Pages/AeromedicalReferenceandWaiverGuide.aspx

Navy waiver requests are submitted to NAMI Aeromedical Qualifications Department for review and appropriate endorsement. All physicals and AMSs should be submitted through the web-based program AERO (https://vfso.rucker.amedd.army.mil/). For assistance, you may also contact NAMI Psychiatry Department at 340 Hulse Road, Building 1954, Pensacola, FL 32508-1092; (850) 452-2783.

References

American Psychiatric Association. 2000. *Diagnostic and Statistical Manual of Mental Disorders, Fourth Edition, Text Revision (DSM-IV-TR)*. Washington, DC: American Psychiatric Publishing.

Bureau of Medicine and Surgery. 1992. *BUMED Instruction 5300.8. Disposition of Rehabilitated Alcohol Dependent Or Abuser Aircrew, Air Controllers,*

Hypobaric Chamber Inside Observers and Instructors. Available at: http://www.med.navy.mil/directives/ExternalDirectives/5300.8.pdf [accessed 25 August 2012].

Bureau of Medicine and Surgery. 2005. *Manual of the Medical Department*, Chapter 15, Medical Examinations. Available at: http://www.med.navy.mil/directives/Documents/NAVMED%20P-117%20(MANMED)/Chapter%2015,%20Medical%20Examinations%20(incorporates%20Changes%20128,%20130,%20135-140%20below).pdf [accessed 25 August 2012].

Department of Defense. 1997a. *DoD Directive 6490.1: Mental Health Evaluations of Members of the Armed Forces*. Available at: http://www.dodig.mil/HOTLINE/Documents/DODInstructions/DOD%20Directive%206490.1.pdf [accessed 25 August 2012].

Department of Defense. 1997b. *DoD Instruction 6490.4: Requirements for Mental Health Evaluations of Members of the Armed Forces*. Available at: http://www.dodig.mil/HOTLINE/Documents/DODInstructions/DOD%20Instruction%206490.4.pdf [accessed 25 August 2012].

Ernsting, J. and King, P. (Eds). 1988. *Aviation Medicine*. London: Butterworths.

Ireland, R. 2002. Pharmacologic considerations for serotonin reuptake inhibitor use by aviators. *Aviation Space and Environmental Medicine*, 73(5), 421–429.

Modell, J.G. and Mountz, J.M. 1990. Drinking and flying—the problem of alcohol use by pilots. *New England Journal of Medicine*, 323(7), 455–461.

Naval Aerospace Medical Institute. 2010. *Aeromedical Reference and Waiver Guide*. Available at: http://www.med.navy.mil/sites/nmotc/nami/arwg/Pages/AeromedicalReferenceandWaiverGuide.aspx [accessed 25 August 2012].

Rayman, R., Hastings, J. and Kruyer, W.B. 2006. *Clinical Aviation Medicine*. 4th Edition. New York: Professional Publishing Group, Ltd.

United States Air Force. 2009. *Air Force Instruction 48-123 Medical Examinations and Standards*. Available at: http://www.e-publishing.af.mil/shared/media/epubs/AFI48-123.pdf [accessed 25 August 2012].

United States Air Force. 2012. *Air Force Waiver Guide. 2012*. Available at http://airforcemedicine.afms.mil/waiverguide [accessed 25 August 2012].

United States Army. 2007. *Army Regulation 40-501, Standards of Medical Fitness*. Available at: http://www.apd.army.mil/pdffiles/r40_501.pdf [accessed 25 August 2012].

United States Army. 2008. *Army Aeromedical Policy Letters*. Available at: http://usasam.amedd.army.mil/dl/Flight%20Provider%20Refresher/References/APLs%20Guide%20Mar08_v4.pdf [accessed 25 August 2012].

Chapter 7
The Motivation to Fly and Fear of Flying

Chris M. Front

When you have once tasted flight, you will forever walk the earth with your eyes turned skyward, for there you have been, and there you will always long to return.

Leonardo Da Vinci

Introduction

The purpose of this chapter is to address some of the complex issues involved in determining the flight status of those with cockpit duties who develop a fear of flying. In the interest of simplicity and clarity, the term flier will refer to pilots, flight officers, navigator/weapon systems operators, and others whose training and duties place them in the cockpit with responsibilities related to the safe and effective operation of the aircraft.

Much has been written about the fear of flying. Indeed, the topic has been of interest ever since fliers have been a focus of study. As early as 1919, the employment of aircraft in combat operations in World War I (WWI) led Anderson to write the first book on aviation medicine, including a chapter on the psychology of aviation which addressed his observations of the fear of flying. A generation later, the extensive use of aircraft in World War II (WWII) produced nearly a million fliers and led to other seminal works (for example, Armstrong 1939; Bond 1952; Grinker and Spiegel 1945). Although fliers and aircraft have continued to play a central role in every conflict since that time, and commercial aviation has evolved into an enormous industry, most contemporary examinations of the fear of flying have been limited to a few brief paragraphs in various chapters on aerospace psychiatry (for example, Jones 1995; 2008) or articles in professional journals (for example, Bucove and Maioriello 1970; Jones 1986; Strongin 1987), often with a number of years between publications. Therefore, a secondary goal of this chapter is to trace the historical development of the most important and cogent contributions to the literature on fear of flying. The final goal is to consolidate and integrate those various contributions in order to provide a comprehensive but concise reference on the motivation to fly and fear of flying.

The chapter will begin with an historical overview, including the predominant theoretical explanations of the motivation to fly and fear of flying. In order to discuss the determination of flight status of those who develop a fear of flying, it is essential to first understand the complex dialectic between the flier's motivation to fly and his or her fear of flying. The chapter will provide a detailed examination of the concept of motivation to fly, summarize the different types of motivation, and discuss the importance of a "healthy" motivation to fly as well as the adaptive defenses necessary to prevent fear of flying. Fear of flying will be examined, including prominent theoretical formulations and the consequent treatment emphases that result from them. Given the consistency of clinical presentations over a century of military flight operations, typical symptomatic presentations will be provided, as will a discussion of treatment options, and issues relevant to medical and administrative disposition. Those familiar with the fear of flying literature are accustomed to the acronym FOF. That convention will be adopted here and the term motivation to fly will hereafter appear as MTF.

Historical Overview

Anderson (1919), a Royal Air Force (RAF) surgeon who himself learned to fly before flight training was a requirement for military flight surgeons, made the following prescient case for the role of aviation psychology:

> From the point of view of medical interest there is perhaps no more important subject than the study of the psychology of flying, in that the practical issues at stake are so great … Mention need only be made of its value in selecting the best type of man suitable for aviation duties, in advising and helping the pupil aviator during his period of instruction, in noting any change in his mental attitude towards flying, in intervening where loss of confidence is beginning, and preventing the establishment of a definite aeroneurosis, in detecting the malingerer, in co-operating with the instructors, and finally in the treatment and disposal of those who have broken down through stress of flying. (67)

Anderson described in detail the physical requirements used in selecting candidates for the Air Service (see Chapters 1 and 2, this volume for more on history and military selection). It is noteworthy that he took the position that "…next to vision, and most important of all in obtaining the best aviator, is the question of temperament…" (19). Anderson had become familiar with the variety of threats and stressors facing aviators, including those associated with wartime missions (anti-aircraft fire, air-to-air attack), those endemic to the machines themselves (such as engine or structural failure), the environment (for example, conditions at altitude, inclement weather), and those that originated in the aviator (for example, becoming disoriented, lost, or falling prey to the particular stresses associated with mission and aircraft type). Anderson noted that, at the outset of WWI, aviators were tasked with a variety of missions in an "odd job" manner, but by the end of the war, missions and the aircraft associated with them had become specialized. He provided descriptions of the particular psychological challenges associated with the various types of aviation missions of the era, and his observations regarding those mission- and aircraft-centric stressors, and the aviator temperament issues associated with them, are still relevant in the modern era (for example, Boyd, Patterson and Thompson 2005).

Anderson noted that the stresses of learning to fly, the psychological trauma that results from mishaps and accidents, and/or the psychological and physical toll of combat flying all can contribute to a condition that he designated Aero-Neurosis. Aeroneurosis, simply, was a reluctance to fly due to fear or anxiety, manifested by a variety of symptoms. He advocated early intervention when pilots experienced "unhappiness in the air, introspection, morbid thoughts, and the feeling that the dangers connected with flying are developing into an obsession … danger signs that the psychologist alone can discern and can take action accordingly" (95).

Anderson also presented what he had learned regarding aeroneuroses seen in flight students and Gotch (1919) of the Royal Navy addressed the aeroneuroses of war flying. Gotch wrote that disease, disorders of conduct, traumatic physical experiences, mental and physical exhaustion, and malingering could all lead to an aeroneurosis. The two writers posited a psychodynamic etiology that would prove durable over the remainder of the twentieth century. They asserted that early negative experiences gave rise to unconscious motivations that interfered with previously adequate functioning, manifesting in an emotional breakdown.

Notably, Gotch's recommendations for treatment also proved to be quite durable: prompt and thorough examination, followed by "sympathetic conversation" with the doctor and rest/convalescent leave, including, if indicated, medication to promote rest. These same treatment techniques were employed a generation later by flight surgeon psychiatrists in WWII (for example, Bond 1952; Grinker and Spiegel 1945). Gotch urged the separation of pilots with good prognoses from those whose presentation made it clear that they would not be returning to flying status, since the latter, if placed in convalescence among the former, would have a deleterious effect on their recovery by acting as a sort of psychological contagion.

Pilmore (1919), an American contemporary, was in agreement with Anderson regarding the importance of temperament in selection. Both of these early flight surgeons held the opinion that, after basic physical qualifications were met, it was temperament that was the most important factor in a student flier's success. Pilmore's monograph is interesting in its diversity, addressing a number of topics such as identifying a variety of both physical and mental conditions that should be considered disqualifying, advocating for a permanent medical specialty for medical officers serving in aviation billets, and the importance of including medical officers on the teams investigating aviation accidents. The astuteness of his observations is evidenced by the fact that most of his recommendations became US Navy policy.

Strongin (1987) observed that, between the two world wars, "There were no longer vast numbers of men training to fly and there was no war to stimulate anxieties or professional interest. The literature was unconcerned with psychological problems of fliers" (264). Strongin observed further that, during the interwar period, the profession of psychiatry was still maturing, and psychologists were devoting their studies to theories of learning, child development, and the development of various psychological tests. Hence, there was virtually no professional writing concerning the subjects of MTF or FOF until Armstrong's *Principles and Practice of Aviation Medicine* in 1939.

Armstrong's (1939) book contained a chapter devoted to the psychology of flight. He retained the term aeroneurosis and provided the following definition: "Aeroneurosis is a chronic functional nervous disorder occurring in professional aviators, characterized by gastric distress, nervous irritability, fatigue of the higher voluntary mental centers, insomnia, and increased motor activity" (453). Armstrong differentiated between ordinary neuroses and aeroneuroses, arguing that, in contrast to ordinary neuroses, aeroneuroses did not respond to "psychoanalytic maneuvers" and required instead the simple interventions of prevention or rest. Like Anderson before him, Armstrong proposed that an aeroneurosis could be precipitated by routine exposure to the real dangers associated with aviation, psychologically traumatic events such as experiencing, witnessing, or hearing of mishaps, and/or chronic physical and/or emotional stressors leading to fatigue. The fact that Armstrong observed this spectrum of precipitants during a time when the various air forces were not at war is a testament to the hazards associated with any type of flying during that era. Additionally, Armstrong recognized that changes in the flier's personal situation could precipitate an aeroneurosis. This observation may seem like a minor addition, but it proved to be important, as evidenced by the inclusion of personal life circumstances, especially romantic/marital relationship complications, in lists of observed FOF precipitants in subsequent papers over a period of decades.

The enormous scope of WWII resulted in nearly a million men being selected for flight training and becoming fliers in a variety of aircraft designed for a wide array of missions. The war itself further accelerated technological advances in airframes and powerplants. When the war began, biplanes were still in service in some roles. By war's end, the first fighter jet aircraft had become operational and flight operations from aircraft carriers had resulted in the first decisive naval battles prosecuted by aircraft alone, with opposing ships no longer even within sight of one another. These improvements in aircraft capabilities, combined with the vast number of aircraft engaged in aerial combat, resulted in even greater physical, cognitive, and psychological demands being placed on the flier. Once again, there was keen interest in MTF and FOF.

Strongin (1987) later made the observation that a common theme emerged in writers of the WWII era, with two emphases: (1) the development of classification categories based upon the extent that anxiety experienced by fliers interfered with their flying duties; and (2) the development of treatment approaches that matched different clinical presentations. Closer examination reveals a third focus: maximizing operational readiness by efficiently salvaging those fliers considered treatable. As Davis (1945) observed, "Early recognition and active therapy in cases presenting phobias would eliminate the tremendous waste of time that has been so evident in the past in pilots who have been grounded for an indefinite period of time because the nature of the condition was not recognized and therefore no definite action could be taken. If Flight Surgeons were more aware of the nature of these problems they could, in a large measure, decrease this loss of time by using the proper psychotherapy" (111).

The reader will note that aeroneurosis had by this time been replaced by the term phobia. This reflects developments in the psychoanalytic formulations of that era. Formulations for flying phobia were also presented by Grinker and Spiegel (1943; 1945) and by Bond (1952), a flight surgeon and psychiatrist in the 8th Air Force. Bond (1952) credited the contributions made by Grinker and Spiegel, but found less efficacy in the use of pentathol interviews for the processing of repressed traumatic experiences. Strongin (1987) asserted that Bond's "most important contributions were the demonstration of a very close association between the number of combat losses and psychological breakdown, and the detailed psychoanalytic descriptions of the dynamics underlying these breakdowns" (265).

Regarding the first part of Strongin's assertion, Grinker and Spiegel (1945) also made this dose-response observation, but Bond presented compelling data showing that there was no correlation between anxiety reactions and number of missions flown; that correlation coefficient was actually negative. In contrast, the correlation coefficient between anxiety reactions and the danger in the air as measured by the number of aircraft shot down on the mission was .75.

As for Strongin's second point, it cannot be overstated. Grinker and Spiegel (1945) presented a well-developed and artfully conveyed analysis of the psychodynamics of men in combat, including aerial combat, and their explication of the development of flying neurosis is both detailed and comprehensive. But Bond's articulation of the psychoanalytic dynamics associated with MTF and FOF had the most influence as it persists today.

The Korean conflict produced new twists in the typical FOF presentation. While the large majority of fliers in Korea served honorably and met their challenges with good adaptation, there was a clear subset of fliers who developed genuine FOF, feigned FOF symptoms, or frankly refused to fly. Schulz (1952) wrote about the challenges of identifying fliers with legitimate FOF or other psychological disorders in the midst of others who were frankly malingering. He observed that only three out of 186 pilots who refused to fly were regular US Air Force (USAF) pilots; the remaining 183 were reservists who had been re-called to active duty. Schulze also noted that 90 percent of those who attempted to avoid flying had received their training during WWII and already had combat experience. A key factor was that most had since married, started families, and gained civilian employment, never expecting to again be called to combat duty. Schulze judged that many of those who presented with FOF did so out of resentment that they had been returned to hazardous duty while there appeared to be an abundance of younger men without family responsibilities who had not yet served in combat.

Jones (1995) retrospectively noted some of the key differences between fliers in WWII and those in Korea, including the fact that virtually all WWII fliers were relatively young, unmarried volunteers who, in a spirit of national unity, perceived the war as an endeavor necessary for the preservation of freedom and democracy. The Korean "police action," in contrast, did not generate such national unity of purpose, and many reservists who were recalled to active duty in Korea resented having their civilian lives and careers interrupted to fight in a war that they did not perceive to carry a clear threat to their nation. Gatto (1954a) also noted that fliers serving in combat in Korea were more likely than their WWII counterparts to be married and settled into family life. He posited that they were, therefore, more susceptible to stress reactions due to the inherent conflict between their duty as providers for their families and their duty to serve. Gatto and his contemporary, Lifton (1953), both noted common FOF presentations in their Korean-era flier cohorts.

Tempereau (1956) arrived in-theater just four months prior to the end of hostilities in Korea, and remained for another 13 months. Hence, he had the opportunity to observe fliers' reactions during the transition from combat to peace-time flying, with fliers presenting with FOF consequent to precipitants other than combat, which likely resulted in his attention to maturational factors. Like Armstrong before him, Tempereau observed an apparent and predictable developmental sequence in the fliers' aviation life cycle that, occasionally, included the development of FOF.

Unfortunately, there were relatively few contributions to the corpus of literature on MTF and FOF in the years after WWII and Korea. Jones observed that, "Most of the literature concerning the effects of combat stress on fliers derives from WWII; little was written on this subject during the Korean conflict, and there is essentially no psychiatric literature on the USAF experience in Southeast Asia during the Vietnam conflict. No significant publications on this topic have emerged from recent USAF operations in Libya, Grenada, Panama, or Kuwait and Iraq" (1995, 179).

The relative lack of monographs examining FOF since the Korean conflict seems especially mystifying given the US Navy's transition to jet aircraft for carrier operations and the carnage that resulted. Rubel's (2010) examination of this transition provides a poignant testimony to the courage (and MTF) of Naval Aviators from the introduction of the Navy's first operational jet in 1947 to 1988, when the Navy's accident rate finally was reduced to a rate comparable to that of the USAF:

> [T]he statistics for the F-8 Crusader, a supersonic fighter designed by Vought in the late 1950s, provide a good illustration of the problem. The F-8 was always known as a difficult airplane to master. In all, 1,261 Crusaders were built. By the time it was withdrawn from the fleet, 1,106 had been involved in mishaps. Only a handful of them were lost to enemy fire in Vietnam. While the F-8 statistics might have been worse than those for most other models, they make the

> magnitude of the problem clear: whether from engine failure, pilot error, weather, or bad luck, the vast majority (88 percent!) of Crusaders ever built ended up as smoking holes in the ground, splashes in the water, or fireballs hurtling across a flight deck. This was naval aviation from 1947 through about 1988. Today, the accident rate is normally one or less per hundred thousand hours of flight time, making mishaps an unusual occurrence. This is in stark contrast to the landmark year of 1954, when naval aviation (that is, Navy and Marine combined) lost 776 aircraft and 535 crew, for an accident rate well above fifty per hundred thousand flight hours—and the rate for carrier-based tactical aviation was much higher than that … During this extended transition period, naval aviation participated in three major wars and numerous crises, and, of course, many planes and crews were lost to enemy fire. However, the vast majority of aircraft losses over this period were due to mishaps… (51–52).

Recalling Bond's (1952) data demonstrating that the incidence of FOF was a function of the dangerousness of the missions, it is inconceivable that accident rates and fatalities at the levels noted by Rubel did not result in FOF among a subset of Navy fliers. Hence, it is astonishing that there were relatively few significant contributions regarding FOF from the 1960s through the 1980s. US Navy psychiatrists Sours, Ehrlich and Phillips (1964) provided an explanation, noting that the nascent space program resulted in a keen interest in topics relevant to astronautics. The result, they argued, was a neglect of psychological issues related to normal Naval flight operations, including FOF.

Reinhardt (1967) examined 46 experienced and successful Naval Aviators who subsequently became unable to continue flying high-performance aircraft and were referred for psychiatric evaluation. His explication of the psychodynamic categories into which those failing fliers best fit both replicated and extended the classic categories. He explicitly aligned himself with Tempereau, Gatto, Sarnoff, and Sours et al. in critiquing the usefulness of FOF as a discrete syndrome, noting that it is misleading when used as a diagnostic term. Reinhardt argued that the diagnosis Adult Situational Reaction best described the majority of the fliers he studied. Notably, his summary of Bond's psychoanalytic explanations and the FOF syndrome produced a published response from Bond 20 years after Bond's original work. It is also noteworthy that Bond's response clarified what he perceived to be a widespread misunderstanding of his intended use of the term FOF, which he never intended to label a syndrome.

Strongin (1987) also lamented the lack of contributions from the Vietnam era. He described a presentation given by McGuire at Brooks Air Force Base in 1969. Strongin wrote that McGuire described five cases that were remarkably similar to reports from previous wars, but that McGuire discouraged the use of FOF as a diagnosis, suggesting instead that it be used only to describe a presenting symptom.

Bond's and McGuire's exhortations to consider FOF a symptom rather than a diagnosis may have contributed to the perspective endorsed by Bucove and Maioriello (1970). They described the cultural milieu of fighter pilots and the taboo nature of psychiatric complaints or admitting to FOF within that culture. Their observations of FOF symptom presentation in rated pilots (USAF terminology) are very consistent with those of Sours, Ehrlich and Phillips (1964) regarding designated aviators (US Navy terminology). They argued that, while operating within those cultural constraints, somatic symptoms provide fliers experiencing FOF with a method of communicating which they otherwise cannot admit, either to themselves or to others: "I don't want to fly jet fighters anymore." Hence, they asserted, such symptoms are better understood as a form of communication than as an illness.

Jones (for example, 1986; 1995; 2000; 2008) has been the most prolific writer on MTF and FOF in the modern era. Jones clarified and modernized the classic psychoanalytic explanations for MTF and FOF. He also addressed the psychiatric evaluation process for both and clarified various issues relevant to case dispositions.

After over half a century of focus on MTF and FOF in military fliers, Leimann Patt (1988) examined the phenomenon from the standpoint of civil aviation. He discussed the origins and role of healthy and unhealthy types of aeronautical motivation, then illustrated the interplay between MTF, healthy and unhealthy defense mechanisms, and aeronautical anxiety. In this manner, he accounted for both positive outcomes, which he labeled Flying Adaptation Syndrome—The Right Stuff, and pathological outcomes, or Flying Disadaptation Syndrome—The Wrong Stuff. His formulations resulted in ten different discrete syndromes, including FOF. Other writers (for example, Dyregrov, Skogstad, Hellesøy and Haugli 1992) have also examined FOF in civil aviation personnel, but writers from within the international military community continue to be the main contributors.

Perhaps the most significant contribution to the FOF literature in recent years has been by writers reflecting the contemporary shift from a purely psychoanalytic formulation of FOF, and the consequent emphasis on uncovering therapy to address the neurotic roots of the problem, to the application of behavior therapy techniques. Behavioral interventions such as systematic desensitization (Wolpe 1958) appear to have first been applied to FOF cases by a group of British clinicians treating fliers in the RAF (see Aitken, Daly, Lister and O'Conner 1971; Aitken, Daly and Rosenthal 1970; Daly, Aitken and Rosenthal 1970; Goorney and O'Conner 1971; O'Connor 1970). Noting many of the same symptoms and circumstances of development, these writers explained the phenomenon from the cognitive–behavioral theoretical perspective. They reported on the use of both imagery and *in vivo* methods, with effective resolution of symptoms and return to sustained adapted flying in a number of cases.

The shift to the application of cognitive–behavior theory is not surprising. Cognitive Behavior Therapy (CBT) has enjoyed contemporary pre-eminence among military clinicians due to an extensive literature demonstrating rapid efficacy with a variety of psychological disorders. In addition, the etiology of anxiety disorders from a cognitive–behavioral perspective is both straightforward and accessible to lay persons, and the clear cause–effect relationships inherent in explanations of etiology and treatment would seem to be a particularly good match for aviation. Even writers known for their psychodynamic training and acumen (for example, Jones 2008) have advocated CBT as a treatment option. Joseph and Kulkarni (2003) provided a useful summary of various CBT techniques and studies in which such techniques were applied to FOF, and Greco (1989) provided a practitioner's guide for CBT with FOF. Despite the recent emergence of CBT in the FOF literature, the classical psychoanalytic formulations for FOF still appear in the modern literature, and the psychodynamic conceptualization of defense mechanisms has proven useful and enduring.

The Motivation to Fly

In order to discuss the determination of an aeromedical disposition for fliers who develop FOF, it is essential to first understand the complex dialectic between the flier's MTF and his or her FOF. Adams and Jones asserted that, "Greater insight into what constitutes the normal, healthy MTF will help those who make judgments regarding the return of grounded aviators to flying duty" (1987, 350). The term MTF seems to imply a unitary concept, but a variety of factors contribute to an individual flier's motivations.

After more than a century of powered flight, and with airline travel having become a rather mundane activity, it is now easy to overlook the powerful symbolism inherent to flight. Flying is now experienced by the vast majority of people as an activity most similar to riding a bus or a train. It is, to most, simply another form of long-distance transportation; one that enjoys the added benefits of speed and safety. But there is an enormous difference between flight as it is experienced by the passenger and flight as experienced by the pilot. For the pilot, flying is an experience so deeply satisfying that it seems to defy description. For the flier, the experience includes a potent blend of freedom, joy, and mastery that are not attainable in any other activity. It is the reward of this experience that produces the ongoing MTF. Yet, with most fliers, the physical act of flying obtained during the first flight lesson is the fulfillment of a yearning to fly that has endured for years, usually since early childhood.

In their focus on the unconscious aspects of MTF, psychoanalytic writers have begun by emphasizing the symbolic meaning of flight. Armstrong (1939) pointed out that, "Except for religion, flying has more profoundly fired the imaginations of great numbers of people than any other thing in history. Strangely enough it is probably our religious teachings which account, to a certain extent at least, for the average person's fascination with aviation since almost all religions depict wings and ascent to the heavens as life's supreme spiritual reward" (442).

In Bond's (1952) detailed examination of the various unconscious motivations to fly, he called attention to the sky's symbolic content representing aspiration, freedom from the restrictions of the earth or of reality, and supernatural achievement. The sky, he observed, represents the dwelling place of the gods and, to most children, is synonymous with their ideas of heaven.

Against this backdrop of the symbolic meaning of the sky, Bond asserted, the activity of flying provides an unparalleled avenue for fantasy. Furthermore, it is precisely this facilitation of the individual flier's fantasy that produces an especially potent experience of fulfillment:

> Over and over one hears this same refrain from among flyers—the separateness of themselves from others; the unity of those who fly against those who do not; the feeling that among them there exists some inexpressible bond, as though they had shared some secret experience together which united them and set them apart; that only in the air are they whole; that there they find something long sought which allows the supreme fulfillment of themselves. It is, of course, the encouragement of each man's separate fantasy that creates this impression of unity. Each flyer in his own individual way has been free to experience his secret pleasure and has become addicted to it. It is the freedom in fantasy that is the uniting force, rather than the separate fantasies themselves. (18)

To this freedom in fantasy are added the themes of power and erotic attachment. Bond observed that the flier's interest in aircraft and flying exceeds the level of interest and pleasure that is evident in other common hobbies or sporting activities. He argued that "The aircraft itself becomes an object of erotic love … No one associated with fliers for any length of time can escape being impressed by the way in which the airplane is personified. Fliers become attached to one particular type of aircraft and will defend it jealously against all others" (21–23).

Bond considered the flier's tendency toward personification of the airplane to be a natural extension of erotic attachment. Such personification is perhaps best illustrated by the ubiquitous practice, which reached its zenith during WWII, of assigning names (usually feminine) to aircraft and decorating them with nose art which most commonly consisted of paintings of scantily-clad women, often in suggestive poses. It is important to note that, although this practice has been prohibited by military regulations in more recent years, it continues among some general aviation aircraft owners today, some of whom are devoted to restoring, maintaining, and flying the surviving military aircraft (warbirds) of the previous century in their original livery, including the nose art. These modern examples of the flier's devotion may be understood as testaments to the universality of Bond's assertions regarding the central role of fantasy in MTF, as well as the erotic attachment and personification of aircraft that was observed throughout the last century.

Bond's observations indicating the erotic love and sexual transference that occurred among the fliers of the 1940s is also still evident in the language of modern aviation. Sexual metaphors and innuendo continue to be commonly employed by fliers in their attempt to describe their MTF, as well as the joy experienced in flight —"Flying is the most fun you can have with your clothes on!" is a well-worn aphorism.

The experience of separateness noted by Bond is also still evident in the modern aviation community. Prior to WWII, there were very few pilots per capita. Recognition of the role of airpower in the attainment of strategic goals during WWII resulted in the training of nearly a million pilots, navigators, and flight engineers during the 1940s. Some of those fliers remained in the military as career aviators and many others left the military in the post-war years but continued to fly in general aviation. Yet, despite their large numbers relative to the pre-war years, flying remained a relatively extraordinary experience. For a number of reasons, the training of new pilots, both military and civilian, never again resulted in the vast number trained for WWII. As a result, over the past few decades, as the cohort of WWII fliers has aged and left the ranks of active pilots, the overall population of fliers has steadily declined. As of this writing, the population in the US is just over 312.5 million. There are fewer than 600,000 certificated civilian pilots, and fewer than 30,000 military pilots. Thus, general aviation pilots constitute only 0.2 percent of the population and military pilots less than 0.01 percent. Membership in such an exceptionally small group naturally equates to an elite status—a feeling that is intensified by the uniqueness of the experience of flight.

Bond's psychoanalytic explanation, perhaps predictably, culminated in his consideration of the act of flying as a "flaunting of phallic power." He posited that the "bodily significance of the aircraft," the "symbolism of flight as intercourse," and the "anthropomorphization of death into a living, threatening father who dwells appropriately in the sky" (30–31) combined to indicate that the Oedipus conflict is acted out in flight. The sublimation of basic drives most fully developed by Bond remained the predominant theme in the understanding of MTF for a number of years.

In discussing the role of mastery in MTF, Reinhardt (1967) credited Schmiedberg (1937) with the initial recognition of the importance of mastery achievement in activities such as flying and competitive driving. The observation was later advanced by Christy (1975) and, while Jones (1986) adopted Bond's psychoanalytic formulations on MTF, he also emphasized the central role of the need for mastery and the experience of joy that results from its attainment.

Studies of military pilots have provided empirical evidence for the central role of mastery in MTF. A study of USAF pilots by Fine and Hartman (1968) revealed that, "Flying is a sublimated, adult extension of early interests in activity, mastery and achievement" (51). Similarly, Reinhardt's (1970) examination of the qualities and characteristics of the most successful US Navy jet pilots revealed that, in a large subset, a strong need for mastery led to the career choice. Reinhardt found that, even among those for whom the pursuit of mastery did not play a role in career selection, the eventual attainment of mastery provided the reward that maintained their motivation: "After many cockpit hours, however, both groups found a sense of mastery of, and unity with, the complex man-aircraft unit of great versatility, maneuverability, and speed. This feeling is deeply gratifying, and many want to be alone during flight. One pilot said, 'This is the best and fastest fighter in the world, and the beauty of it is that no one can take hold of the controls except me, the pilot'" (33).

Whether one ascribes to the purely psychoanalytic drive-based explanations posited by Bond and others, or to the more developmental notions of joy associated with mastery as described by Reinhardt, Christy, and Jones, it is clear that for the prototypical flier, MTF is experienced at a deep level that defies adequate verbal description. Although hypotheses regarding the potential neurocognitive underpinnings of MTF do not appear in the extant literature, one might posit that MTF involves limbic system activation in addition to cortical thought processes. This proposal is consistent with a succinct summary of MTF presented by Jones and Marsh (2001, 131): "Briefly, a healthy MTF contains some elements of rational choice (it's a good job), and some of emotional attraction (I've wanted to fly as long as I can remember)."

Unhealthy Motivation to Fly

The discussion thus far has focused on healthy motivations to fly. They are considered to be healthy simply because they contribute to successful adaptation as a flier; success being indicated by the attainment of safety, proficiency, and longevity (Berry 1961). It is important, however, to understand that there are other motivations that have been associated with training failures, unsafe flying practices, psychological decompensation, and other career-ending maladaptive outcomes. For this reason, such motivations to fly are considered to be unhealthy. As Christy (1975, 309) observed:

> ...[I]t appears that an interest in flying may involve a need for mastery, for prestige, for control, the need for aggressive expression, or the need for symbolic expression in competition with siblings (peers) or with father or other authority figures. Other adjectives which are descriptive of expressed desires and feelings by those who love to fly include: thrill, freedom, excitement, power, speed, escape from earth, independence, competition, omnipotence and, more relevant for others, the obverse themes such as defiance, doing the forbidden, denying fear or defying death. The latter tend to be counter-phobic motivations for flying and therefore are potentially dangerous at some future date to safety of flight or to the resistance of the individual to development of a flying phobia.

Christy's use of the term counter-phobic originated with Morganstern (1966). Essentially, the counter-phobic flier seeks to fly in an attempt to overcome a FOF that is fundamentally unacceptable to him or her due to rigid defenses. For such a flier, the act of flying is an avenue for proving to one's self (and, perhaps, a domineering parent) that one is capable and can overcome the denied, but present, underlying anxieties. Due to this compulsion to prove their fearlessness, these fliers exhibit a tendency to engage in unsafe practices, such as unauthorized maneuvers and stunts. They usually become training failures but, if they manage to contain their tendencies until fully qualified, their disregard of operational risk management is eventually recognized by their more adaptive peers, who consider them "accidents in search of a grid coordinate" and refuse to fly with them. Hence, either through training attrition, mishaps, or administrative actions, their flying careers are chaotic and brief.

Similar to the counter-phobic are those who enter flying "due to an exaggerated need for power or attention or omnipotence and become unsafe fliers by their impulsive, over-compensating, danger-seeking flying" (Christy 1975, 310). The underlying motivation is different from that of the counter-phobic, but the disregard for safety, rules, and established procedures produces overt behavior that is essentially the same, with similar outcomes.

The selection of an aviation career in response to parental pressures is another unhealthy MTF. This may be more difficult to recognize, since trainees with a healthy MTF frequently come from families with a flying tradition. In the twenty-first century, it is not uncommon for a trainee to have a father and/or a grandfather who was a military flier. (With current trends, this will soon include mothers.) This usually contributes to a healthy MTF through early introduction to aircraft, to flight experience, and to flying culture via acquaintance with military squadron mates or general aviation flying activities. Furthermore, studies have demonstrated a correlation between strong positive paternal relationships and good adaptation to the demands of military flying (Perry 1971; Reinhardt 1970). But the presence of a family tradition can be an unreliable indicator of healthy MTF. For cases in which the tradition is seen as creating a foregone conclusion regarding career path, the flier may be choosing an aviation career so as to fulfill either overt or tacit pressure to continue the tradition, despite the lack of a desire to fly. Alternatively, it may be an attempt to surpass the accomplishments of a dominant father or older sibling who was a successful flier. Finally, this may also occur under circumstances in which a parent who yearned to fly, but was unable to due to medical disqualification or other circumstances, pressures the child to satisfy his/her unfulfilled goals. In any case, motivation that arises out of an attempt to satisfy, placate, or surpass parental or other authority figures is usually insufficient to sustain a career.

The film *Top Gun* (1986) provided a convenient and succinct euphemism to clinicians at the Naval Aerospace Medical Institute (NAMI) to describe another form of unhealthy MTF. Top Gun Syndrome describes an applicant for aviation training who is motivated by the status and prestige attributed to Naval Aviators. Such applicants lack the healthy MTF that is described above and are drawn to aviation only by the mystique associated with the military flier. This is an unhealthy twist to the sense of separateness noted by Bond. Rather than experiencing the sense of separateness as a natural by-product of membership in an elite group that is incidental to the fulfillment of a deeply experienced desire to fly, these applicants actively seek the elite status itself in their pursuit of narcissistic gratification. They do not care about flying; they care simply about being identified as a flier. This unhealthy motivation typically results in either training attrition due to a lack of healthy motivation, or to career dissatisfaction and/or adaptive difficulties.

Recognition of the negative influence on flight performance of such maladaptive personality traits led to the US Navy's classification of such trainees or fliers, once identified, as Not Aeronautically Adaptable (NAA) (see Chapter 6, this volume) and resulted in empirical study of such traits, as well as their more adaptive counterparts (Berg and Moore 1997; Berg, Moore, Retzlaff and King 2002; Christen and Moore 1998; King and McGlohn 1997; Maschke 2004; Moore and Ambrose 1998; Moore, Berg and Valbracht 1996; Paullin, Katz, Bruskeiwicz, Houston and Damos 2006; Picano and Edwards 1996; Retzlaff and Gibertini 1987; Siem and Murray 1994). For additional discussion of maladaptive personality characteristics and flying, please see Chapter 14, this volume.

Motivation to Fly and Successful Adaptation to Flying

As noted above, Berry (1961) used the three indicators of safety, proficiency, and longevity to define success in high-performance aircraft and space flight. While it is relatively easy to recognize that unhealthy MTF actively interferes with success, the role of healthy MTF is more subtle. Those familiar with the rigors of military/professional flying, however, recognize MTF as one of the most essential components of enduring success. In his observations of WWI RAF student pilots, Anderson (1919) was the first to note the importance of motivation, observing it to be a crucial component both in training and thereafter.

Anderson's position has continued to be supported throughout the decades since. O'Connor (1970) noted the small number of applicants to the RAF who eventually qualified as a military flier (one out of 50) and observed, "The aim of aircrew selection is to identify and reject those applicants who do not possess the necessary aptitude to learn to fly in a reasonable time, who lack the dedication required for military aviation, or whose personality and temperament seem ill-suited to the stresses of aircrew life" (877).

Even the most advanced airplanes of WWI were primitive, unstable, fragile, and highly unreliable by today's standards. The flier of the day needed a strong MTF in order to persist through the crashes and emergency landings that resulted from the airplane's inevitable engine and structural failures. Those who were not seriously injured or killed certainly lost friends and fellow fliers to the mishaps of the era. As decades passed, airframes and powerplants became more sophisticated and reliable, but they also became faster and able to withstand maneuvers that placed more demands on the pilot, with the result that crashes and mishaps continued to be commonplace. In the modern era, although aircraft are highly reliable, they are also highly sophisticated systems requiring a combination of physical fitness, motor, and neurocognitive abilities in order to operate them safety and effectively, an outcome that can only be achieved after a great deal of very challenging study and training. Hence, despite the reduction in mishaps achieved in the modern era, the flier still requires a strong MTF in order to complete training and sustain a successful career. It is notable that motivation was among the personality traits identified by Picano, Williams, and Roland (2006) in their meta-analysis as predictive of successful performance not only in military flying, but also in other military professions involving high-risk operations. In such professions, adequate motivation sustains the operator through the rigors of training and, importantly, it also enables perseverance through the ongoing adversity inherent to the profession.

Motivation and the Flier's Developmental Stages

Armstrong (1939) appears to have been the first to identify common developmental stages experienced by fliers in the course of their flying careers, which he described as four psychologic periods. About a generation later, Tempereau (1956) described the five stages that he repeatedly observed during the development of FOF. Although the two authors delineate their phases differently, there is significant overlap between their observations.

Armstrong observed that, "The first few years of flying are glamorous ones and there are few inhibitions due to age or experience. Usually during this period there are no serious domestic or financial responsibilities to be concerned with and the dangers of flying exist in the mind of each individual only for the other fellow who is not quite so capable as he" (448). Armstrong's first psychologic period contains the first two stages described by Tempereau, which he labeled Stage 1: The Initial Thrill and Stage 2: The Hot Pilot. Stage 1 occurs during the first few training flights and is noted to be closely akin to the thrill one experiences when riding a rollercoaster. Tempereau noted that this thrill occurs among passengers as well as among new pilots. He viewed it as masked fright and posited that it requires the new pilot to engage the defense mechanisms of denial and rationalization in order to cope with the innate fear of height and separation from the earth's surface (and, therefore, FOF) that was described by Gatto (1954a). Tempereau's Stage 2 occurs after the pilot has engaged in several initial flights and has successfully defended against the innate fear such that it is now completely denied or repressed. When this occurs, "the awesome, uncanny, 'wonderfully-terrifying' thrill passes, to be supplanted by a stage in which flying quite obviously acts to satisfy an intense emotional need. It is here that the symbolism of flight seems to become important. The 'hot pilot' or 'tiger,' boisterous and flamboyant, is most at ease in the cockpit. On the ground he is restless; while he is flying his body and his plane through the sky he is supremely happy" (220).

The second psychological period described by Armstrong typically begins after two to four years of flying. He noted that entry into that period may be abrupt or gradual, depending on circumstances. Assuming no unusual issues, the pilot grows more conservative, recognizes his limitations and exhibits an appreciation for the hazards of flying. In other cases, this change toward realism and conservatism may occur abruptly due to one of two causes. The first is exposure to narrow escapes or crashes; either by personally experiencing one or more close calls or by witnessing or learning of the deaths of other pilots known to be at least as capable and competent as the flier. The second cause is rooted in marriage and the establishment of a family, with the attendant concerns and responsibilities produced by this change in status. Armstrong observed that, during this period, "the accumulated experience and judgment of the individual increases his flying ability and he becomes a better and safer pilot" (448). This period is consistent with the beginning of Tempereau's Stage 3: The Airplane Driver, which he noted was a term "coined by pilots to define an analogy between the career flier and the truck or bus driver." This stage is characterized by the more experienced pilot having settled down. Flying is experienced as less stimulating, is now more of a job to perform. Tempereau noted that this stage is reached by most older pilots and represents the normal end point in the developmental stages.

Armstrong's third psychologic period would seem to contain Tempereau's Airplane Driver stage, as it is noted to extend "over a period of about 10 years and occupies roughly that part of the life span between the ages of 27 and 37" in the career of the typical flier. He observed this period to be characterized by a continued trend toward conservatism and phasing out of tendencies toward recklessness. Armstrong hypothesized that this stage requires the flier to engage in a series of emotional adjustments, noting that failure to accomplish these emotional developmental tasks will result in adjustment problems such as aeroneurosis. Hence, Armstrong's third psychologic period may or may not contain Tempereau's Stage 4: Emergence of Anxiety. In Tempereau's model, the emergence of anxiety can be precipitated by a stressful event during any of his first three stages.

If all continues as it should, however, Armstrong's fourth and final psychologic period in the life of the professional flier is achieved. He noted that this period occupies the time from about age 37 to the end of a flier's career, is the safest period of flying, and is achieved if the flier has successfully achieved the prior stages. Tempereau's Stage 3: Airplane Driver appears to also span this period. Both writers make the observation that advancing age, conservatism, and the necessary corresponding emotional adjustments are associated with increased safety. This observation is evident in the well-known aviation aphorism: "There are old pilots and there are bold pilots, but there are no old and bold pilots."

Armstrong observed that this final period may continue indefinitely or it may evolve further due to changing personal priorities. During this later period of the flight career, the flier may opt to stop flying due to a variety of reasons, to include hazardous duties, increased administrative duties, introduction of new aircraft or flying techniques or increasing conservatism. It is important to note that, regardless of whether the flier in Armstrong's final psychologic period continues to fly or, due to evolving personal priorities, decides to make a career change that includes a cessation of flying, there is no manifest adjustment problem; using Armstrong's nosology, there is no aeroneurosis. It is a reasoned career decision rather than an aeromedical issue; the flier is moving toward a new developmental goal rather than fleeing from maladaptive anxiety. In contrast, if an adaptive failure occurs during

Armstrong's third psychologic period, in a manner equivalent to Tempereau's Stage 4, then the condition requires aeromedical evaluation and disposition.

It is also important to note that, consistent with Armstrong's later periods, Tempereau did not hypothesize that all fliers continued into his fourth and fifth stages. Most enjoy the relative banality of completing their careers as Stage 3 Airplane Drivers. The final two stages in Tempereau's model were included in order to accommodate the full range of observed developmental courses, including those fliers that experienced the emergence of anxiety that is the hallmark of his fourth stage.

Tempereau labeled his final stage, following the emergence of anxiety, Stage 5: Defense Formation. It is characterized, simply, as "the erection of psychological defenses against developing anxiety, whether in the form of phobic reactions, hysterical symptoms, depression, or behavioral disturbances" in direct response to the experienced anxiety. He hypothesized that, "The type of defense will depend upon the subject's pre-morbid personality and his earlier conflicts" (220–221). He hypothesized further that the effectiveness of the developed defenses determines the longevity of the experienced anxiety. Tempereau endorsed and promoted the explanation of the role of defenses as explicated by Gatto (1954a), who argued that the very nature of the activity of flying requires the flier to develop psychological defenses in order to repeatedly engage in the activity without anxiety.

The *Motivation–Defense–Fear Balance* and Aeronautical Adaptation

In order to assess a flier, it is necessary to examine the complex relationship between MTF and FOF. Essentially, MTF can be conceived as providing an opposing drive or force which balances out the natural and instinctive FOF. This balance exists in the psychic life of all fliers, whether or not they are consciously aware of it. Importantly, the balance is dynamic in that the psychic defenses that are employed in the management of the fear evolve over time, which accounts to a great extent for the different stages in the flier's development as described above. The balance is also dynamic in that it may tilt more or less in one direction or the other as the flier experiences increased skill and confidence in his or her flying ability, which is then tempered by close calls and the recognition that, due to a host of uncontrollable variables, risk remains regardless of skill level. When the defenses are structured such that the balance is clearly tilted toward MTF, then the flier is more able to experience the thrill and joy noted in the earlier developmental stages. The more mature, conservative flier has a more balanced homeostasis. The flier who experiences circumstances that challenge his or her ability to develop adaptive defenses finds the balance tipped toward FOF and, unless new and/or more effective defenses evolve, will develop anxiety. Hence, FOF may be conceived as not a new phenomenon experienced only by a subset of disturbed fliers, but rather as an aspect of flying that is fundamental to the activity of flying, and therefore must be managed by all fliers through healthy defenses, but which may break through at various points in the flying career, depending upon the circumstances and the effectiveness of the flier's defenses.

In Leimann Patt's (1988) creative explication of this concept, he drew a comparison between the critical components of the *Motivation–Defense–Fear Balance* noted here and a simplified version of the aerodynamic formula for Lift. He noted that Lift = [Speed x Wing Performance]/Weight and proposed that a healthy psychological adjustment to flying (which he termed Flying Adaptation Syndrome) could similarly be represented as follows:

$$\text{Flying Adaptation Syndrome} = \frac{\text{Aeronautical Motivation x Defense Mechanisms}}{\text{Aeronautical Anxiety}}$$

Leimann Patt argued that, "Healthy aeronautical motivation is as important to carrying out a successful flying activity as adequate defense mechanisms are to counteract aeronautical anxiety" (956). He used variations in the relative contributions of the constituent elements in this basic formula to illustrate the resulting impact on different types of Flying Disadaptation Syndrome, including FOF.

The fundamental role of MTF in good adaptation to flight was present as early as 1919, in Anderson's (93) observation that, "In the psychological study of the aviator one is struck by the importance of the motive in taking up aviation. This gives more or less driving power to the conscious endeavour to overcome the obstacles in learning to fly—and supplies the determination to surmount difficulties throughout the whole flying career. The author places determination, grit, 'guts,' call it what you may, as the most important factor in flying." Despite the ever-present role of the innate FOF in the *Motivation–Defense–Fear Balance*, most clinicians familiar with FOF as it is manifested in experienced fliers have argued that the issue is often more complex than a simple breaking through of the innate fear.

Fear of Flying

FOF might most simply be defined as a reluctance to fly due to fear or anxiety, manifested by a variety of symptoms. The complexity of the phenomenon becomes apparent once one begins to consider the difference between fear and anxiety, the types of psychic defenses that predominate, the variety of possible symptom presentations, whether the flier is a student or experienced flier, and other issues, such as whether one is using the term to refer to a symptom, a syndrome, or an administrative disposition.

A number of writers have pointed out that the term FOF is problematic. Responding to Reinhardt in 1967, Bond noted that the aviation training commands of the late 1930s and early 1940s had taken to using the term "to denote a nonmedical, administrative reason for removal from flying status. It was a term of opprobrium and was first cousin to the term 'lack of moral fiber' of the RAF. In combat in England the Eighth and Ninth Air Force started using this term for more or less punitive removals from flying and the term 'operational exhaustion' for medical or nonpunitive removals." Consequently, Bond began using the term anxiety reactions with FOF representing one symptom of the reaction. Sours, Ehrlich and Phillips (1964) argued that FOF is an imprecise term, comparable with depression in that it is variably used to refer to a symptom, a syndrome, or a nosological entity. Clinicians must be aware of this terminology problem and be both intentional and explicit about their use of the term FOF.

Psychodynamic Formulations

Psychoanalytic formulations of aeroneuroses, flying phobia, and FOF dominated clinical thought for most of the past century. The student flier must have adequate healthy MTF and must quickly develop adaptive defenses in order to cope with the reality that the activity of flying leaves little room for error—at a point in the learning process when the student is acutely aware of the limitations of his or her flying skills—yielding the conclusion that serious injury or death is, therefore, a potential outcome of the endeavor. Hence, Jones (1986) points out:

> Perhaps one should not ask "Why do some fliers become afraid to fly?" but rather, "Why are not ALL fliers afraid to fly?" The answer appears to me to lie in the amount of pure joy the flier derives from flying, the amount of anxiety mixed with natural fear, the extent to which the flier's defenses have been challenged by circumstance, and in the adequacy and maturity of the flier's psychic defenses. Real fear about real danger must be clearly distinguished from a basic, primitive anxiety, i.e., a neurotic component. As an example of the difference, contrast the logical and useful fear that we feel when a rattlesnake is set loose in the room with us with the anxiety felt by someone with a snake phobia when a snake unexpectedly appears on a movie screen. These two emotional responses and their psychophysiologic concomitants may appear to be identical, but fear is instinctive and anxiety is neurotic. The stimuli define the difference: real danger evokes fear, and a symbolic threat evokes anxiety. (131–132)

Jones's points are subtle, but of crucial importance in understanding FOF reactions from the psychodynamic perspective. The student flier is being presented with real danger, which may (depending on the motivation and defenses) result in a fear response or a neurotic response. It is no surprise that this has proven to be an enduring phenomenon. Anderson (1919) argued for making a distinction between flying pupils and qualified pilots. His perspective was that the challenges of flight training quickly exposed the fact that some students were simply not suited for flight duty. He observed that some students honestly admitted that they did not want to continue flying. Others were not so forthright and their difficulty was manifested instead by some form of aeroneurosis. Anderson noted that he initially attempted to treat them and eventually reached the conclusion that this was essentially impossible. Consequently, his practice became to simply reassign such students to other, more traditional, military duties, with no negative career consequences.

It is a testament to Anderson's clinical acumen that nearly a century after he practiced, military specialists in aerospace psychiatry and psychology still distinguish between symptoms of fear or anxiety experienced by flight students versus those experienced by qualified pilots. The USAF policy, for example, is to classify difficulties related to excessive fear or anxiety associated with flight training as Manifestation of Apprehension (MOA) to flying (King 1999). MOA is viewed simply as an indication of unsuitability for flight training, leading to reassignment to other USAF duties, with no career penalties. The same symptom presentation by a rated USAF pilot, in contrast, is likely to result in evaluation by the Department of Neuropsychiatry at the USAF School of Aerospace Medicine.

The reason for completely different responses—administrative for the student and clinical for the qualified flier—is rooted in Jones' distinction between reality-based fear and neurotic anxiety. The student is experiencing fear in reaction to the reality-based risks associated with flying, a reaction that signals inadequate MTF and/or inadequate or incomplete formation of the defenses required to establish an adaptive *Motivation–Defense–Fear Balance*. In contrast, the development of anxiety in an already-adapted flier represents what Jones referred to as a serial change in functioning that signals a failure in the system of previously adequate coping defenses, and thus requires a more sophisticated evaluation. This perspective is supported by the findings of Sours, Ehrlich and Phillips (1964) that student and early-career fliers were more likely to present with Manifest (Overt) FOF reactions, whereas seasoned fliers almost always present with Latent (Covert) reactions. The types and quality of defenses employed by the flier play an integral role in the *Motivation–Defense–Fear Balance* and are an essential component in psychodynamic formulations of FOF.

Coping defenses

An understanding of the differences between adaptive and maladaptive coping defenses is fundamental to the psychodynamic formulation of FOF. Noting the need for coping defenses in order to adequately adjust to the reality-based fear inherent to flying, Jones (1986) observed that fliers employed a combination of the defense mechanisms of denial, humor, suppression, intellectualization, and rationalization in response to near misses, deaths of fellow fliers and other hazards of aviation. Leimann Patt (1988) expanded on Jones' observation, describing and drawing a distinction between defense mechanisms that typically contribute to good adaptation and those that lead to disadaptation. He argued that denial, repression or suppression, psychophysiological habituation (for example, in cases of motion sickness), rationalization, and identification are normal and effective.

Although not noted by Leimann Patt, based upon the observations of Jones and others, humor, intellectualization, sublimation, and thought suppression (ability to compartmentalize) are also included among those defense mechanisms that contribute to good adaptation for fliers.

In discussing defenses that contribute to disadaptation, Leimann Patt focused on reaction formation, evasion, displacement, and isolation, noting that they are "usually tragic … the first leads to substandard operational behaviors, and the other three to severe psychopathologic downfalls" (956). Regression, acting out, and somatization were not noted by Leimann Patt, but are represented prominently in clinical cases presented by numerous psychoanalytic writers on FOF (for example, Bond 1952; Gatto 1954a; Gatto 1954b; Grinker and Spiegel 1945; Jones 1986; Phillips and Sours 1963; Sours, Ehrlich and Phillips 1964).

If the flier employs defenses that are adequate and appropriately developed, then the *Motivation–Defense–Fear Balance* is adequately maintained and the flier's career progresses normally. If, on the other hand, the defenses are maladaptive, or the flier's circumstances change such that previously adaptive defenses are no longer adequate to maintain the *Motivation–Defense–Fear Balance*, then problems ensue. The various constellations of physical, emotional, and/or behavioral symptoms that result from this disruption to the *Motivation–Defense–Fear Balance* have been described variously as aeroneuroses, flying phobias, FOF, and with the more general rubric FOF Syndrome.

Neurotic roots of Fear of Flying

Some writers (Bond 1952; Jones 1986) focus on the special aspects of flight that make FOF a neurosis unique to fliers. Others (Armstrong 1939) propose that one type of flying phobia is a version of the types of neuroses that are common to all patients, while another is specific to fliers. Still others (Gatto 1954a) argue that the neurotic roots of FOF are essentially the same as those for non-fliers.

Bond noted that there is an important difference between those neuroses that develop spontaneously and those that develop as the result of a catastrophe not the fault of the pilot. It is important to recall that, at the time of Bond's writing, Acute Stress Disorder and Post-Traumatic Stress Disorder (PTSD) did not yet exist in the clinical nosology. Bond's handling of such cases noted their particular presentation while keeping with the clinical thought of that era. He observed that traumatic experiences led to a more rapid but circumscribed onset of phobic symptoms, whereas cases with non-traumatic precipitants had a slower rate of development, but spread progressively into broader areas of phobic response. He also noted that while the developing phobia is still unconscious and has yet to become a clear phobia, it may be expressed in physical symptoms, the most common of which included airsickness, pseudo bends, headaches, and vertiginous attacks. A final symptom common to the latter group, "is a growing cautiousness in the air, a symptom that can be very dangerous, for it hampers direct action and decisions. Often this symptom takes the form of obsessively following one safety precaution to the exclusion of others." Importantly, Bond also observed that the most common type of phobia that occurs during flight training is the type that develops as the result of an aircraft accident not of the pilot's making.

Armstrong (1939, 453) also argued for only two types of neuroses in fliers, but differed from Bond in his distinctions. One type, he proposed, is a true aeroneurosis, which he defined as "a chronic functional nervous disorder occurring in professional aviators, characterized by gastric distress, nervous irritability, fatigue of the higher voluntary mental centers, insomnia, and increased motor activity." He noted that this type "develops in relatively stable individuals, is not necessarily disabling, and is an entity seen only among aviators." He argued that the second type of neurosis observed in fliers "develops in relatively unstable pilots or in relatively stable pilots under conditions of unusual stress, is disabling, and is identical with those neuroses seen in general practice." Unlike the true aeroneurosis, this type is undifferentiated from typical clinical neuroses but simply has flying as the cause.

Gatto (1954a) either did not recognize or did not concur with the importance of the symbolic meaning of flight advocated by Bond; he did not even address the issue of MTF in his extensive coverage of FOF. This likely led to his view that FOF arose due to conflicts over instinctual needs, superego forces, and external reality, noting that the clinician who understands the military aspects of these factors will have an advantage.

Gatto proposed that conflicts over instinctual needs result from disturbances in personal relationships and libidinal ties that interfere with the need to express affection, anger, aggression, and so forth. He noted that this is especially true in the military flight environment where expressions of any type of significant emotion are generally not considered consistent with the culture. As a result, such situations are more challenging for those with especially strong superegos (a psychic configuration typical of fliers). Gatto argued that, "When men with strong superegos are bombarded constantly by internal and external conflicts, severe somatizations may occur." He described cases involving gastrointestinal reactions, suspected hyperthyroidism, and musculoskeletal reactions.

Regardless of their differences, common to all of the psychodynamic writers is the belief that the expressed symptoms are merely the overt manifestations of underlying conflicts that exist at varying levels of consciousness. Therefore, identification of the underlying neurotic conflict is the key to understanding the particular symptomatology and determines the most useful treatment and/or disposition.

Fear of Flying Syndrome and clinical presentations

The variety of presentations that stem from disruption of the *Motivation–Defense–Fear Balance* led clinicians to hypothesize a broad category or syndrome that would encapsulate the entire array of clinical presentations. Gotch (1919) was the first to propose such a classification system, with six categories: fatigue, neurasthenia, cases in which confidence has not been lost but in which a toxic element caused the breakdown, such as the flu or syphilis, psychopaths, physical causes, such as lack of oxygen at high altitudes, and malingerers. Gatto (1954a) described the following seven patterns of behavior: obsessive over-concern related to the functioning of the plane; phobias (that is, flying, claustrophobia); psychosomatic disturbances; behavior disturbances revealing inadequacy or delinquency, such as alcoholism or malingering; neuroses; pseudopsychoses; and psychoses. In their reappraisal of the extant FOF literature, Sours, Ehrlich and Phillips (1964) distilled all of the various types of clinical presentations into two overarching groups: Manifest (Overt) versus Latent (Covert) FOF reactions. They included among the Manifest (Overt) FOF reactions: acute situational reactions, aesthenic psychophysiological reactions, neurotic character disorders, gross stress reactions, and sociopathic reactions (malingering).

Acute situational reactions were clearly predominant in flight students with poor motivation. The authors also described this group as failing to develop adequate coping defenses, as evidenced by their over-focus on the risk of death while flying. Their symptoms, which included depression and anxiety, as well as sympathetic nervous system overactivity resulting in various psychophysiological reactions, did not derive from long-standing neurotic conflicts and did not simulate organic disorders. These individuals simply did not want to be fliers and felt stuck in the role.

The aesthenic psychophysiological reactions group was characterized by marked obsessional traits and rigid perfectionism. Their symptoms typically included lassitude, fatigability, and multiple somatic symptoms that progressed until they were in marked distress whether flying or on the ground. Nevertheless, they "doggedly continued to fly after recognizing that they were not aeronautically adapted. They tried to substitute rituals for flight procedures in order to control fear of the aircraft" (158–159). Like the first group, their fears were based upon realistic appraisals of the risks associated with flight, but they attempted to manage their difficulties through flying, resulting in significant symptomatic periods and distress.

The neurotic character disorder group also experienced realistic FOF, but their histories revealed underlying neurotic character disorders and frequently included unhealthy MTF and a lack of enjoyment of flight. Despite the fact that they were fully conscious of their FOF, what distinguished them from the acute situational reaction group was a relative lack of insight.

Gross stress reactions were so-called due to the extremity of symptoms exhibited consequent to aircraft accidents. The description offered by Sours, Ehrlich and Phillips (1964) is entirely consistent with the modern diagnostic category of PTSD: "Typically, these aviators withdrew from flying after the accident. They appeared tense and anxious to family and friends. Sleep was usually interrupted by nightmares which recapitulated details of the traumatic event. Occasionally startle reactions would occur if explosions had been associated with the accident. Although resembling in some respects acute anxiety states, these reactions were the sequelae of gross stress experiences." They noted that these fliers did not have histories indicative of neurotic tendencies; rather, they were stable fliers whose symptoms were clearly consequent to the traumatic experience. Had PTSD been part of the diagnostic nosology of the time, there is little doubt that Sours, Ehrlich and Phillips would have applied that diagnosis to this group.

The final group of manifest reactions was labeled sociopathic reactions due to the conscious simulation of organic symptoms characteristic of this group. This group was noted to claim feigned symptoms related to vision and hearing in particular, and experienced complete resolution of symptoms once removed from flying. It was also noted that subsequent investigations into their histories revealed other examples of manipulation and deception.

The Latent (Covert) FOF reactions noted by Sours, Ehrlich and Phillips included: chronic recurrent airsickness, syncopal reactions in flight, anxiety reactions, and hysterical reactions of two subtypes: somato-sensory and somato-motor and special senses. Importantly, they noted that three of the four types of latent reactions were observed predominantly in student fliers (mean flight time = 60 hours). In contrast, the anxiety reactions group was composed of seasoned, previously adapted fliers.

The chronic recurrent airsickness reactions common to students differed somewhat from the transient airsickness experienced by many students upon initiation of aerobatic training. In contrast, the students in this group often began experiencing airsickness even before entering the aircraft, and in some cases symptoms persisted after the flight and into the evening. Some experienced no airsickness if they were in control of the aircraft, but grew airsick if another pilot was in control. Sours, Ehrlich and Phillips noted that this group often had an overdeveloped need for control and were very aware of their bodily sensations. In an interesting corollary finding, McMichael and Graybiel (1963) found that Rorschach measures associated with rigidity and emotional lability were significantly correlated with susceptibility to motion sickness.

Syncopal reactions were noted to most commonly occur during aerobatic maneuvers when G forces in the range of 3–4 Gs were experienced. Notably, these reactions oftentimes occurred after the flier had already developed tolerance for Gs, and they could not be reproduced in the human centrifuge with Gs experienced on the ground. Citing several studies, Sours, Ehrlich and Phillips asserted that, "Psychophysiological studies of low G tolerance have demonstrated that anxiety, fear of loss of control of the aircraft and depression correlate with low 'blackout' level and epinephrine production. On the other hand, expressed anger, minimal anxiety and a sense of well-being in flight are associated with high G tolerance and norepinephrine secretion. The interplay of neurotic latent FOF and altered physiology is most clearly seen in syncopal reactions."

Hysterical reactions were characterized by various types of conversion reactions. They included in this group fliers presenting with unfounded somato-sensory and somato-motor impairments, as well as conversion reactions involving the special senses of sight and hearing. Sours, Ehrlich and Phillips asserted that these were hysterical reactions using bodily impairment as a means of defending against anxiety. They noted that the somato-sensory and somato-motor impairments were easily recognized due to their flagrant qualities, but the conversion reactions involving the special senses were more difficult to evaluate, requiring repeated and extensive medical evaluation before the true nature of the problem could be identified.

The final group of Latent presentations, anxiety reactions, was typical only of previously adapted and well-seasoned Naval Aviators (most had accumulated over 1,000 flight hours). These fliers were generally described as motivated to fly, closely identified with Naval aviation, and had previously effectively used denial and repression to manage fears of flying. Sours, Ehrlich and Phillips asserted that the key to this group was an underlying neurotic conflict. Examination of their past histories revealed some evidence of overtly or covertly conflicted relationships with parents, and many were noted to be experiencing anxiety and frustration secondary to physical separation from their wives.

The symptom of acute anxiety in these fliers was transient and circumscribed to flying, yet they attributed their symptoms to "insignificant medical illnesses, poor flying conditions, inadequate oxygen equipment, defective anti-gravity suits and the eccentricities of their particular aircraft." These fliers failed to establish the connection between their anxiety and flying. This conflict was frequently evidenced by prodromal behavioral changes such as requests for flying lower performance aircraft, unnecessary abortion of missions, and overly cautious behavior. They noted that, with careful scrutiny, subtle changes in motivation could be detected as the result of such precipitating factors as career assignment disappointments (for example, failure to be selected for command, assignment to undesirable billets), poor squadron records, marital difficulties, and accident near-misses. Sours, Ehrlich and Phillips also cited a few occurrences of a decline in motivation in conjunction with Tempereau's stage progression from hot pilot to airplane driver. They observed that it was apparent that these aviators now derived more pleasure from other areas of their lives, but could not admit any decline in MTF and reacted negatively when that conclusion was proffered to them. This observation is consistent with the description of the double-bind later observed by Bucove and Maioriello (1970); namely, that the culture of fliers, particularly the squadron milieu of military fliers, provides a strong disincentive for fliers to admit a decline in MTF, contributing to latent expressions.

Fear of Flying: A synthesis model

Joseph and Kulkarni (2003) found the Manifest and Latent categorization system advanced by Sours, Ehrlich and Phillips (1964) to be compelling enough that, in their attempt to integrate and synthesize the FOF construct, they organized their model around this distinction in presentations. They also adopted the information presented by Sours, Ehrlich and Phillips (1964) regarding the influence of underlying neurotic personality styles and motivation on such reactions. Sours, Ehrlich and Phillips (1964) had used the work of Bond (1952) and Gatto (1954a) for their theoretical underpinnings, and their work is also consistent with the later work by Jones (for example, 1986). Hence, the model proposed by Joseph and Kulkarni provides a useful summary to the psychodynamic formulations of FOF.

Joseph and Kulkarni (2003) acknowledged the role of coping defenses in their review of the FOF literature, but they did not make the role of coping defenses explicit in their model depicting the interrelationships of the various components. The proposed role of defenses presented here will extend the work of Joseph and Kulkarni. As portrayed in Figure 7.1, a stable personality combined with adequate ability, good motivation, and the development of adaptive coping defenses leads to adequate adjustment to flying.

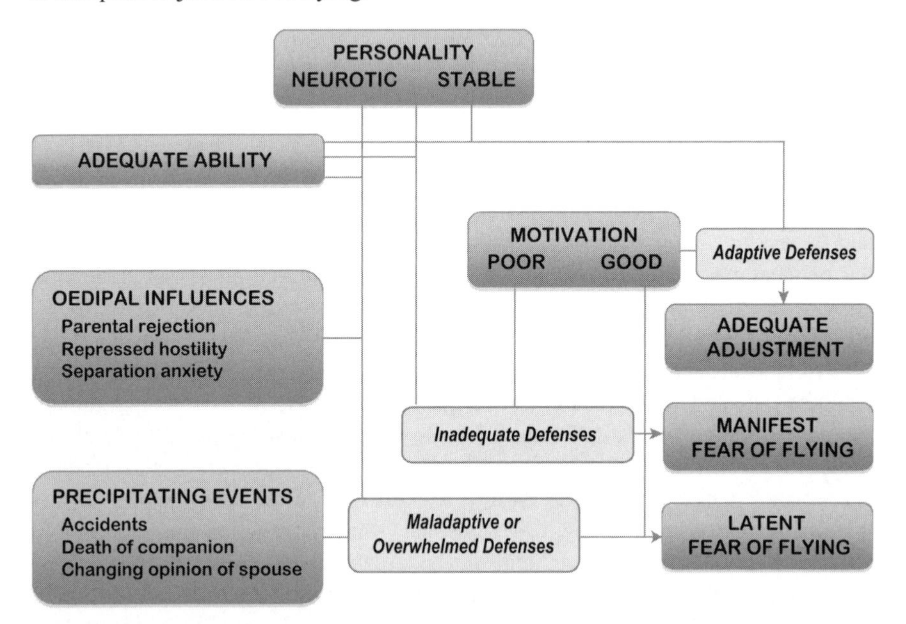

Figure 7.1 The Motivation–Defense–Fear Balance: Variables Resulting in Adjustment versus Manifest FOF versus Latent FOF

Source: Adapted from Joseph & Kulkarni (2003)

In contrast, a manifest FOF may result in two ways. In the first, a stable personality combined with poor motivation and inadequate defenses leads to one of the manifest presentations of FOF. In such cases, the role of ability may vary. For instance, a perception of inadequate ability on the part of the student flier could easily impede the development of adequate defenses. It is important to note that, in such a case, this could signal good judgment on the part of the student, who recognizes and simply acknowledges that inadequate ability constitutes a realistic hazard. Adequate ability, on the other hand, is no guarantee of adequate defense development. Thus, manifest FOF can still result despite adequate ability.

The latent types of FOF can also result from two different courses. A neurotic personality combined with a strong MTF and maladaptive defenses is the first course. In this scenario, the maladaptive defenses will not be sufficient to sustain the flier and a latent FOF will result. This is consistent with the latent presentations observed in student and early-career fliers. Latent presentations in previously adapted fliers are explained by an underlying neurotic personality combined with strong ability, strong motivation, and coping defenses that are initially adequate but not robust. Under these conditions, a period of adapted flying can ensue, but a precipitating event of some kind triggers the underlying neurotic conflict to overwhelm the previously adequate defenses. This results in the serial change observed by numerous writers.

The Cognitive–Behavioral Perspective

The clinical usefulness of behaviorism was catapulted by Wolpe's (1959) demonstration of the efficacy of the use of reciprocal inhibition in the systematic desensitization to anxiety-provoking stimuli, as experienced in phobias. Systematic desensitization had already become an established behavior therapy for anxiety-provoking stimuli when treatments based on modifying maladaptive cognitions also began to gain acceptance due to their demonstrated efficacy in the treatment of mood disorders and anxiety (for example, Beck 1976). Initially, psychotherapists who practiced behavior modification and those who relied on altering cognitions were sharply divided, but the merits of both approaches eventually led to a rapprochement and integration of the two approaches into what is now termed the cognitive–behavioral approach.

From the cognitive–behavioral perspective, every experience of excessive anxiety is rooted in the following two maladaptive cognitive appraisals: (1) over-estimation of the perceived threat in a situation; and (2) under-estimation of the subject's ability to cope with the threat (low self-efficacy). This combination of maladaptive appraisals leads to excessive sympathetic nervous system arousal, which causes uncomfortable physical sensations, distressing emotions, and reduced cognitive performance (for example, attention/concentration), as well as the overall performance decrements that ensue (Yerkes–Dodson Law). The distressing emotions and performance decrements that result from the over-arousal are perceived as validation of the subject's appraisal that the situation was excessively threatening and exceeded his or her capabilities, which leads to avoidance of the situation in the future. Avoidance of similar situations is experienced as rewarding since it results in a removal of the perceived threat and its concomitants (negative reinforcement in classic behaviorism), and is thus likely to be adopted as a "useful" strategy. Persistent avoidance leads to an absence of opportunity to discover that the threat is, in fact, over-estimated and/or that the subject is, in fact, able to adequately cope with it. This completes a cycle that results in entrenchment of the maladaptive cognitions and behavior.

From the cognitive–behavioral perspective, anxiety disorders that are related to flying are viewed the same as those embedded in other contexts. The root of the problem is the maladaptive cognitions, but in order to effectively alter the maladaptive cognitions the attendant physiological arousal and related behaviors must also be addressed. Hence, in addition to cognitive psychotherapy and its focus on addressing maladaptive cognitions, behavior therapy to teach skills such as deep muscle relaxation and tactical breathing to maintain optimal levels of autonomic arousal, as well as treatment techniques that integrate these skills, such as exposure therapy, are included in the treatment approach.

CBT has enjoyed contemporary pre-eminence, particularly among military clinicians, due to an extensive literature demonstrating rapid efficacy with a variety of psychological disorders, its straightforward and accessible approach, and its clear cause–effect relationships inherent in explanations of etiology and treatment which are a particularly good match for aviation. The mechanistic language that can be used to describe and explain the physiological and emotional responses to various cognitions, and the practical focus of CBT interventions ("If you do X, then Y will result") correspond perfectly to the pragmatic, procedural approach to trouble-shooting preferred by fliers. Indeed, CBT for anxiety disorders related to flying can be construed as similar to the training instruction provided for coping with G forces and other flight physiology issues, and thus enjoys face validity with the flier who is otherwise skeptical of "shrinks" and their "psychobabble."

Evaluation of Fear of Flying: An Integrative Approach

In order to discuss the evaluation of FOF, the issue of diagnostic clarification must be addressed. Psychiatric diagnostic categories have evolved a great deal since Anderson and Gotch published their observations of aeroneuroses in 1919. The variety of symptom presentations included under the rubric FOF syndrome by several writers of the twentieth century included a host of presentations that, in the current nosology, would be diagnosed as discrete Axis I or Axis II disorders, such as Acute and Posttraumatic Stress Disorder, Alcohol Abuse or Alcohol Dependence, Personality Disorder, or Malingering (American Psychiatric Association 2000). Within the *DSM* classification system, FOF does not enjoy a discrete Axis I disorder designation, although the phobic FOF presentation is included among the environmental subtypes of a Specific Phobia. When one of the above-noted discrete disorders fully explains the flier's presenting symptoms, then the evaluation poses no unusual difficulties since all well-trained clinicians are capable of evaluating such conditions; attention can then be turned to aeromedical disposition and treatment.

But the *DSM* fails to account for other types of FOF presentation. One can argue that the modern usefulness of the FOF construct is in explaining those latent clinical cases in which the somatic symptoms presented by the flier either are unfounded or do not adequately account for the whole picture. But there is no FOF subtype among the *DSM* Somatoform Disorders that can be invoked. This situation is understandable, given the small number of clinicians practicing clinical psychology and psychiatry in the aerospace environment. Nevertheless, clinicians are constrained by the current nosology when making diagnoses. The result is that clinicians evaluating a flier with a classic presentation of latent FOF find themselves armed with a century's worth of accumulated clinical knowledge, but unable to apply a precise diagnosis because one does not exist in the currently accepted nosology!

One adaptation is to accept this unfortunate reality and simply choose a *DSM* category that is most closely aligned with the aspects of the case, such as Psychological Factors Affecting a General Medical Condition or Somatoform Disorder Not Otherwise Specified. Regardless of the diagnostic category selected, however, the more fundamental task of determining whether the condition is rooted in FOF requires training that is not included in residencies in psychiatry or clinical psychology due to the applicability only to practice in the aviation environment. The accurate evaluation of such cases requires a familiarity with the FOF literature and, preferably, mentoring by an experienced aerospace clinician. Jones (1986) and Strongin (1987) provide a useful grounding in this endeavor.

Strongin (1987) recommended that clinicians begin by asking themselves three questions that will provide a focus to their evaluation: "First, ask yourself whether the symptoms stem from a preexisting disorder. Second, ask yourself whether the ego was overwhelmed by situational stress, or simply physically and emotionally exhausted from overwork. Finally, you should question whether changes in life circumstances have temporarily altered the flier's motivational and defensive structure. The factors can occur singly or in combination with each other." (267)

Jones's three questions

Jones (1986) also provided "three questions," but his were designed for use during the clinical interview with fliers suspected of presenting psychosomatic symptoms due to a repressed anxiety, consistent with latent expressions of FOF. The questions are designed to expose critical disruptions in the flier's *Motivation–Defense–Fear Balance*.

Question 1: "What do you think will happen if you continue to fly with this problem?" One of the first lessons for clinicians learning to assess fliers is that fliers with a healthy motivation tend to minimize or deny symptoms in order to avoid being grounded, even to the extent of positive or reverse malingering. A flier with a true disease process will respond to this first question in terms of concern that the disease is getting worse; concern that it may cause him or her long-term health problems and/or result in permanent grounding. In contrast, "If the answer is along the lines that the symptom makes flying dangerous, as 'I'm not safe to fly feeling this way,' or 'I might get killed,' the flier is seeing his disorder in terms of flying rather than of his or her own health, and this is a clue as to where the real anxiety lies" (135).

Question 2: "Will you fly when we get you well?" Jones emphasizes to listen not only to the substantive content of the answer, but also to the manner in which it is stated. This question will seem ridiculous to a motivated flier, who is likely to respond with something akin to "That's a dumb question, Doc! Of course I'll fly when you guys fix this!" They may then seek information about the most expeditious treatment, seeking to avoid delay, since they have a heavy flying schedule to keep and their squadron mates are covering for them. In contrast, a flier who does not respond in this manner is providing information about his or her lack of MTF: "If the answer is given with hesitation, or with any equivocation at all, suspect anxiety. If the answer is 'No,' or if other symptoms are quickly introduced, the underlying anxiety is clearly manifested" (135).

Question 3: "What do you think about these symptoms?" Jones notes that, "If the answer to the question is indifferent or unworried, and yet he has just told you that he is too sick to fly, or has some doubts that you can cure him, he may well be telling you that he doesn't want to fly anymore." A flier who believes that he has significant physical symptoms should be concerned about the flight surgeons accurately determining the etiology and providing effective treatment. They are likely to seek reassurance that physicians with proper expertise will be engaging in effective evaluation so that they can get back to normal functioning in order to return to flying. Jones points out that a flier who knows subconsciously that the symptoms arise primarily from anxiety is less likely to have this as his or her primary concern, perhaps even exhibiting *la belle indifference.*

It is important to recognize that even those using CBT for overt Specific Phobias related to flying have provided evidence regarding the usefulness of considering neurotic components to the disorder. Daly, Aitken and Rosenthal (1970), for instance, described a cohort of 14 British Air Force fliers referred for CBT for a flying phobia. They noted initial precipitants to the development of the phobias such as hypoxia or cockpit contamination and flying conditions in which visual cues were reduced, such as flying over an escarpment, in clouds or at night. They also noted, however, that 11 of the 14 patients had sexual or marital problems and in only four of those did the difficulty appear after the development of the phobia. Importantly, they also observed that childhood experiences may also have been determinants of adult problems, to include an overprotective mother, a stern father, death of a key family member, and/or aircraft mishaps experienced by family members. The descriptions of childhood determinants could easily have been extracted from one of the psychodynamic writers of the last century, yet they were descriptions of patients included in a CBT course of treatment (including *in vivo* desensitization in flight).

Daly, Aitken and Rosenthal (1970) provided a cogent integration of the cognitive–behavioral and psychodynamic aspects of phobia development, demonstrating that rigid adherence to only one framework for understanding such disorders is ineffective. Even writers known for their psychodynamic training and acumen (for example, Jones, 2008) have advocated CBT as a treatment option. In their synthesis of the psychodynamic perspective of FOF, Joseph and Kulkarni

(2003) provided a summary of various CBT techniques and studies in which such techniques were successfully applied. An integrative theoretical approach appears to be best suited for evaluating FOF.

Finally, psychological testing can be helpful in the accurate evaluation of various FOF presentations. But this must be accomplished by a clinical psychologist who has both advanced expertise in the interpretation of the tests used and thorough familiarity with the pilot norms and typical score patterns for those tests. The interpretive remarks provided by various computer narrative report interpretive programs, for instance, are based on general population norms, and are frequently unreliable and inaccurate when applied to fliers.

Disposition and Treatment

A complete discussion of the considerations required for disposition of the variety of FOF presentations is beyond the scope of this chapter. Jones (2008, 413) summarizes the issue concisely by noting that, "Whatever its genesis and presentation, symptomatic FOF should medically disqualify any aviator from active flying until the causes are delineated and the underlying disorder has been successfully treated. Such treatment may involve meticulous psychiatric history taking and some psychodynamic exploration, as well as brief pharmacotherapy and behavioral modification techniques."

Decisions regarding treatment for FOF are simplified after consideration of the point made above; namely, that the majority of symptom presentations included under the rubric FOF syndrome by writers of the twentieth century can now be diagnosed as discrete Axis I or Axis II disorders. Virtually all of those disorders now have well-validated treatments available. An added benefit is that, when MTF is strong, positive treatment outcomes commonly exceed those for non-fliers, both in military and civil aviation. This is even evident in disorders that are typically difficult to treat without frequent and/or sustained relapses, such as Alcohol Dependence. Data obtained from the Human Intervention and Motivation Study (HIMS), the civil aviation program for coordinating the treatment and monitoring the recovery of airline pilots, indicates sustained successful recovery and safe flying in over 90 percent of the pilots treated (see Chapter 5, this volume for more on substance abuse and pilots).

Similarly, "patients with flying phobia differ from those with most other phobias in that their livelihood is completely in jeopardy immediately" (Daly, Aitken and Rosenthal 1970, 881). Hence, the motivation to break through avoidance and to endure the initial discomfort inherent in such techniques as exposure therapy is higher than for the non-flier with a flying phobia. There is evidence for the efficacy of CBT in this area, especially if the exposure therapy component incorporates *in vivo* exposure, which can be accomplished with virtual reality technology or in flight in dual-control aircraft with a therapist who is also a qualified pilot (Aitken,

Daly, Lister and O'Conner 1971; Aitken, Daly and Rosenthal 1970; Wiederhold et al. 2002).

In contrast, when MTF is weak, treatment is likely to be ineffective. However, if latent FOF is properly diagnosed, a number of clinicians have reported good treatment outcomes via psychodynamic psychotherapy aimed at addressing the underlying neurotic conflict. Sours, Ehrlich and Phillips (1964) attributed their positive outcomes with this group to the *Motivation–Defense–Fear Balance* contrasts with the manifest group:

> On the other hand, the latent FOF patients had employed strong defenses against anxiety associated with flying. Their defenses and motivation had been high. Their symptoms express their instinctual—often narcissistic—need to fly. Strengthening denial and repression through therapy and providing them with superficial insight into what had weakened their motivation and denial brought about remission of symptoms and enabled them to remain in flying status.
>
> We believe, therefore, that manifest FOF reactions—except gross stress reactions—do not respond to short-term psychotherapy. On the other hand, latent FOF reactions can often be sufficiently worked through in therapy to enable the aviator to return to flying. The results of this study suggest the following criteria for treatment: (1) good motivation for flying, (2) latent as opposed to manifest FOF reactions, (3) effective defense mechanisms and the ability to use counterphobic maneuvers to allay anxiety, (4) the ability to understand symptomatology in terms of the ongoing life situation and its relation to flying. (164)

The issue is more complicated with the latent expressions of FOF, however, since the accurate diagnosis of that presentation requires clinical acumen that is available in only a limited number of clinicians. In both military and civil aviation, but especially in civil aviation, the presentation of somatic symptoms may result in the FOF root of the somatic presentation going unrecognized, leading to unnecessary, expensive, and potentially invasive diagnostic medical procedures (Jones 2008). If placed on medical disability status, this unfortunate outcome can then be further complicated by secondary gain.

On the positive side, the number of true, latent FOF cases observed is extremely small. The clinical psychologists and psychiatrists practicing in the Aerospace Psychiatry consultation services for the US Navy and Air Force, where such cases are most likely to occur, count the number of cases per year in single digits.

Jones (1986) observed an overarching or common factor in the treatment of fliers regardless of the specific diagnosis. He asserted: "If any one personality factor symbolizes successful fliers, it is their self-confidence, their absolute faith that they can always depend upon themselves. Events which shake this faith, that cast doubt on their self-control, may lead to disproportionate anxiety about flying. Restoring that faith is therapeutic, at times quite rapidly so" (134). This point should be a prominent consideration of the clinician who is treating fliers, irrespective of the diagnosis and specific treatment utilized.

Summary

The purpose of this chapter was to address some of the complex issues involved in determining the flight status of pilots and flight officers with cockpit duties who develop a FOF. MTF and FOF have occupied a prominent place in the evolution of aeromedical psychology and must be addressed by flight surgeons, aeromedical psychiatrists and psychologists in a concerted effort that combines clinical acumen with expert knowledge of the flight environment and an understanding of the flier's *Motivation–Defense–Fear Balance*. The importance of these considerations in the effective evaluation, disposition, and treatment of the various clinical presentations of FOF in pilots or flight officers cannot be overstated.

References

Adams, R.R. and Jones, D.R. 1987. The healthy motivation to fly: No psychiatric diagnosis. *Aviation, Space, and Environmental Medicine*, 58(4), 350–354.

Aitken, R.C., Daly, R.J., Lister, J.A. and O'Conner, P.J. 1971. Treatment of flying phobia in aircrew. *American Journal of Psychotherapy*, 25(4), 530–542.

Aitken, R.C.B., Daly, R.J. and Rosenthal, S.V. 1970. Treatment of flying phobia in trained aircrew. *Proceedings of the Royal Society of Medicine*. 63(9), 882–886.

American Psychiatric Association. 2000. *Diagnostic and Statistical Manual of Mental Disorders, 4th Edition, Text Revision*. Washington, DC: Author.

Anderson, H.G. 1919. *The Medical and Surgical Aspects of Aviation.* London: Oxford University Press.

Armstrong, H G. 1939. *Principles and Practice of Aviation Medicine*. Baltimore, MD: Williams and Wilkins Co.

Beck, A.T. 1976. *Cognitive Therapy and the Emotional Disorders*. London: Penguin Press.

Berg, J.S., Moore, J.L., Retzlaff, P.D. and King, R.E. 2002. Assessment of personality and crew interaction skills in successful naval aviators. *Aviation, Space, and Environmental Medicine*, 73(6), 575–579.

Berry, C.A. 1961. Human qualifications for and reactions to jet flight, in *Human Factors in Jet and Space Travel*, edited by S.B. Sells and C.A. Berry. New York: Ronald Press.

Bond, D.D. 1952. *The Love and Fear of Flying*. New York: International Universities Press.

Boyd, J.E., Patterson, J.C. and Thompson, B.T. 2005. Psychological test profiles of USAF pilots before training vs. type aircraft flown. *Aviation, Space, and Environmental Medicine*, 76(5), 463–468.

Bucove, A.D. and Maioriello, R.P. 1970. Symptoms without illness: Fear of flying among fighter pilots. *Psychiatric Quarterly*, 44(1), 125–141.

Christen, B.R. and Moore, J.L. 1998. A descriptive analysis of "not aeronautically adaptable" dispositions in the US Navy. *Aviation, Space, and Environmental Medicine*, 69(11), 1071–1075.

Christy, R.L. 1975. Personality factors in selection and flight proficiency. *Aviation, Space, and Environmental Medicine*, 46(3), 309–311.

Daly, R.J., Aitken, R.C.B. and Rosenthal, S.V. 1970. Flying phobia: Phenomenological study. *Proceedings of the Royal Society of Medicine*. 63(9), 878–882.

Davis, D.B. 1945. Phobias in pilots. *The Military Surgeon: Journal of the Association of Military Surgeons of the United States*, 105, 105–111.

Dyregrov, A., Skogstad, A., Hellesøy, O.H. and Haugli, L. 1992. Fear of flying in civil aviation personnel. *Aviation, Space, and Environmental Medicine*, 63(9), 831–838.

Fine, P.M. and Hartman, B.O. 1968. *Psychiatric Strengths and Weaknesses of Typical Air Force Pilots.* US Air Force School of Aerospace Medicine, Aerospace Medical Division (AFSC). Brooks Air Force Base, Texas. (SAM-TR-68-121).

Gatto, L.E. 1954a. Understanding the "fear of flying" syndrome I: Psychic aspects of the problem. *United States Armed Forces Medical Journal*, 5(8), 1093–1116.

Gatto, L.E. 1954b. Understanding the "fear of flying" syndrome II: Psychosomatic aspects and treatment. *United States Armed Forces Medical Journal*, 5(9), 1267–1289.

Goorney, A.B. and O'Conner, P.J. 1971. Anxiety associated with flying: A retrospective survey of military aircrew psychiatric casualties. *British Journal of Psychiatry*, 119(549), 159–166.

Gotch, O.H. 1919. The aero-neurosis of war pilots, in *The Medical and Surgical Aspects of Aviation*, edited by H.G. Anderson, M.W. Flack and O.H. Gotch. London: Oxford University Press, 109–149.

Greco, T.S. 1989. A cognitive-behavioural approach to fear of flying: A practitioner's guide. *Phobia Practice and Research Journal*, 2(1), 3–15.

Grinker, R.R. and Spiegel, J.P. 1943. *War Neurosis in North Africa: The Tunisian Campaign (January–May 1943).* New York: Josiah Macy, Jr. Foundation.

Grinker, R.R. and Spiegel, J.P. 1945. *Men Under Stress*. Philadelphia, PA: Blakiston.

Jones, D.R. 1986. Flying and danger, joy and fear. *Aviation, Space, and Environmental Medicine*, 57(2), 131–136.

Jones, D.R. 1995. US Air Force combat psychiatry, in *Textbook of Military Medicine: War Psychiatry, Part I*, edited by F.D. Jones, L.R. Sparacino, V.L. Wilcox, J.M. Rothberg and J.W. Stokes. Washington, DC: Office of the Surgeon General of the United States, 177–210.

Jones, D.R. 2000. Fear of flying—No longer a symptom without a disease. *Aviation, Space, and Environmental Medicine, 71*(4), 438–440.

Jones, D.R. 2008. Aerospace psychiatry, in *Fundamentals of Aerospace Medicine,* 4th Edition, edited by J.R Davis, J. Stepanek, R. Johnson and J.A. Fogarty. Baltimore, MD: Lippincott, Williams & Wilkins, 406–424.

Jones, D.R. and Marsh, R.W. 2001. Psychiatric considerations in military aerospace medicine. *Aviation, Space, and Environmental Medicine*, 72(2), 129–135.

Joseph, C.J. and Kulkarni, J.S. 2003. Fear of flying: A review. *Indian Journal of Aerospace Medicine*, 47(2), 21–31.

King, R.E. 1999. *Aerospace Clinical Psychology*. Brookfield, VT: Ashgate.

King, R.E. and McGlohn, S.E. 1997. Female United States Air Force pilot personality: The new right stuff. *Military Medicine*, 162(10), 695–697.

Leimann Patt, H.O. 1988. The right and wrong stuff in civil aviation. *Aviation, Space, and Environmental Medicine*, 59(11), 955–959.

Lifton, R.J. 1953. Psychotherapy with combat fliers. *U.S. Armed Forces Medical Journal*, 4, 525-532.

Maschke, P. 2004. Personality evaluation of applicants in aviation, in *Aviation Psychology: Practice and Research*, edited by K.L. Goeters. Burlington, VT: Ashgate, 141–151.

McGuire, T.F. 1969. A second look at fear of flying. *Proceedings of the Sixteenth Annual Conference of Air Force Behavioral Scientists*, Brooks AFB, Texas, 84–93.

McMichael, A.E. and Graybiel, A.G. 1963. Rorschach indications of emotional instability and susceptibility to motion sickness. *Aerospace Medicine*, 34, 997–1000.

Moore, J.L. and Ambrose, M.R. 1998, May. *Personality Clusters Among Experienced Naval Aviators*. Paper presented to the Aerospace Medical Association Scientific Meeting, Seattle, WA.

Moore, J.L., Berg, J.S. and Valbracht, L.E. 1996, May. *Prediction of Aeronautical Adaptability Using NEO-PI-R Facet Scores.* Paper presented to the Aerospace Medical Association Scientific Meeting, Reno, NV.

Morganstern, A.L. 1966. Fear of flying and the counter-phobic personality. *Aerospace Medicine*, 37, 404–407.

O'Connor, P.J. 1970. Phobic reaction to flying. *Proceedings of the Royal Society of Medicine*. 63(9), 877–878.

Paullin, C., Katz, L., Bruskeiwicz, K.T., Houston, J. and Damos, D. 2006. *Review of Aviator Selection (Tech Rep. No. 1183).* United States Army Research Institute for the Behavioral Sciences, Arlington, VA.

Perry, C.J.G. 1971. Aerospace psychiatry. In *Aerospace Medicine*, edited by H.W. Randel. Baltimore: Williams & Wilkins, 534-549.

Phillips, P.B. and Sours, J.A. 1963. Pseudo-organic illness in the failing flyer. *Journal of the Florida Medical Association*, 50, 127–130.

Picano, J.J. and Edwards, H.F. 1996. Psychiatric syndromes associated with problems in aeronautical adaptation among military student pilots. *Aviation, Space, and Environmental Medicine*, 67(12), 1119–1123.

Picano, J.J., Williams, T.J. and Roland, R.R. 2006. Assessment and selection of high risk operational personnel, in *Military Psychology: Clinical and Operational Applications*, edited by C.H. Kennedy and E.A. Zillmer. New York, NY: The Guilford Press, 353–370.

Pilmore, F.U. 1919. The nervous element in aviation. *Naval Medical Bulletin*, 13, 458–478.

Reinhardt, R.F. 1966. The compulsive flyer. *Aerospace Medicine*, 27, 411–413.

Reinhardt, R.F. 1967. The flyer who fails: An adult situational reaction. *American Journal of Psychiatry*, 124(6), 48–52.

Reinhardt, R.F. (1970). The outstanding jet pilot. *American Journal of Psychiatry*, 127(6), 732–735.

Retzlaff, P.D. and Gibertini, M. 1987. Air force pilot personality: Hard data on the "right stuff." *Multivariate Behavior Research*, 22(4), 383–399.

Rubel, R.C. 2010. The US Navy's transition to jets. *Naval War College Review*, 63(2), 49–59.

Siem, F.M. and Murray, B.S. 1994. Personality factors affecting pilot combat performance: A preliminary investigation. *Aviation, Space, and Environmental Medicine*, 65(5 Suppl), A45–A48.

Schmideberg, M. 1937. On motoring and walking. *International Journal of Psychoanalysis*, 18, 42–53.

Schulz, H.A. 1952. Fear of flying. *USAF Medical Services Digest*, Nov, 29-31.

Scott, T. (Director, Motion Picture). 1986. *Top Gun*. Los Angeles: Paramount.

Sours, J.E., Ehrlich, R.E. and Phillips, P.B. 1964. The fear of flying syndrome: A reappraisal. *Aerospace Medicine*, 35, 156–166.

Strongin, T.S. 1987. A historical review of the fear of flying among aircrewmen. *Aviation, Space, and Environmental Medicine*, 58(3), 263–267.

Tempereau, C.E. 1956. Fear of flying in Korea. *American Journal of Psychiatry*, 113, 218–223.

Wiederhold, B.K., Jang, D.P., Gervitz, R.G., Kim, S.I., Kim, I.Y. and Wiederhold, M.D. 2002. The treatment of fear of flying: A controlled study of imaginal and virtual reality graded exposure therapy. *IEEE Transactions on Information Technology in Biomedicine*, 6(3), 218–223.

Wolpe, J. 1958. *Psychotherapy by Reciprocal Inhibition*. Stanford, CA: Stanford University Press.

Chapter 8
Airsickness and Space Sickness

Erik Viirre and Jonathan B. Clark

Air and space sickness are extremely important topics when discussing ability to perform a variety of flight duties, as these can be debilitating conditions and typical pharmacologic treatment is often contraindicated. This chapter will explore both the physical and psychological components of airsickness, will review the operational importance of motion sickness for flight, will give some guidelines for flight crews susceptible to airsickness and will outline the psychological and pharmacological interventions used to eliminate or ameliorate motion sickness. Special cases of motion sickness related to aviation including space motion sickness, high-acceleration environments and virtual aviation systems will be described.

Motion Sickness

Motion sickness is a condition that can be provoked by movement of the human head in a gravity field or by moving visual fields. If a person keeps their head stationary relative to the earth's 1G gravity vector and their eyes closed, motion sickness will not arise. However, once they begin to move, especially with linear or angular accelerations greater than normal self-propelled locomotion, motion sickness may occur. Being resident in a non-1G environment, such as weightlessness in space or on the surface of a non-1G object, may also induce motion sickness, even if there is no relative movement within the local gravity field. Also, moving visual stimuli, even with the head held stationary, can provoke motion sickness.

Importantly, there is a wide range of susceptibility to motion sickness in humans. Particularly with disorders of the inner ear vestibular apparatus, there may be rapid onset of motion sickness provoked by even small head movements. In contrast, some individuals can experience high G maneuvers, up to the point of loss of consciousness, and not experience motion sickness. Ironically, motion sickness does not occur in those with non-functional inner ears, so-called labyrinthine defectives.

Motion sickness results in feelings of nausea, headache, fatigue, and incipient or actual vomiting. Sometimes, following the nausea and vomiting, there is a common concomitant condition that is named sopite syndrome, where subjects are fatigued and cognitively impaired (Graybiel and Knepton 1976). Many sufferers are highly avoidant of conditions that may provoke their motion sickness, especially air or sea travel.

Interestingly, sea sickness is perhaps more prevalent that air sickness. It is infrequent that one observes a person taking oral medication or wearing a skin patch for air sickness, but it is very common to see people with transdermal skin patches for motion sickness on sea voyages, such as ocean cruises. The fact that motion sickness may be more common on sea-going vessels is related to the triggering or potentially causative stimuli for motion sickness and how the body reacts to motion.

Theories of Motion Sickness

Theories of motion sickness have to link two seemingly disparate phenomena: gastric distress, with its concomitant nausea and stomach peristalsis, and motion imparted to the body and particularly to the head. Why would movement be linked to vomiting? There might be a direct connection, in that movements of the body could cause shifting of internal organs as they are suspended from the thorax and abdomen by gravity. The shifting could stimulate stretch receptors in the walls of the gastrointestinal (GI) viscera and through a reflex loop cause a GI contraction. However, that would not explain motion sickness provoked by head movement alone or even by pure visual stimuli without head movement.

The ecological theory of motion sickness (Gibson 1966, 1979; Stoffregen and Riccio 1991) says that a comparison of visual and vestibular inputs that results in a measured mismatch, that is, a visual–vestibular mismatch, is the trigger for motion sickness. The teleological reason for this comparison is to determine if the vestibular sensors are malfunctioning. There are several lines of evidence for this concept.

Motion sickness usually occurs in conditions with visual–vestibular mismatch:

- Riding in the back seat of a car while reading, being below decks in a boat or not looking out of the window in a plane. The visual stimuli provide no indication of body motion, even though there are acceleration stimuli. Motion sickness is triggered with prolonged viewing inside the vehicle and is mitigated or avoided by looking outside the vehicle, such as at the horizon.
- Viewing of rapidly varying moving fields of visual scenes. The common situation for this would be in a movie theatre watching images recorded from a hand-held camera. Indeed, this sort of stimulus can be used in experiments where motion sickness is induced (Strychacz, Viirre and Wing 2005).
- The motion stimuli that are most provocative for motion sickness are in the frequency range for visual–vestibular comparison. Griffin (1990) found, by polling people for sea-sickness on ships with motion acceleration recording

systems, that the most provocative stimulus was vertical heave (up and down) motion at about 0.2 hz. These frequencies are un-natural motion accelerations for ambulatory humans. Until the development of boats, humans did not normally experience motion accelerations that vehicles produce.

- Biomarkers for the presence of motion sickness in humans include electro-encephalographic signals in the region of Ventral Intra-Parietal Region (VIP), where visual–vestibular comparison and sensorimotor integration occurs. The spectral content of VIP region signals changes with the presence and absence of reported motion sickness (Strychacz, Viirre and Wing 2005; Lin et al. 2007).

Anatomy of Motion Sensation

Movement of images on the retina can be used to calculate the movement of the head in a fixed spatial environment. As fixed walls or other objects are approached linearly or obliquely, motion estimates occur in the visual brain. The common experience of *vection* tells us this is so: when we are sitting stationary beside a large moving object, say a truck beside us in traffic, movement of the truck can induce a strong feeling of motion even though we are sitting still. Movement of a large portion of the visual field is a trigger for motion sensation: a fact that movie makers frequently exploit.

Of course, even with our eyes closed, we can sense movement. Pressure sensors throughout the body can impart motion sensation, but the primary sensors for motion are in the vestibular system. The otoliths are large calcium carbonate crystals that rest on hair cells in the utricle. They are sensors for tilt or linear acceleration of the head. Also in each inner ear are the semi-circular canals that are rotational accelerometers, good for detecting rapid rotations of the head. The semi-circular canals can detect motion in three dimensions and are optimized for detection of rapid onset and high-frequency movements of the head. They are directly connected to centers in the brainstem that drive vision stabilizing movements of the eyes. Signals from the canal also rapidly drive balance reflexes and go to high-cortex regions to deliver self-motion information.

The motion information from the body sensors goes to a variety of centers to help drive motion computations and create the conscious percept of motion. Importantly, the acceleration information from the vestibular sensors goes to motion integration regions of the brain, such as the VIP (Bremmer et al. 2002). In VIP information from a variety of sources is compared and integrated, presumably to give accurate updates of motion of the body. Each motion sensor system is used for the stimuli they are optimized for. Vision is optimal for detecting slow-onset, long-duration motion such as slow walking in a straight line. In contrast, the vestibular sensors are good for rapid onset movements, such as head turns or slips or falls. Vision sensors are too slow to detect a rapid head turn and vestibular sensors are insensitive to slow accelerations, so the two systems complement each other (Robinson 1981). Importantly, there is some overlap in the motion-sensing capabilities of each system. In this overlap the brain can compare the outputs of the two systems and with the comparison optimize the motion sensation. The comparison region is centered around the 0.1 to 1.0 Hz frequency range. Both systems can detect accelerations in these frequencies. Figure 8.1 below shows the frequency responses of the acceleration detection systems and their neuro-anatomic linkages.

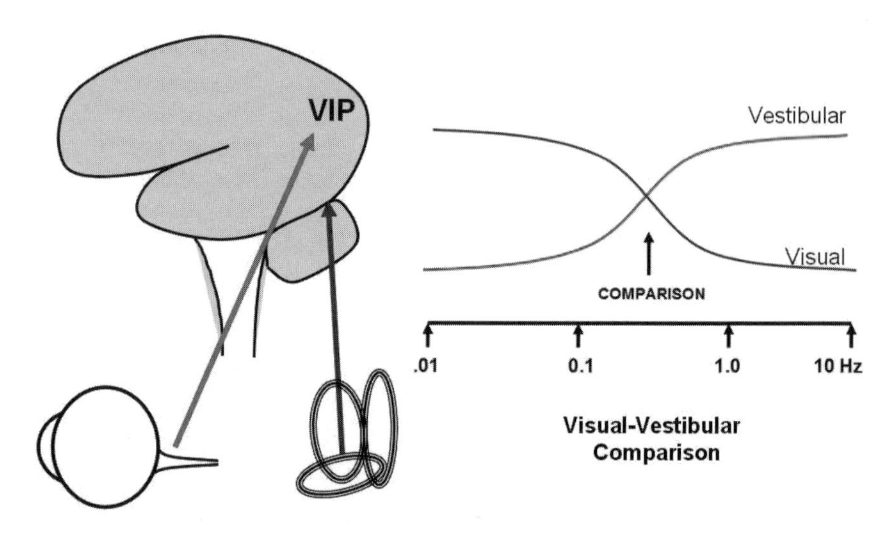

Figure 8.1 Visual Vestibular Conflict and Motion Sickness

Why Vomit with Motion Stimuli? The Ecological part of The Ecological Theory of Motion Sickness

But why nausea and vomiting with motion sickness (Yates, Miller and Lucot 1998)? Perhaps the clue is the other initiator for nausea and vomiting: GI toxins or poisons. The common human experience of alcohol intoxication shows us the two phenomena simultaneously. Alcohol rapidly diffuses into the fluid of the vestibular apparatus: the endolymph. There alcohol measurably changes the specific gravity of the endolymph and induces malfunction of the semi-circular canals by altering the relative movements of the acceleration sensors. As many a college student can attest, two things can occur: the bed-spins and vomiting. The malfunctioning inner ear induces the spinning sensation and then motion sickness, nausea, and vomiting occur. Interestingly, by consuming tritiated (also called "heavy") water, the reduced specific gravity of the endolymph can be reversed to normal and motion sickness from drinking alcohol can be averted (Money and Myles 1974). Unfortunately, this is a very expensive hang-over cure.

Vomiting is presumed to rapidly expel toxins (and everything else) from the stomach. Nausea is a highly aversive stimulus that is teleologically believed to result in a conscious avoidance of the food associated with poisoning. Some people will develop an aversion to alcohol after drinking and vomiting. However, unusual food avoidances appear to confirm the role of aversion and vomiting. Seligman and Hager (1972) describe Sauce-Béarnaise syndrome where he developed a strong distaste for his previously-beloved Sauce Béarnaise after vomiting it up. One of the authors of this chapter (EV) had the same reaction to coveted donuts after being sick with them. Thus it is believed that visual-vestibular mismatch is a signal interpreted by the body as vestibular labyrinth poisoning and thus *inappropriately* induces nausea and vomiting. The fact that most disease conditions of the inner ear, including viral infection (or labyrinthitis) or abnormal fluid pressure (endolymphatic hydrops) or altered endolymphatic fluid dynamics (canalithiasis), strongly induce vertigo with nausea and vomiting also supports the labyrinth poisoning explanation. The observation that motion sickness can be attenuated with motion stimulation experience also lends some credence to this concept. Seeing someone be sick is a stimulus for nausea and vomiting, perhaps because we think that we ate what they ate.

Thus as mentioned above, it is accelerations that humans never experienced in evolution, riding in vehicles, experiencing large visual field motion, such as movies or Virtual Reality (Draper, Viirre, Furness and Gawron 2001), or moving outside the earth's gravitational field, that are the stimuli for motion sickness.

Operational Importance of Motion Sickness

Beyond its great unpleasantness, motion sickness has obvious operational relevance. People who are motion sick have relative or complete incapacitation for completing tasks requiring quick action, concentration, or strength. Reflexes are slowed, cognition is impaired and strength is reduced during motion sickness (Griffin 1990). Control of vehicles will be impaired and emergency responses slowed. Further, as mentioned above, observation of vomiting (or thinking about it, like with this chapter) will induce nausea and thus potentially propagate motion sickness through a flight crew or passengers.

Airsickness and Flight Crew Fitness-For-Duty

It would seem obvious, but not all people who are prone to motion sickness will avoid a career in aviation. Occasional stories of WWII aces are told where the pilot would prevail over severe motion sickness. In fact, flight training itself will result in a reduced sensitivity to motion sickness. However, in the current era, repeated nausea and vomiting during flight are an adverse selection criteria for advancement in flight training and repeated incidents of vomiting during flight will be disqualifying (for example Naval Aerospace Medical Institute 2010). The over-abundance of people desiring to be pilots means that the motion-sick susceptible may not be selected. However, adaptation training can result in motion sickness control in some individuals. Adaptation can be carried out by flight experience or by stimulation of motion sickness in a "Barany Chair." The subject is rotated in the chair around an earth-vertical axis and motion sickness can be induced by tipping the head. Repeated sessions in the chair delay the onset of symptoms, reduce their severity and can eliminate them to the point that personnel can fly. Some military flight training programs have such chair training. Now, civilian programs with centrifuge training and other technologies may offer the "Barany Chair." Medications are not allowed for any long-term use for motion sickness as the drugs themselves affect cognition and performance and induce sleepiness (Paul, MacLellan and Gray 2005). An excellent overview and Clinical Practice Guideline has been prepared by the American Society for Aerospace Medicine (Rogers and Van Syoc 2011).

An important situation is the development of a vestibular disorder in someone already trained to fly. What are the criteria for fitness-for-duty? Can labyrinthine disorders be a medical disability for flight or ocean crew? Repeated incidents of vomiting during flight training can be disqualifying from further flight operations, a condition called Chronic Intractable Motion Sickness (CIMS). Some labyrinthine disorders will remain static after their active phases and personnel can be rehabilitated to control motion sickness.

Factors Influencing Airsickness

The motion environment itself can be more or less provocative. As mentioned above, low-frequency vertical motion is the most provocative stimulus. Modern civilian airlines carefully select flight altitudes and flight paths to avoid turbulence, which is a strong negative passenger experience factor. The view can also make a difference. For example, for crash safety in some military aircraft, flight passengers sit facing aft and there is a higher incidence of motion sickness. A view out a window or to the horizon on the deck of a ship can mitigate motion sickness. As is well known to many with motion sickness susceptibility, being in active control of accelerations is better than being a passive recipient of the motion. Thus being the driver or the pilot-in-control will reduce motion sickness. Indeed, many pilots would be well advised to be sympathetic to their passengers lest they be put in the back seat.

Anxiety can exacerbate motion sickness and thus become a perpetuating or inciting factor. Anticipating a repeat of a motion sickness experience will make many passengers fearful and even more prone to the problem. Conversely, repeated controlled exposures can reduce motion sickness incidence. Flight crew can help or exacerbate motion sickness in their passengers. Unawareness of a passenger's susceptibility can set off a nasty chain of events. Dismissing fears or even jokingly provoking them ("Does this check-list tell me how to fly the plane?") and being unaware of the provocativeness of flight maneuvering is unforgiveable. Passengers, especially anxious ones, should be spared flight experiences such as check-rides where emergency procedures, such as spin recovery, are demonstrated. Conversely, a professional demeanor, with calm but clear instruction on emergency procedures, will increase passenger confidence in their flight crew and reduce anxiousness and motion sickness.

Heat, bright lights, and unpleasant smells are also provocative for motion sickness. Thus a sunny afternoon on an airport ramp with avgas or jet-fuel refilling can be a set-up for motion sickness in flight. Passengers and flight crew should be provided cool ventilation with strong but comfortable air flow. They should avoid strong smells such as fuel, lubricants and of course, vomit. Shade, on the ground and in the vehicle, is helpful. Commercial operators sometimes forget the importance of these environmental factors as passengers and crew are trapped on the tarmac waiting for a departure clearance.

Food is an important trigger of gastric sensitivity. An empty stomach may actually be somewhat provocative for motion sickness. Moderate amounts of clear fluids and light carbohydrates will reduce motion sickness. Heavy meals of protein and especially fats, such as dairy products, should be avoided, as should carbonated beverages. Interestingly, ginger, in a variety of forms is a good antiemetic agent, delaying the onset and reducing the severity of motion sickness. Ginger ale, served cold, in small sips is a reasonable first line of defense. Various forms of ginger have been found to mitigate motion sickness (Grontved and Hentzer 1986; Grontved, Brask, Kampskard and Hentzer 1988).

A variety of demographic factors are related to motion sickness susceptibility. Women are about twice as susceptible to it as men. Very young children are moderately resistant, but later age children and adolescents are very susceptible. As age progresses, susceptibility gradually reduces (Park 1998). Motion sickness is particularly predominant in people with migraine. As migraine is a genetic condition, whole families can be prone to the problem.

Mitigations for Airsickness

Aircrew

In aircrew, experience is perhaps the most important factor. Those susceptible to motion sickness will usually be self-selected out. With flight training and practice, motion sensitivity goes down. Importantly, a prolonged hiatus from flying, especially for a medical problem, might herald a return of susceptibility and some flight experience may be necessary to return to status. Anti-motion sickness medications are strongly prohibited for flight crew as the drugs can have a profound influence on performance and cognition.

Passengers

Calm understanding is perhaps the most important mitigant of motion sickness. People prone to motion sickness will usually be keenly aware of their condition and avoid provocative situations. Avoidance of provocative conditions: turbulent air on hot, sunny afternoons, warm temperatures, lack of hydration, and provocative foods should be paramount.

Medications

For passengers without flight critical duties, anti-motion sickness medications may be appropriate (Wood 1979). These include: scopolamine (Transderm Patches or Scopace), meclizine (Antivert), dimenhydrinate (Dramamine), and promethazine (Phenergan). All of these drugs have potential side-effects, the most common of which is sedation (Wood et al. 1984). Indeed, scopolamine is so sedating in oral form, it is often given with a stimulant, including powerful controlled substances such as d-amphetamine. Other side-effects are possible, such as anti-cholinergic effects of scopolamine including angle closure glaucoma or urinary bladder retention.

Case Example: Civilian Flights with Parabolic Flight Paths for Weightlessness

In the modern era, civilians are seeking and are being offered adventure experiences such as weightlessness in flight. Such operations now are offered in the United States under FAR Part 121 regulations by appropriately licensed operators, such as Zero G Corporation. The parabolic flight paths generally include 0.4G, 0.16G, and 0G weightless intervals followed by hyperG periods up to 1.8G. The flight experiences are designed to maximize the enjoyment of passengers and minimize the motion sickness. A number of features are generally incorporated including:

- Provision through the program advertising emphasizing the adventure nature of the experience to signal to all passengers what the flights are like. Particularly for the motion-susceptible, the advertising description indirectly indicates that they should carefully consider, but not necessarily avoid, the experience. Passengers are given recommendations to review their medical conditions with their physician and, if appropriate, receive prescriptions for anti-nausea medication. Often a prescription anti-emetic, such as oral scopolamine, plus a mild stimulant such as caffeine, are used. Sedation may occur during the pre and post-flight intervals, but passengers do not sleep during the parabolas.
- A well-designed pre-flight program with careful control of flight timing, control of food and beverages, and pre-flight training and counseling.

Avoidance of turbulent flight conditions, especially prior to the parabolas, is very helpful.
- Parabolic programs with frequent breaks and an appropriate number of parabolas.
- Minimization of movement, especially head movement, during hyper-G intervals.
- Trained cabin crew who recognize and counsel anxious passengers and complement an environment of adventure and anticipation.
- Controlled cabin environmental conditions such as cool air flow and reduced smells.
- Rapid identification and segregation of ill passengers.

Through these mitigations, hundreds of people have had enjoyable experiences, even those with partial or complete paralysis (for example, Mackenzie, Viirre, Vanderploeg and Chilvers 2008).

Space Motion Sickness

Incidence

Space Motion Sickness (SMS) is the constellation of symptoms which include stomach awareness (a pre-nausea sensation of fullness or discomfort in the epigastric region), loss of appetite, nausea, vomiting, drowsiness, headache, and dizziness associated with the spaceflight microgravity environment (Matsnev et al. 1983). Sweating and pallor are not as prominent in SMS as in other forms of motion sickness, which may relate in part to the cephalad fluid shifts (Oman, Lichtenberg, Money and McCoy 1986). SMS is a subset of the maladaptive physiologic effects termed Space Adaptation Syndrome (SAS) (Homick et al. 1984) and is the sensorimotor system's maladaptive response to microgravity. Gastric motility and bowel sounds are diminished (Thornton, Linder, Moore and Pool 1987). Vomiting often occurs following orbital insertion and may come on suddenly without prodromal symptoms (that is, projectile vomiting) often followed by a symptom-free period of hours (Thornton, Moore, Pool and Vanderploeg 1987). The sopite syndrome is a component of SMS, which includes malaise, loss of initiative, irritability, decreased attention, and lack of concentration, and can degrade crew performance (Graybiel and Knepton 1976). SMS is not correlated with prior motion sickness on sea, air, rotating chairs, or parabolic flight. SMS is not protective against post-flight motion sickness, as nine of 25 Apollo astronauts became seasick in their capsule after landing. Skylab astronauts who took motion sickness medication prior to landing did not get seasick. SMS incidence is lower on subsequent space flights.

The Russian cosmonaut Gherman Titov was the first to report vomiting when he tried to eat on the sixth orbit during the second manned spaceflight (Vostok

2) on 6 August 1961. There were no reports of SMS in the Mercury and Gemini programs. Although SMS was not reported in the Gemini program, caloric intake was low for many astronauts, and might be attributable to mild SMS-associated loss of appetite (Homick 1979). The first reported American experience occurred during Apollo 8 where all three crewmembers experienced SMS (Hawkins and Zieglschmid 1975). The first mission impact due to SMS occurred on Apollo 9 (Homick and Miller 1975). The Apollo 9 Lunar Module Pilot vomited suddenly after donning his pressure suit for an Extravehicular Activity (EVA) 40 hours into the mission without any prodromal motion sickness symptoms. About four hours later, he suddenly vomited again then felt much better. The spacewalk was delayed for two days and EVA activities were curtailed. Post-flight vestibular function tests were normal. SMS occurred in 35 percent of Apollo astronauts and 60 percent of Skylab astronauts (Davis et al. 1988). The low incidence of SMS reported early in manned spaceflight may be due to smaller cabin volumes, which reduce freedom of movement, particularly head movement. The incidence of SMS in the Russian program has been about 50 percent (Gorgiladze and Bryanov 1989). Russian capsules had slightly larger habitable volumes than the US capsules. The incidence of SMS in the Space Shuttle Program is over 70 percent in rookie astronauts (Davis et al. 1988; Jennings 1998). See Table 8.1 for SMS incidence in various space missions. SMS is graded as mild, moderate, or severe based on the criteria in Table 8.2. During the Shuttle program about half of the SMS cases were mild, one-third were moderate, and one-sixth were severe (Davis et al. 1988).

Table 8.1 Space Motion Sickness in American and Russian Space Programs

American				**Russian**			
Program	**Total Crew**	**Crew with SMS**	**% SMS**	**Program**	**Total Crew**	**Crew with SMS**	**% SMS**
Mercury	6	0	0%	Vostok	6	1	17%
Gemini	20	0	0%	Voskhod	5	3	60%
Apollo	33	11	33%	Soyuz	38	21	55%
Skylab	9	5	56%	ASTP	2	2	100%
ASTP	3	0	0%	Salyut 5	6	2	33%
Shuttle (pre-1986)	85	57	67%	Salyut 6	27	12	44%
Shuttle (1988–1998)	315	252	80%				
Totals	471	325	70%		84	41	49%

ASTP—Apollo Soyuz Test Project

Table 8.2 Space Motion Sickness Grading Criteria

None(0):	No signs or symptoms reported with the exception of mild transient headache or mild decreased appetite.
Mild(1):	One to several symptoms of a mild nature; may be transient and only brought on as the result of head movements; no operational impact; may include single episode of retching or vomiting; all symptoms resolved in 36–48 hours.
Moderate(2):	Several symptoms of a relatively persistent nature which may wax and wane; loss of appetite; general malaise, lethargy and epigastric discomfort may be most dominant symptoms; includes no more than two episodes of vomiting; minimal operational impact, all symptoms resolved in 72 hours.
Severe(3):	Several symptoms of a relatively persistent nature that may wax and wane; in addition to loss of appetite and stomach discomfort, malaise and/or lethargy are pronounced; strong desire not to move head; includes more than two episodes of vomiting; significant performance decrement may be apparent; symptoms may persist beyond 72 hours.

(Davis, et al. 1988)

Resolution of symptoms usually occurs within 48 hours after exposure to microgravity although there are reports of SMS not fully resolved after two weeks (Matsnev et al. 1983). Twenty-five percent of Russian cosmonauts have symptoms lasting 14 days or more, and 17 percent of cosmonauts on long-duration missions (83–365 days) may develop symptoms during their flight (Kornilova, Yakoleva, Tarasov and Gorgiladze 1983). Age, gender, crew qualification (pilot versus mission specialist), and aerobic fitness do not correlate with incidence or severity (Jennings, Davis and Santy 1988). SMS on previous spaceflights is predictive of SMS on a subsequent flight, but repeat fliers often have less severe symptoms than their previous flight (Davis et al. 1988). A high-protein/low-carbohydrate diet correlated with SMS symptoms on Skylab missions (Simanonok, Kohl and Charles 1993). Head motion can elicit SMS symptoms (Oman and Shubentsov 1992) with pitch head movements being the most provocative, followed by roll and yaw (Lackner and Graybiel 1987). Soyuz and Shuttle crew bend forward to doff their pressure suit within an hour of orbital insertion, so this was a vulnerable period.

Mitigations

Rapid head movements, particularly in pitch and roll, should be avoided during the early spaceflight. Extra caution is recommended during suit-doffing procedures. Astronauts tend to move the head *en bloc* with the body until adaptation to microgravity has occurred. Maintaining a lower cabin temperature generally is better tolerated in SMS-affected crew. Astronauts often fluid restrict to reduce headward fluid shifts early in the mission, but adequate hydration is recommended. Many crewmembers avoid high-fat and protein diets. Emesis bags should be easily accessible in case of sudden vomiting. If SMS develops, symptomatic individuals should stay on clear liquids until symptoms improve. Intravenous fluids are used for those with severe symptoms.

Training

Pre-flight training to enhance adaptation may be an effective countermeasure for SMS (Reschke, Parker, Harm and Michaud 1988). Training includes education on perceptual illusions and novel sensory inputs, and avoidance of provocative factors. Training devices and procedures designed to pre-adapt astronauts to the novel sensory and perceptual conditions similar to those of weightlessness are based on the concept of training crews with different gravity stimuli or visual–vestibular conflicts on earth and rapidly switching between microgravity and earth gravity sensorimotor states. The pre-adaptation trainer projects unusual visual scene orientations to induce orientation illusions such as unusual body attitude in relation to the spacecraft. Unexpected or unusual body attitudes related to the visual scene, like being upside down, have triggered acute SMS in astronauts in space. These conflicting visual scenes may attenuate or eliminate SMS (Harm and Parker 1994).

Virtual reality training may simulate specific effects of microgravity and may prove to be an effective countermeasure against SMS. A study of 17 men and 13 women using the NASA Device for Orientation and Motion Environments (DOME) virtual reality system to examine the effects of navigation and switch activation tasks in a virtual space station demonstrated sensory conflicts similar to that seen in space (Stroud, Harm and Klaus 2005). This type of pre-flight virtual environment training may be an effective countermeasure by enhancing adaptation and habituation. The down side to this is crew time in the pre-launch phase and the duration of effect of the dual adapted state.

Another non-pharmacologic modality that may reduce SMS is autogenic feedback training exercise (AFTE). AFTE is a combination of biofeedback and autogenic therapy, where subjects are trained to voluntarily control several physiological responses within six hours (Cowings 1990). Previous research has demonstrated that AFTE significantly improves tolerance to motion sickness induced by cross-coupled Coriolis acceleration for individuals suffering from intractable airsickness (Cowings and Toscano 2000).

The Russians use pre-flight vestibular adaptation training which involves cross-coupled Coriolis acceleration utilizing active head movements while on a rotating chair. The incidence of SMS symptoms in Russian and American space crews is similar despite this pre-flight vestibular adaption training, and cross-coupled Coriolis acceleration does not duplicate the sensory conflicts encountered in weightlessness (Gorgiladze and Bryanov 1989).

Mission critical timing

During the Space Shuttle Program flight rules restrict critical activities, such as EVA, within three days of reaching orbit and lighten crew activities during the first few days because of known decreased performance (Davis, Jennings, Beck and Bagian 1993). SMS is not a disqualifying medical standard, but Shuttle commanders had to demonstrate no/minimal SMS on their previous flight as a Shuttle pilot prior to qualifying for commander.

Mediation

An on orbit modality that might reduce SMS is the use of stroboscopic illumination or liquid crystal display shutter glasses to reduce the severity of motion sickness symptoms where retinal slip is a significant factor. Preliminary studies have shown promise in this methodology (Reschke, Somers, Leigh and Jones 2006).

Medications

Pharmacologic intervention remains the most efficacious method of preventing and managing SMS. Oral treatment after symptom development is complicated by bioavailability and decreased GI motility and absorption (Davis, Jennings, Beck and Bagian 1993). Scopolamine/Dexedrine (0.4mg/5mg) prophylaxis was used in the early space program as it had been efficacious in sea sickness and other types of motion sickness (Wood and Graybiel 1968). Scopolamine compounds prevented SMS but symptoms recurred, as scopolamine may delay the normal adaptation to microgravity (Davis, Jennings and Beck 1993). These scopolamine effects are consistent with ground-based research (Wood et al. 1986). Many antimotion sickness medications have been tried with varying degrees of success (Graybiel 1980). Oral Phenergan/Dexedrine (25mg/5mg) combination (known as Phen/dex) has been used prophylactically hours before launch and efficacy has been acceptable, despite concerns about performance effects (Hordinsky et al. 1982). Phenergan (promethazine) does not seem to delay adaptation (Davis, Jennings, Beck and Bagian 1993) and may actually hasten adaptation to provocative motion (Lackner and Graybiel 1994). The use of prophylactic medications is most appropriate for those susceptible to SMS, hence SMS prophylaxis is not recommended for first time fliers. Additionally, primary Shuttle flight crew (commanders, pilots, and flight engineers) are not allowed to take anti-motion sickness medications prior to launch due to potential adverse performance effects. The most efficacious in-flight treatment is parenteral promethazine, usually intramuscular (IM) at doses of 25–50 mg (Davis, Jennings, Beck and Bagian 1993; Jennings 1998). The first IM administration of promethazine in the US space program occurred on STS

29 in March 1989 (Bagian 1991). The suppository form is also effective when symptoms preclude oral administration (Davis, Jennings, Beck and Bagian 1993). SMS medication is often administered in the pre-sleep period to reduce the impact of drowsiness or lethargy on crew activities (Davis, Jennings, Beck and Bagian 1993; Jennings 1998), although drowsiness has been reported infrequently (Bagian and Ward 1994; Davis, Jennings, Beck and Bagian 1993). The focus on critical crew tasks may counteract the soporific effects of medication (Lackner and Graybiel 1994). Less than 1 percent have persistent symptoms (Davis, Jennings, Beck and Bagian 1993). A study in humans of SMS medication side-effects on working memory suggest that, at clinically useful doses, the rank order of the drugs with best cognitive profiles is meclizine (H_1 antihistamine) > scopolamine (muscarinic anticholinergic agent) > promethazine (H_1 antihistamine) > lorazepam (GABAergic) (Paule, Chelonis, Blake and Dornhoffer 2004).

In summary, SMS can be considered a variant of motion sickness that occurs in microgravity (Lackner and DiZio 2006). It has persisted in human spaceflight and is still experienced by a large proportion of space travelers. SMS is often treated with medication with undesirable side-effects and it is hard to predict who will experience it and to what extent it will affect performance. New approaches to preventing or treating SMS need to be developed as well as further work to elucidate the underlying mechanisms (Heer and Paloski 2006).

Special Case: Motion Sickness in Virtual Operators

An important new flight crew category has recently developed: the operator of unmanned aerial systems (UAS). Being outside a remotely operated vehicle takes away the acceleration stimuli of the vehicle, but the visual–vestibular conflict still exists. The operator viewing displays from video cameras on-board a UAS is experiencing a visual–vestibular conflict in that the operator is still and the view is moving. With small field-of-view displays, visual–vestibular conflict is minimized. The large area of peripheral vision signals visually that the operator is still. However, wide field-of-view displays either fixed on the ground with the operator, or head mounted on the operator (Head Mounted Display: HMD) can give rise to motion sickness. Configuration of video imagery, such as instability, delay, or magnification or minification can provoke motion sickness with prolonged viewing (Draper, Viirre, Furness and Gawron 2001). A UAS operator in a moving environment, such as the back of a truck, might be doubly exposed to mismatched motion signals and have both motion sickness symptoms and reduced performance.

Beyond UAS operators, an even larger contingent of people are viewing computer-generated flight simulations on fixed or head mounted displays. As with the video camera images, simulator scene configuration can be very provocative for motion sickness. For example, minification (zoom-out) to allow a viewing of a large angular extent of a scene is actually the most provocative for motion sickness compared to correct magnification or a "zoom-in" condition (Draper, Viirre, Furness and Gawron 2001). Trainees and operators of UAS systems or simulators need to be monitored for motion sickness or even mild cognitive effects that may be present with the sopite syndrome (Draper, Viirre, Furness and Gawron 2001).

The Future

Importantly, modern understanding of the physiology of motion sensation is leading to better design of means of preventing motion sickness. For example, virtual environments may be used to pre-adapt personnel to reduce their motion sickness (Dai, Raphan and Cohen 2011; Stroud, Harm and Klaus 2005; Trendel et al. 2010). Neurophysiologic signals have been developed that indicate the presence of visual attention and minimization of motion sickness. Such biomarkers can be used to assess people and systems in training and operational environments to maintain optimal performance. Advanced techniques, such as mild galvanic stimulation of the vestibular nerve (Dilda, MacDougall and Moore 2011) and new classes of anti-emetic drugs without sedation, such as ondansetron or intra-nasal scopolamine (Simmons et al. 2010) hold promise that motion sensitivity, nausea, and reduced performance can be eliminated.

References

Bagian, J.P. 1991. First intramuscular administration in the US space program. *Journal of Clinical Pharmacology*, 31(10), 920.

Bagian, J.P. and Ward, D.F. 1994. A retrospective study of promethazine and its failure to produce the expected incidence of sedation during space flight. *Journal of Clinical Pharmacology*, 34(6), 649–651.

Bremmer, F., Klam, F., Duhamel, J., Ben Hamed, S. and Graf, W. 2002. Visual-vestibular interactive responses in the macaque ventral intraparietal area (VIP). *European Journal of Neuroscience,* 16(8),1569–86.

Cowings, P.S. 1990 Autogenic-feedback training: a preventive method for motion and space sickness, in *Motion and Space Sickness*, edited by G. Crampton. Boca Raton, FL: CRC Press, 354–372.

Cowings, P.S. and Toscano, W.B. 2000 Autogenic feedback training exercise is superior to Promethazine for the treatment of motion sickness. *Journal of Clinical Pharmacology,* 40(10), 1154–1165.

Dai, M., Raphan, T. and Cohen, B. 2011. Prolonged reduction of motion sickness sensitivity by visual-vestibular interaction. *Experimental Brain Research*, 210(3–4), 503–513.

Davis, J.R, Jennings R.T. and Beck, B.G. 1993. Comparison of treatment strategies for space motion sickness. *Acta Astronautica*, 29(8), 587–591.

Davis, J.R., Jennings, R.T., Beck, B.G. and Bagian, J.P. 1993. Treatment efficacy of intramuscular promethazine for space motion sickness. *Aviation, Space, and Environmental Medicine*, 64(3, Pt 1), 230–233.

Davis, J.R., Vanderploeg, J.M., Santy, P.A., Jennings, R.T. and Stewart D.F. 1988. Space motion sickness during 24 flights of the space shuttle. *Aviation Space, and Environmental Medicine*, 59(12), 1185–1189.

Dilda, V., MacDougall, H.G. and Moore S.T. 2011. Tolerance to extended galvanic vestibular stimulation: optimal exposure for astronaut training. *Aviation, Space, and Environmental Medicine*, 82(8), 770–774.

Draper, M.H., Viirre, E.S., Furness, T.A. and Gawron, V.J. 2001. The effects of virtual image Scale and system time delay on simulator sickness within head coupled virtual environments. *Human Factors*, 43(1), 129–146.

Gibson, J.J. 1966. *The Senses Considered as Perceptual Systems*. Boston, MA: Houghton Mifflin.

Gibson, J.J. 1979. *The Ecological Approach to Visual Perception*. Boston, MA: Houghton Mifflin.

Gorgiladze, G.I. and Bryanov II. 1989. Space motion sickness. *Kosmicheskaya Biologiya I Aviakosmicheskaya Meditsina*, 23(3), 4–14.

Graybiel, A. 1980. Space motion sickness: Skylab revisited. *Aviation, Space, and Environmental Medicine*, 51(8), 814–822.

Graybiel, A. and Knepton, J. 1976. Sopite syndrome: A sometimes sole manifestation of motion sickness. *Aviation, Space, and Environmental Medicine*, 47(8), 873–882.

Griffin, M.J. 1990. *Handbook of Human Vibration*. London: Academic Press.

Grontved, A., Brask, T., Kampskard, J. and Hentzer, E. 1988. Ginger root against seasickness: A controlled trial on the open sea. *Acta Otolaryngologica*, 105(1–2), 44–49.

Grontved, A. and Hentzer, E. 1986. Vertigo reducing effects of ginger root: A controlled clinical study. *Journal of Otorhinolaryngology and its Related Specialties*, 48(5), 282–286.

Harm, D.L. and Parker, D.E. 1994. Preflight adaptation training for spatial orientation and space motion sickness. *Journal of Clinical Pharmacology*, 34(6), 618–627.

Hawkins, W.R. and Zieglschmid, J.F. 1975. Clinical aspects of crew health, in *Biomedical Results of Apollo (NASA SP-368)*, edited by L.F. Dietlein, R.S. Johnston and C.A. Berry. Washington, DC: US Government Printing Office, 43–81.

Heer, M. and Paloski, W.H. 2006. Space motion sickness: Incidence, etiology, and countermeasures. *Autonomic Neuroscience,* 129(1–2), 77–79.

Homick, J.L. 1979. Space motion sickness. *Acta Astronautica*, 6(10), 1259–1272.

Homick, J.L. and Miller, E.F. 1975. Apollo flight crew vestibular assessment, in *Biomedical results of Apollo, NASA SP 368*, edited by R.S. Johnston and L.F. Deitlein, 323-340.

Homick, J.L., Reschke, M.F. and Vanderploeg, .JM. 1984. *Space Adaptation Syndrome: Incidence and Operational Implications for the Space Transportation System Program*. NATO AGARD Symposium on Motion Sickness: Mechanisms, Prediction, Prevention and Treatment, Williamsburg, VA, Paper 36, AGARD CP-372.

Hordinsky, J.R., Schwertz, E., Beier, J., Martin, J. and Aust, G. 1982. Relative efficacy of the proposed space shuttle antimotion sickness medications. *Acta Astronautica*, 6(7), 375–383.

Jennings, R.T. 1998. Managing space motion sickness. *Journal of Vestibular Research*, 8(1), 67–70.

Jennings, R.T., Davis, J.R. and Santy, P.A. 1988. Comparison of aerobic fitness and space motion sickness in the space shuttle program. *Aviation, Space and Environmental Medicine*, 59(5), 448–451.

Kornilova, L.N., Yakovleva, I.Y., Tarasov, I.K. and Gorgiladze, G.I. 1983. Vestibular dysfunction in cosmonauts during adaptation to zero-g and readaptation to 1g. *Physiologist*, 26, S35–S40.

Lackner, J.R. and DiZio, P. 2006. Space motion sickness. *Experimental Brain Research*, 175, 377–399.

Lackner, J.R. and Graybiel, A. 1987. Head movements in low and high force environments elicit motion sickness: Implications for space motion sickness. *Aviation, Space and Environmental Medicine*, 58(9, Suppl.), A212–217.

Lackner, J.R. and Graybiel, A. 1994. Use of promethazine to hasten adaptation to provocative motion. *Journal of Clinical Pharmacology*, 34(6), 644–648.

Lin, C.T., Chuang, S.W., Chen, Y.C., Ko, L.W., Liang, S.F. and Jung, T.P. 2007. EEG effects of motion sickness induced in a dynamic virtual reality environment. *Conference Proceedings — IEEE Engineering in Medicine and Biology Society*, 2007, 3872–3875.

Mackenzie, I., Viirre, E., Vanderploeg, J. and Chilvers E. 2007. Effects of zero g and high g in a patient with advanced Amyotrophic Lateral Sclerosis. *The Lancet*, 370(9587), 566.

Matsnev, E.I., Yakovleva, I.Y., Tarasov, I.K., Alekseev, V.N., Kornilova, L.N., Mateev, A.D. and Gordiladze, G.I. 1983. Space motion sickness: Phenomenology, countermeasures, and mechanisms. *Aviation, Space, and Environmental Medicine*, 54(4), 312–317.

Money, K.E. and Myles, W.S. 1974. Heavy water nystagmus and effects of alcohol. *Nature*, 247(5440), 404–405.

Naval Aerospace Medical Institute (2010). *Aeromedical reference and waiver guide*. Available at: http://www.med.navy.mil/sites/nmotc/nami/arwg/Pages/AeromedicalReferenceandWaiverGuide.aspx [accessed 26 August 2012].

Oman, C.M., Lichtenberg, B.K., Money, K.E. and McCoy, R.K. 1986. MIT/Canadian vestibular experiments on the Spacelab-1 mission: 4. Space motion sickness: Symptoms, stimuli, and predictability. *Experimental Brain Research*, 64(2), 316–334.

Oman, C.M. and Shubentsov, I. 1992. Space sickness symptom severity correlates with average head acceleration, in *Mechanisms and Control of Emesis*, edited by A.L. Bianchi, L. Grolot, A.D. Miller and G.L. King. John Libbey Eurotext, Montrouge: 185–194.

Park, A.H. 1998. Age differences in self-reported susceptibility to motion sickness. *Perceptual and Motor Skills*, 87(3 Pt 2),1202.

Paul, M.A., MacLellan, M. and Gray, G. 2005. Motion-sickness medications for aircrew: impact on psychomotor performance. *Aviation, Space, and Environmental Medicine* 76(6), 560–565.

Paule, M.G., Chelonis, J.J., Blake, D.J. and Dornhoffer, J.L. 2004. Effects of drug countermeasures for space motion sickness on working memory in humans. *Neurotoxicology and Teratology*, 26(6), 825–837.

Reschke, M.F., Parker, D.E., Harm, D.L. and Michaud, L. 1988. Ground-based training for the stimulus rearrangement encountered during space flight. *Acta Oto-Laryngologica*, Suppl. 460, 87–93.

Reschke, M.F., Somers, J.T., Leigh, R.J., Jones, G.M. 2006. *A Countermeasure For Space Motion Sickness*. NASA Johnson Space Center Technical Report Server. Available at: http://ntrs.nasa.gov/search.jsp?R=20060025995 [accessed 26 August 2012].

Robinson, D.A. 1981. The use of control systems analysis in the neurophysiology of eye movements. *Annual Review of Neuroscience,* 4(1), 463–503.

Rogers, D. and Van Syoc, D. 2011. *Clinical practice guideline for motion sickness*. Available at: http://www.asams.org/guidelines/Completed/NEW%20Motion%20Sickness.htm [accessed 26 August 2012].

NASA/Space Biomedical Research Institute USRA/Division of Space Biomedicine.

Seligman, M.E.P. and Hager, J.L. 1972. Biological boundaries of learning. The sauce-bearnaise syndrome. *Psychology Today*, 6, 59–61, 84–87.

Simanonok, K.E., Kohl, R.L. and Charles, J.B. 1993. The relationship between space sickness and preflight diet. *Physiologist*, 36(1), S90–S91.

Simmons, R.G., Phillips, J.B., Lojewski, R.A., Wang, Z., Boyd. J.L. and Putcha, L. 2010. The efficacy of low-dose intranasal scopolamine for motion sickness. *Aviation Space, and Environmental Medicine*, 81(4), 405–412.

Stoffregen, T.A. and Riccio, G.E. 1991. An ecological critique of the sensory conflict theory of motion sickness. *Ecological Psychology,* 3(3), 159–194.

Stroud, K.J., Harm, D.L. and Klaus, D.M. 2005. Preflight virtual reality training as a countermeasure for space motion sickness and disorientation. *Aviation, Space, and Environmental Medicine,* 76(4), 352–356.

Strychacz, C., Viirre, E. and Wing, S. 2005. The use of EEG to measure cerebral changes during computer-based motion sickness-inducing tasks. *Proceedings of the SPIE*, 5797, 139–147.

Thornton, W.E., Linder, B.J., Moore, T.P. and Pool, S.L. 1987. Gastrointestinal motility in space motion sickness. *Aviation, Space, and Environmental Medicine*, 58 (9 Pt 2), A16–A21.

Thornton, W.E., Moore, T.P., Pool, S.L. and Vanderploeg, J.M. 1987. Clinical characterization and etiology of space motion sickness. *Aviation, Space, and Environmental Medicine*, 58 (9 Pt 2), A1–A8.

Trendel, D., Haus-Cheymol, R., Erauso, T., Bertin, G., Florentin, J.L., Vaillant, P.Y. and Bonne, L. 2010. Optokinetic stimulation rehabilitation in preventing seasickness. *European Annals of Otorhinolaryngology, Head and Neck Diseases*, 127(4), 125–129.

Wood, C.D. 1979. Antimotion sickness and antiemetic drugs. *Drugs*, 17(6), 471–479.

Wood, C.D. and Graybiel, A. 1968. Evaluation of sixteen anti-motion sickness drugs under controlled laboratory conditions. *Aerospace Medicine*, 39(12), 1341–1344.

Wood, C.D., Manno, J.E., Manno, B.R., Odenheimer, R.C. and Bairnsfather, L.E. 1986. The effect of antimotion sickness drugs on habituation to motion. *Aviation, Space, and Environmental Medicine*, 57(6), 539–542.

Wood, C.D., Manno, J.E., Manno, B.R., Redetzki, H.M., Wood, M. and Vekovius, W.A. 1984. Side effects of antimotion sickness drugs. *Aviation, Space, and Environmental Medicine*, 55(2), 113–116.

Yates, B.J., Miller, A.D. and Lucot, J.B. 1998. Physiologic basis and pharmacology of motion sickness: an update. *Brain Research Bulletin,* 47(5), 395–406.

Chapter 9
Fatigue and Aviation

J. Lynn Caldwell and John A. Caldwell

Fatigue from inadequate sleep has been a problem for many years, possibly beginning with the advent of the electric light bulb when the ability to be active at night as well as during the day became possible. Modern transportation technology contributed further when air travel enabled time-zone transitions at a rate that severely outpaced the body clock's physiological rate of adaptation. In recent years, the problem of sleepiness/fatigue has been a focus in aviation due to the safety hazards associated with impaired operator alertness.

Pilot fatigue is often cited as a major contributor to severe aviation mishaps. In the US Navy and Marine Corps for example, fatigue was identified as the leading cause of Class A mishaps (that is, damage exceeds one million dollars and involves fatality or permanent disability or destroyed aircraft) between 2000 and 2006 (Davenport 2009). Aviator fatigue has been blamed at least in part for such incidents and accidents as the 2009 crash of a Continental Connection flight in which 50 people were killed (National Transportation Safety Board 2010), the 2004 crash of Corporate Airlines flight 5966 in which 13 people perished (National Transportation Safety Board 2006), the 1997 crash of Korean Air flight 801 in which 228 people died (National Transportation Safety Board 1999), the 1985 near crash of China Airlines flight 006 in which 24 people were injured (National Transportation Safety Board 1986), and the 1999 mishap involving American Airlines Flight 1420 in which 11 people were killed (National Transportation Safety Board 2001). For a play-by-play look at fatigue and an aircraft mishap, please see Wesensten and Balkin (2010) who outline the role of sleep loss and consequent cognitive dysfunction in the 1993 crash of American International Airways Flight 808, which resulted in serious injuries to all three crewmembers.

While it was once thought that pilot fatigue could be overcome simply through the exertion of sufficient willpower, a plethora of recent evidence has established that drowsiness on the flight deck is a physiologically-based reality that has little to do with motivation, professionalism, or training. In the 1920s, Charles Lindbergh noted in his eighteenth hour of flight, "I've lost command of my eyelids. They shut and I shake myself and lift them with my fingers. I stare at the instruments, wrinkled forehead, muscles tense. I've got to find some way to stay alert" (Lindbergh 1953, 354). Clearly, Mr Lindbergh possessed the mental drive to stay alert and vigilant, but his physiological drive for sleep was not to be deterred by his strong will and good intentions. Advances in aviation have led to improved aircraft, but pilots continue to fight the problems of sleepiness in flight, due mainly to the long duty periods and circadian disruptions that are common in both civilian and military flight operations (Samel, Wegmann and Vejvoda 1995; Nevelle et al. 1994).

In this chapter, the physiological underpinnings of pilot fatigue will be reviewed along with the factors that are currently thought to influence fatigue vulnerability as well as the rate at which individuals recover from fatigue. In addition, strategies for optimizing crew schedule design, tools for monitoring the sleep crewmembers actually obtain on their work/rest schedules, and methods for ensuring that each sleep opportunity is maximally restorative will be highlighted. The concluding text will provide a brief overview of several on-duty counter-fatigue techniques that will help compensate for insufficient pre-duty sleep and/or circadian disruptions. An integrated consideration of the information presented here will help ensure optimal alertness, performance, and safety in the aviation context.

Effects of Inadequate Sleep on Alertness and Performance

Sleepiness on the job (or conversely, the level of alertness on the job) is a function of two major factors—the homeostatic drive for sleep and the circadian rhythm, or body clock. In people who are working consistent daytime schedules, these two factors work together to maintain consolidated sleep at night and stable alertness during the day; however, in people who are sleep restricted or engaged in rotating work/rest schedules, the homeostatic and circadian factors can combine to cause significant problems.

The homeostatic drive

The homeostatic drive for sleep is primarily a function of the amount of sleep recently and/or routinely obtained and the amount of time between the end of the last sleep period and the beginning of the duty period. Healthy adults need between 7.5 and 8.5 hours of sleep per day to perform optimally (Reynolds and Banks 2010), and the failure to obtain this amount rapidly impairs both alertness and performance. Furthermore, remaining awake for longer than 16 continuous hours significantly degrades performance (Williamson and Feyer 2000), especially when the latter part of this time span coincides with the late night hours. Remaining awake on any single occasion for more than 24 continuous hours produces a variety of acute adverse effects such as degradations in reaction time, poor attention, memory loss, and impaired decision-making (Lim and Dinges 2010). Chronically shortened sleep spanning several consecutive days produces similar effects (Van Dongen, Maislin, Mullington and Dinges 2003). Two classic studies in which the effects of chronic sleep deprivation were assessed indicated that chronic sleep loss produced a "dose response" effect, with performance decrements progressively increasing as time in bed (and thus time asleep) was systematically reduced (Belenky et al.

2003; Van Dongen, Maislin, Mullington and Dinges 2003). Sleep durations under six hours per night were particularly problematic in both investigations—a finding confirmed by other reports (Banks, Van Dongen, Maislin and Dinges 2010; Rupp, Wesensten, Bliese and Balkin 2009; Van Dongen, Maislin, Mullington and Dinges 2003).

The circadian drive

The circadian clock, or time of day according to the body's clock, is the second important contributor to operator alertness. As diurnal animals, humans are programmed to be active during the day and to be asleep at night. Whenever this natural order of wake and sleep is disrupted, problematic consequences ensue. Performance deficits are noteworthy from as early as 02:00 to as late as 10:00, particularly if no sleep occurred during the night (Goel, Van Dongen and Dinges 2011). Aviation personnel who work at night, or who fly across multiple time zones, often find they are required to be alert at a time when the body is physiologically prepared for sleep. Research shows that during the circadian trough (the low-point in the body's circadian rhythm that generally occurs between 02:00 and 06:00) alertness is lower, reaction time is slower, and accuracy is poorer than during the circadian peak (that is, during daytime hours) (Folkard and Tucker 2003). Studies from the aviation arena have shown that microsleeps, short episodes of brain activity associated with attentional lapses, are up to nine times more likely during the nighttime compared to the daytime (Wright and McGown 2001; Samel, Wegmann and Vejvoda 1995), and psychomotor vigilance lapses are five times greater (Dinges et al. 1990). In one nighttime flight simulation study, nine of 14 pilots experienced actual sleep episodes during the flight (Neri et al. 2002), corroborating Moore-Ede's (1993) assertion that pilots nod off at the controls significantly more frequently in the dark than during daylight. Keep in mind that, when multiple time zones are rapidly crossed, aircrews may experience serious performance decrements during local daytime hours in a new time zone because these daytime hours on the local clock correspond to nighttime hours in terms of the body's internal circadian clock. Generally speaking, adaptation to a new time zone occurs at an average of one day per time-zone crossed in the westward direction and 1.5 days per time zone crossed in the eastward direction (World Health Organization 2010).

The combined effect

When considering the impact of the homeostatic and circadian drives on alertness and performance, it is important to remember that these two drives interact with one another. Thus, cognitive performance which occurs during the circadian trough after many hours of continuous wakefulness will be far more impaired than performance during the circadian trough following a recent sleep period. This is because, in the former case, there would be high sleep pressure stemming from both the circadian and the homeostatic factors whereas in the latter case, sleep pressure would largely be a function of only the single circadian factor. Research involving flight performance from pilots undergoing total, acute sleep deprivation from 37 hours to 64 hours during which the performance periods spanned both the daytime and nighttime illustrates this interactive pattern (Caldwell, Caldwell, Brown and Smith 2004; Caldwell, Smythe, LeDuc, and Caldwell 2000; Caldwell, Caldwell, Smythe and Hall 2000). As can be seen in Figure 9.1, there is an overall decline in performance which begins after the last sleep period due to increased homeostatic sleep pressure; this decline is accentuated during the circadian troughs due to elevated circadian-driven sleep pressure.

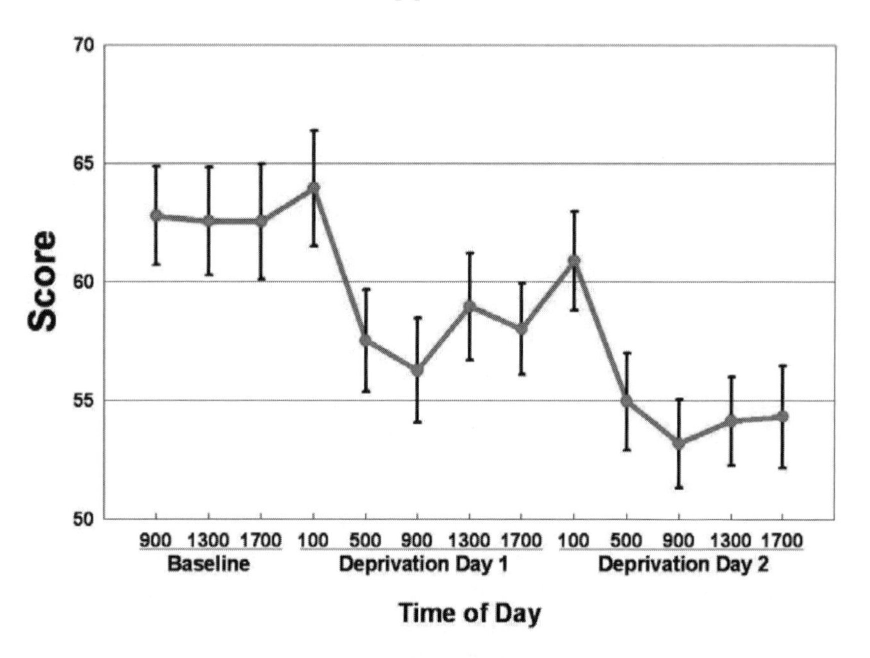

Figure 9.1 Illustration of flight-performance scores across a period during which no sleep was obtained for 64 continuous hours

Recovering from Fatigue Attributable to Sleep Loss

After sleep restriction, people frequently fail to fully recover after only a single eight-hour night of sleep. In fact, as demonstrated by two studies cited in the previous section (Belenky et al. 2003; Van Dongen, Maislin, Mullington and Dinges 2003), people were not fully recovered even after three consecutive days during which eight-hour recovery sleep episodes were provided. Generally, recovery from acute, total sleep deprivation appears to be faster than recovery from chronic sleep restriction. Results from most acute sleep-deprivation studies show that performance returns to baseline levels within two nights of recovery sleep of at least eight hours each (Balkin, Rupp, Picchioni and Wesensten 2008). However, this is not the case with chronic sleep restriction, where return to well-rested baseline performance can take several days--even as long as a week (Axelsson et al. 2008). Banks, Van Dongen, Maislin and Dinges (2010) suggested that a prolonged sleep period may speed recovery, but even so, subjects who had experienced five nights of severe sleep restriction were not fully recovered after one night of ten hours in bed.

In an effort to determine if prior sleep deficits affected recovery, Rupp and colleagues (Rupp, Wesensten, Bliese and Balkin 2009; Rupp, Killgore and Balkin 2010) assigned subjects to either ten hours time in bed (extended group) or their habitual time in bed (mean of seven hours) for one week after which they were tested across seven nights of sleep restricted to three hours time in bed, followed by five nights of recovery sleep of eight hours time in bed. The results indicated that for subjects in the extended sleep group, performance on the Psychomotor Vigilance Task (PVT), a simple sustained-attention, reaction time test, returned to pre-restriction levels after the first recovery night, but for subjects in the habitual sleep group, recovery was delayed. Overall, reduction of sleep debt prior to sleep restriction allowed faster recovery from sleep restriction; thus, operational personnel should be careful to optimize their sleep whenever possible so that they can "bounce back" from unavoidable episodes of sleep loss more quickly.

Factors Affecting Vulnerability to Fatigue Attributable to Sleep Loss

While sleep deprivation causes performance deficits in everyone, research indicates individual variation in the magnitude of decrements observed. These individual differences are stable, trait-like inter-individual characteristics in susceptibility to fatigue (Van Dongen, Baynard, Maislin and Dinges 2004; Rupp, Wesensten, Bliese and Balkin 2009; Van Dongen, Vitellaro and Dinges 2005). Research over the past decade has investigated possible explanations for why people respond differently to inadequate sleep.

Central nervous system effects

Brain activation studies indicate there may be differences in basic physiological profiles that influence the response to sleep deprivation. Caldwell and colleagues (2005) showed that F-117 pilots who displayed high cortical activity while not sleep-deprived performed better on tasks during 37 hours of continuous wakefulness than did individuals with lower levels of pre-sleep-deprived cortical activity. Mu and colleagues (2005) found similar results in a study during which ten fatigue-resilient subjects showed more brain activation during a rested functional Magnetic Resonance Imaging (fMRI) session than did fatigue-vulnerable subjects. Chuah, Venkatraman, Dinges and Chee (2006) indicated that subjects who were more vulnerable to fatigue demonstrated poorer inhibitory efficiency on a go/no-go task as well as higher levels of cortical activation during 24 hours of continuous wakefulness.

Personality effects

Outgoing (extroverted) individuals may have a disadvantage over introverted people in terms of resisting the effects of sleep loss, and this is of particular interest in the present context because there is reason to believe pilots are more outgoing, energetic, and socially competent (thus, more extroverted) than non-pilots (Caldwell et al. 1993; Campbell, Moore, Poythress and Kennedy 2009). Research with non-pilots has revealed that extroverted subjects suffer greater performance impairment than introverted subjects (Killgore et al. 2007). This individual difference in response to sleep deprivation may be due to greater cortical arousal in introverts than in extroverts (Johnson et al. 1999). Furthermore, there appears to be an interaction between personality and social environment. Rupp, Killgore and Balkin (2010) exposed introverts and extroverts to either a socially-enriched or socially-isolated environment during 36 hours of continuous wakefulness and found that extroverts in the socially-enriched condition were more vulnerable to sleep deprivation than when they were in a socially-impoverished environment, while introverts were unaffected by the social interactions. Social exposure may combine with personality to modulate vulnerability to sleep loss via effects on brain activation.

Age effects

In addition to differences in brain activation and personality, age may also influence the response to extended wakefulness. Generally speaking, older adults may have an advantage over their younger counterparts in terms of resisting the impact of sleep loss, but younger adults seem to have the edge in terms of recovery (Bliese, Wesensten and Balkin 2005; Rupp, Wesensten, Bliese and Balkin 2009). Unfortunately, the younger adults seem to deteriorate faster as a consequence of sleep loss and awareness of their cognitive impairment appears to be less accurate than that of older people (Rupp, Wesensten, Bliese and Balkin 2009).

Recommendations for Countering Fatigue in Aviation

Traditionally, aviation regulators have failed to fully appreciate the physiological complexity of fatigue and have focused instead strictly on regulating hours on duty without sufficient regard to either sleep or circadian factors. The Federal Aviation Association (FAA), however, recently formally recognized that alertness and performance are impacted by the body clock, sleep duration, and sleep efficiency. The new rules reflect our understanding that night duty is more fatiguing than day duty and time-zone crossings require a period of acclimation (Federal Register

2012). In addition, a recognition of the impact of cumulative fatigue led to changes in the amount of time allotted for rest; aircrew are now required to have ten hours of rest between successive duty periods, with 30 continuous hours off after a seven-day duty period. Comprehensive Fatigue Risk Management Systems (FRMS) are beginning to replace prescriptive duty hours and set crew-rotation approaches. The International Civil Aviation Organization (ICAO) recently adopted fatigue management recommendations which include methods of monitoring and managing fatigue based on scientific principles (International Civil Aviation Organization 2011).

Nevertheless, even with new crew rest guidelines, pilots will continue to find many of the more typical aviation schedules challenging from an alertness and performance standpoint. Fortunately, there are a number of countermeasures available to help optimize on-duty alertness as well as off-duty fatigue recovery. When schedule-related fatigue cannot be altogether avoided, it at least can be mitigated via schedule-optimization strategies. The impact of these optimization strategies can be bolstered by the utilization of technologies that permit ongoing monitoring of pre-duty sleep quantity and sleep/wake timing. Sleep difficulties (which are at the heart of on-duty crew fatigue) can be attenuated by educating aircrew members on techniques designed to improve pre-duty and layover sleep. In-flight counter-fatigue measures can be relied upon to improve in-flight alertness despite any scheduling or pre-duty sleep problems that prove difficult or impossible to resolve.

Schedule optimization

As regulatory agencies increasingly recognize the physiological realities of human fatigue they are beginning to require airlines to address these realities in their scheduling practices. For instance, Attachment K Annex 6 to the Convention on International Civil Aviation guidance on dealing with fatigue risks in aircraft operations requires the development and maintenance of processes for fatigue hazard identification. One of these processes is the predictive process, which involves identifying flight hazards by examining scheduling from the standpoint of "factors known to affect sleep and fatigue and their effects on performance" (Section 2.1.1), and a major component of this examination involves the use of bio-mathematical models to help account for the way in which scheduling will affect the circadian and sleep-regulation systems that are primary fatigue determinants (International Civil Aviation Organization 2011).

A bio-mathematical model is basically a set of integrated equations that predict human fatigue based on such factors as recent sleep quantity, sleep quality, and sleep/wake timing; the current time of day (during duty); and workload in terms of things like number of recent takeoffs and landings and/or other factors. Prior to use in operational contexts, models are validated against various types of performance data (such as reaction time or accuracy measures) collected in laboratory sleep-restriction or sleep-deprivation studies, or against accident probability and/or accident severity data collected in real-world environments. Models are typically updated in an iterative process as new data and/or new scientific information become available, and they are quite useful in applied contexts because they translate basic scientific principles established from empirical investigations into generalized predictions that are relevant to operational settings.

At present, airlines can utilize different types of bio-mathematical models to optimize duty schedules from a fatigue risk standpoint. Several models are available to help determine the impact of work/rest schedules on aviator performance. In addition, the models can be used to explore scheduling modifications that will mitigate fatigue factors. The Civil Aviation Safety Authority provided a recent overview of six fatigue models (Civil Aviation Safety Authority 2010). Of these, two were specifically tailored for the aviation environment—the Sleep, Activity, Fatigue, and Task-Effectiveness (SAFTE) model and the System for Aircrew Fatigue Evaluation (SAFE). The UK Civil Aviation Authority (CAA) sponsored the development of the SAFE model for assessing flight-time limitations for operators. It has been validated using operator data and is used by the CAA for fatigue risk assessment of rosters. The SAFE model was not available for commercial use as of 2010, but plans include making it available to companies who develop pilot schedules as well as to individual operators.

The SAFTE model (Hursh et al. 2004) which has been instantiated in the Fatigue Avoidance Scheduling Tool (FAST) software (Hursh et al. 2004) was validated in ground transportation operational studies and has been tailored to accept aviation-specific input. The SAFTE model has been validated as an accurate predictor of sleep restriction on performance (Van Dongen 2004). In addition, it has been shown to accurately predict the impact of scheduling factors on accident risk (Hursh et al. 2006). Although the precision of SAFTE fatigue predictions as well as those from other available bio-mathematical models of fatigue and performance are adversely affected by individual differences and uncertain pre-duty conditions (Van Dongen et al. 2007), model-based optimization of crew schedules represents a step in the right direction toward mitigating operational fatigue risks.

Ongoing sleep monitoring

Bio-mathematical models such as SAFE and SAFTE provide an excellent quantification of the expected fatigue risks associated with crew schedules. However, in most cases the models must rely on *sleep estimates* that are based on the characteristics and timing of off-duty periods rather than *sleep measurements* obtained from an accurate and reliable source. Software instantiations of the SAFTE model include a routine called AutoSleep which estimates an operator's sleep based on what science has determined about the sleep propensity of people who have sleep opportunities of varying durations at different points in the circadian cycle. AutoSleep considers factors such as the commuting time to and from the workplace as well as the influence of body clock factors when determining the estimated duration and quality of off-duty sleep. This is certainly a better approach than assuming that the amount of off-duty sleep is equal to the amount of off-duty time, but it is nonetheless an estimate and not a measurement of sleep. Thus, in cases during which the sleep environment is unknowingly poor, or some unidentified stressor prevents or disrupts sleep, the AutoSleep routine would predict more sleep than would actually be obtained, and the fatigue risk scores subsequently based on this overestimation would be lower (i.e., more favorable) than they actually should be. Thus, when possible, it is always better to measure sleep directly.

The gold standard for sleep assessment is polysomnography, a procedure which involves the measurement and scoring of several physiological parameters to quantify the characteristics and duration of each sleep episode. If feasible, it would be best to monitor the off-duty and pre-duty sleep of crewmembers in an ongoing fashion via polysomnographic techniques since this would provide the most accurate input to the bio-mathematical models that calculate the impact of different sleep patterns on fatigue risk. Unfortunately, the effort and expense involved in continuous physiological sleep monitoring make it impractical for day-to-day real-world application. However, direct, empirical measurements of sleep and sleep/wake timing can be obtained via another, far more feasible, methodology. Wrist-activity monitoring in conjunction with validated "sleep-scoring" classification algorithms has long been used to assess basic sleep characteristics in situations where polysomnography is impractical (Morgenthaler et al. 2007; Sadeh and Acebo 2002). Recent advances in sleep monitoring via actigraphy have made this technique applicable for more routine fatigue assessment applications.

Wrist-activity monitoring basically translates the frequency and time-course of body movements into measures of sleep quantity, sleep quality, and sleep/wake timing. Since these measurements provide the necessary input for bio-mathematical models which consider sleep and the circadian rhythm to be the primary determinants of fatigue-risk, they are well-suited to fatigue management applications in aviation and other operational contexts. An example of one such wrist-activity monitor is the Fatigue Science ReadiBand™ (see Figure 9.2). The accuracy of ReadiBand™ actigraph sleep/wake classifications have been verified in a study of 50 patients undergoing polysomnographic evaluation, with results indicating 92 percent accuracy in comparison to polysomnography. Although, sleep calculations based on activity monitoring are not completely fail-safe since activity monitoring cannot accurately detect relaxed (movement-free) wakefulness or microsleeps, it is better at tracking bed times, wake-up times, and sleep times than subjective sleep logs. Individuals unfortunately are not the best judges of their own sleep, and they certainly are not good judges of the full impact of their sleep patterns on their fatigue status, particularly when a considerable amount of time passes prior to the recording of recent sleep. Thus, activity-based sleep histories are a solid foundation for the calculation of operational fatigue-risk levels attributable to sleep loss and disrupted sleep/wake cycles.

Given the accuracy of actigraphs in predicting sleep/wake activity, they could be used to establish a conservative fitness-for-duty program even without submitting the recorded recent-sleep-history data to a model analysis. Since it is well-known that the average adult needs a minimum of eight hours of sleep in order to be fully rested (Van Dongen, Maislin, Mullington and Dinges 2003), the actigraphy record of pilots reporting to duty could be examined, and those showing significantly less than eight hours of sleep in the preceding 24-hour period could be excluded from upcoming flights (or at least warned about their potential level of impairment).

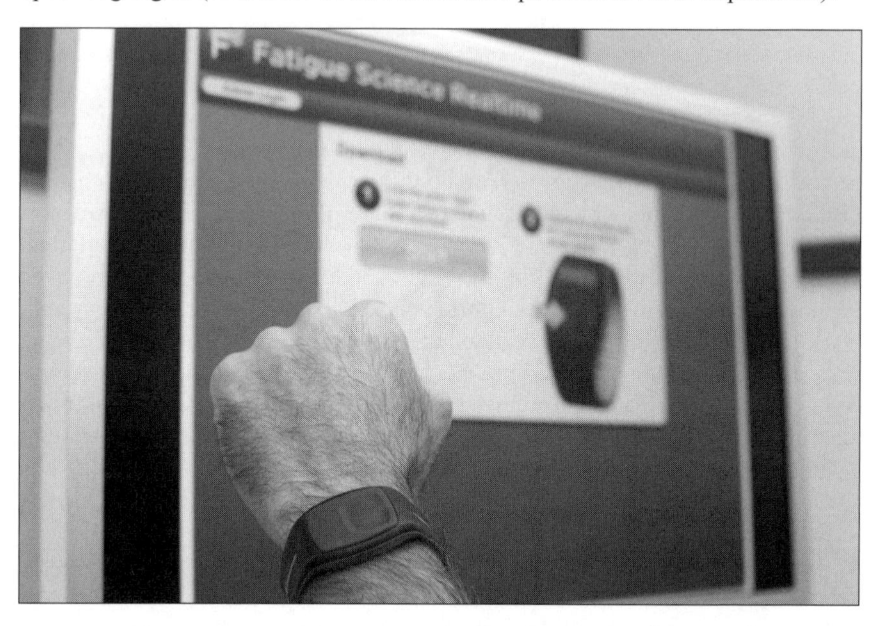

Figure 9.2 A Wrist-Activity Monitor for Sleep and Fatigue Monitoring

Techniques designed to improve pre-duty and layover sleep

As already discussed, the greatest driver of fatigue in the aviation context is insufficient or disrupted sleep. Of course, some sleep problems are unavoidable, but others are amenable to modification, especially if the crewmember has a pre-arranged strategic sleep strategy. There are general sleep-habits that are relevant for any situation (see Table 9.1), and there are tips to deal with the special problems associated with poor layover sleep opportunities or environments. Details on each recommendation are covered in the text that follows. Due to the aviation scheduling context, some recommendations will be more difficult to follow than others, but crewmembers should strive to adhere to as many of the suggestions as possible whether at home or on trips.

Stick to a consistent wake-up and bedtime every day of the week Although adhering to a consistent sleep/wake schedule every day of the week may be impossible for many crewmembers because of changing duty hours and time-

Table 9.1 Sleep-optimization Strategies

Sleep-optimization Strategies
Stick to a consistent wake-up and bedtime every day of the week
Use the bedroom only for sleep and sex
Resolve daily dilemmas outside of the bedroom
Establish a bedtime routine
Establish an aerobic exercise routine and stick to it
Create a comfortable sleep environment (for example, light, noise, temperature, and sleep surface)
Familiarize the sleep environment
Properly manage dietary issues, including caffeine and alcohol consumption
Don't be a clock watcher
Don't take naps during the day (if you have trouble sleeping at night)
Don't smoke cigarettes immediately before bedtime
Get up and go to another room if sleep doesn't come in 30 minutes

zone transitions, variations in bedtimes should be avoided when possible. This is because circadian rhythms have a strong effect on sleepiness levels across the 24-hour period, and constant fluctuations in sleep timing can interfere with these rhythms and/or place sleep initiation attempts at inopportune times. Note that maximum sleepiness occurs near the low-point of the body's temperature rhythm (that is, after midnight—*body clock time*), and it is easiest to fall asleep for a reasonable period of time before or after this point (Lack and Wright 2007). But this period of sleepiness is bracketed by "wake-maintenance zones," and one of these occurs six to ten hours before the core temperature minimum (that is, from around 1700 to 2100 for most adults on consistent day-shift schedules). During this zone, it is quite difficult to fall asleep even under the best of circumstances as anyone who has ever tried to go to sleep earlier than normal will verify. In particular, the evening wake-maintenance zone is problematic for travelers and night workers because, even though they realize the need for sleep during the late-afternoon or early-evening hours, they simply are unable to fall asleep during this period. Unfortunately, people who fail to adhere to a regular bedtime routine also can develop problems going to sleep if they inadvertently time their attempt at sleep onset inside the wake-maintenance zone. While this may not be a problem if it happens only occasionally, it may pave the way for the development of psychological apprehension about sleep initiation, and this could ultimately lead to insomnia. Thus, for this reason and many others, it is best to adhere to a consistent well-placed bedtime as much as possible.

Use the bedroom only for sleep and sex Using the bedroom only for things that are compatible with sleep is important because keeping non-sleep-related activities out of the bedroom avoids the development of associations between the sleep environment and potentially sleep-disrupting thoughts, actions, and habits (Morgenthaler et al. 2006). Few people think about the power of stimulus control in day-to-day life, but a good example is the way in which the sight and smell of good food promotes feelings of hunger even at times when the last meal wasn't so long ago. Another example is how the simple act of entering a health club promotes the willingness to exercise even when feelings of fatigue almost resulted in a workout cancellation moments before. When people make their bedroom a sleep-only area, simply entering this area will begin to promote sleep. However, when people read, watch TV, talk on the phone, play computer/video games, answer emails, and engage in stressful discussions about the day's events while lying in bed, the bedroom becomes a confusing place from a mental-association standpoint.

Resolve daily dilemmas outside of the bedroom Resolving "worry issues" outside of the bedroom can help minimize the time spent lying awake in bed worrying about tomorrow. Prior to bedtime, worriers should make a "worry list" and write a brief action item beside each concern. This can eliminate the process of thinking about something in the bedroom when the focus should be on sleep. Creating a written worry list will not solve the problem itself, but it can bring a sense of closure regarding a reasonable course of action which sometimes can stop the process of re-thinking (that is, cognitively perseverating) the same situation over and over throughout the night. Another technique which can help those who suffer sleep problems due to anxiety and worry is relaxation training. Progressive muscle relaxation techniques or biofeedback strategies can reduce muscle tension, override intrusive thoughts, and set the stage for a comfortable consolidated sleep period (Morgenthaler et al. 2006).

Establish a bedtime routine The value of establishing and maintaining a consistent pre-bedtime routine is based on an operant technique called chaining in which each behavior in a routine series of actions is reinforced by the next action, and all the behaviors in the chain ultimately lead to the final occurrence in the chain, which in this case is sleep. Chaining is a highly effective strategy for creating complex sequences of behaviors, and it can help promote rapid sleep onset if routinely followed. An example of a good pre-bedtime chain of events would be to turn off the TV at 2100, take a bath, lay out a wardrobe for the next day, read a relaxing book for 30–45 minutes, set the alarm clock, and go to bed by 2230. Once this same routine is completed night after night for several nights, the very first action in the chain (turning off the TV at 2100) will signal the brain that sleep-time is near. As each action in the series is completed, the association with "closing in on bedtime" becomes stronger so that by the time the alarm clock has been set for the next morning, the brain is fully prepared for the onset of sleep.

Establish an aerobic exercise routine The importance of exercise to restful sleep has been clearly established in the scientific literature (Uchida et al. 2012). Studies have shown that activities such as running, cycling, and swimming during the day will subsequently make it easier to fall asleep and stay asleep during the night. Weight-lifting exercises might not be as helpful as aerobic exercise, but it appears that any exercise is better than nothing. The only caveat about exercise is that the exercise period should not be too close to bedtime since physical activity is known to have a short-term alerting or stressor effect. As indicated in the review by Uchida et al. (2012), exercise periods close to bedtime result in reduced amounts of the deepest (and presumably most-restorative) stages of sleep. The rule of thumb is that exercise should be performed each day, but it usually should not be performed within three to four hours of bedtime.

Create a comfortable sleep environment The restorative value of every sleep opportunity can be enhanced by attending to several environmental and/or psychological factors. Creating a dark, quiet, cool, comfortable, and familiar sleep environment is important for the best possible sleep both at home and when staying in lodging facilities during layovers.

It is important for the bedroom (or hotel room) to be completely dark since research indicates that light exerts alerting effects on a variety of brain regions in addition to its influence on circadian regulation (Vandewalle, Maquet and Dijk 2011). A good rule of thumb with regard to the recommended level of darkness is that the bedroom should be sufficiently dark to prevent seeing a hand-held object within one or two feet in front of the eyes. To promote this level of darkness in the home bedroom, windows should be covered with blackout shades and light-emitting devices (such as digital clocks) should be hidden from view or removed altogether. Pilots who must sleep in hotels during layovers should carry clothespins or metal clips to help securely close hotel-room curtains; in the event that this strategy fails to create sufficient darkness, they may use a good sleep mask.

The minimization of noise is also important for promoting good sleep. Warwick (2011) reports that road, railway, and aircraft noise can significantly detract from sleep quality. Thus, foam earplugs should be used along with some type of masking-noise device (even a typical box fan is good) to create the least noise-contaminated sleep environment possible. When selecting hotel rooms in which to stay during layovers, rooms closest to roadways, runways, elevators, stairways, vending machines, and housekeeping staging areas should be avoided.

Ensuring an appropriate room temperature is also critical to optimizing each sleep opportunity. Studies show that about 68 degrees Fahrenheit (16–20 degrees Celsius) is most conducive to good sleep, whereas bedrooms that are too hot or too cold will lead to sleep disturbances (Onen, Onen, Bailly and Parquet 1994). Keeping the home sleep environment at the appropriate temperature usually is not a problem, but when considering the layover environment, pilots should check and adjust hotel room thermostats upon arrival so that there is plenty of time for the room to reach the optimal temperature prior to bedtime. If possible, the heating/cooling unit in the room should be set in such a way that the fan runs continuously while the heating/cooling components cycle on and off since the continuous fan noise will keep air circulating while simultaneously masking potentially-disruptive hallway noises.

Regarding the importance of the sleep surface, research has shown that the firmness of the bed surface exerts an impact on both physical comfort and sleep quality (Jacobson, Wallace and Gemmell 2006). Home mattresses should be rotated regularly. They also should be replaced every five to seven years in most cases. When sleeping in hotels, pilots should check the comfort of the bed immediately upon arrival and request another room if the bed is not satisfactory. If another room is not available, pilots should check to see if the other side of the bed is less worn, and when all else fails, they should either request a bed board or actually pull the mattress on to the floor in an effort to promote a satisfactory level of comfort.

Familiarize the sleep environment Creating some degree of familiarity within the sleep environment can be as important as many of the previously-mentioned comfort factors; clinical studies have shown first-night effects such as increased wakefulness, decreased sleep, and reduced sleep efficiency in unfamiliar places (Agnew, Webb and Williomas 1966). The home sleep environment is of course already familiar, but when staying in unfamiliar places during layovers the novelty of the hotel room can be a sleep disrupter even when the physical environment is perfect in other respects. To overcome this problem, pilots can bring along familiar items such as their own pillows, small blankets, or family pictures. They can sleep in the same type of clothing as when at home, and should make every effort to engage in the same type of pre-bedtime routines as those used in the home setting.

Properly manage dietary issues Optimizing the sleep environment can have a surprising effect on the quality and quantity of sleep obtained both at home and when traveling. However, it is likewise important to properly manage dietary habits, particularly when work or travel schedules introduce meal-time and bedtime complications.

Eating too much immediately prior to bedtime or eating the wrong kinds of food can create sleep difficulties; conversely, being hungry can cause awakenings during the night (Urponen, Vuori, Hasan and Partinen 1988). When in the home environment, it is usually easy to plan appropriate meal timing and content, but when traveling things become more difficult due to unpredictable schedules and limited availability of suitable dietary choices. When departing on trips that will involve sleep away from the home setting, pilots should pack light snacks such as dried fruit, granola bars, nuts, and so on in preparation for the possibility of very late hotel arrivals. Such light snacks will help to prevent nighttime hunger pangs while avoiding indigestion from late consumption of large meals. In general, heavy meals should be avoided within two hours of bedtime, and particularly when traveling across time zones in which jet lag will be a factor, unfamiliar and overly-spicy foods should be avoided since these may exacerbate the stomach discomfort that often accompanies rapid schedule changes.

Caffeine consumption should be carefully managed to ensure that caffeine ingestion at the wrong times does not delay sleep initiation or interfere with sleep maintenance. Caffeine is a readily-available stimulant that is sometimes contained in products where it might not be expected (such as in some brands of carbonated orange sodas). Caffeine's alerting properties can help create or maintain wakefulness during duty periods where fatigue becomes an issue; however, it is known to exert a negative effect on sleep quality if taken too close to bedtime, despite protests on the part of chronic caffeine users that this is not the case (Committee on Military Nutrition Research 2001). People should avoid caffeine within four hours of bedtime, and even longer in people who already have sleep difficulties. Note that the impact of caffeine on sleep quality changes with age since sleep architecture becomes more fragile with age, and the amount of caffeine that had little effect at the age of 20 may wreak havoc on sleep quality at age 45. For a discussion of caffeine and the flight environment, please see below as well as Chapter 12, this volume.

Alcohol has long been considered a "sleep promoter" and, while it is true that alcohol reduces sleep-onset latency, it also increases wakefulness after sleep onset and suppresses rapid eye movement (REM) sleep (Ramakrishnan and Scheid 2007). The negative impact of alcohol on sleep quality, combined with its effects on next-day blood-sugar levels, makes it a bad choice as a sleep aid. No more than two drinks should be consumed within four hours of bedtime.

Don't be a clock watcher The caution against being a clock watcher is especially important for someone who is already worried about sleep. Checking the time throughout the night may be one of the most difficult habits to overcome, but it is very important to fight the temptation to check the time constantly throughout the night. Watching the clock sets up a maladaptive pattern of thinking that can destroy the chances of getting enough sleep. Knowing the time will not improve the quality of sleep, it will not make it easier to go back to sleep, and it will not increase the amount of available sleep time. If concern about waking up in time for a meeting or a flight departure prompts the desire (or compulsion) to constantly check the time during the sleep period, this worry can be managed by setting two different alarm clocks and/or by setting clocks and then arranging for a backup wake-up call from the hotel operator during layovers.

Don't take naps as long as you have a full eight-hour sleep period coming up The longer the period of continuous waking, the greater the pressure to fall asleep, and since napping is a form of sleep, it reduces sleep pressure. This is a great feature of naps when they are used to supplement sleep that is unavoidably shortened or disrupted due to work or travel demands. But in people who have satisfactory upcoming sleep opportunities, naps during the day are likely to create sleep initiation problems later on. Pilots who are transitioning to new time zones in the eastward direction should be particularly wary about napping upon arrival because, although they might be quite sleepy due to the prolonged wakefulness necessitated by travel, alleviating this sleepiness with a daytime nap any longer than about 45 minutes will ultimately make it harder to fall asleep when the upcoming bedtime arrives. Earlier in this chapter the wake-maintenance zones were discussed. The reason eastward travelers often have difficulties falling asleep in the new time zone is because the destination bedtime occurs earlier than usual—within one of the wake-maintenance zones. Pilots can partially combat this problem by forcing themselves to remain awake for longer than usual (post-flight) prior to the advanced sleep time so that the sleep pressure from extended wakefulness counteracts the circadian-driven urge to remain awake. In general, either when staying at home or when traveling abroad, naps should be used only when some factor is interfering with normal off-duty sleep and not as a solution to boredom or inactivity.

Don't smoke cigarettes immediately before bedtime Smoking cigarettes right before bedtime is one more action that falls in the category of a chemical sleep disrupter. Tobacco smoke contains nicotine, and because smoke is inhaled into the lungs, its constituents are rapidly absorbed into the bloodstream. Although nicotine is a weak stimulant compared to caffeine, it should be avoided by those experiencing sleep difficulties for obvious reasons. Research has shown that cigarette smokers are significantly more likely than non-smokers to report problems going to sleep and problems staying asleep (Phillips and Danner 1995). Try not to smoke within one hour of bedtime.

Get out of bed and go to another room if sleep does not come in 30 minutes A final helpful sleep habit is to get out of bed and out of the bedroom for a few minutes on those occasions during which sleep onset fails to occur within 30 minutes of lights out. Simply lying in bed fretting about being unable to sleep will not only make things worse for the night on which the sleep problem occurs, but if done excessively it could become a chronic problem leading to longer-term bouts of insomnia. Everyone has problems going to sleep on occasion, and when this happens, it is best to just take it in stride, get out of bed, and go do something relaxing in another part of the house for a while. Reading a book, meditating, and listening to music are good choices since these activities are easy to stop when sleepiness returns. Conversely, working or socializing on the computer, watching TV, playing video games, or watching a movie are less desirable since they tend to be more engaging and are likely to compete with the motivation to return to bed.

In-flight Fatigue Mitigation

Other techniques can be implemented during the duty period to address fatigue issues that remain once the schedules are optimized and sleep is improved. The strategies presented below may not completely resolve the full extent of fatigue-related operational problems, but all have been shown to help sustain performance in aviation settings where non-standard work/sleep schedules and rapid/frequent time-zone crossings are the norm rather than the exception.

On-board sleep

On-board, out-of-cockpit bunk sleep is one of the most important in-flight countermeasures that can be implemented to address sleep loss and circadian disruption during extended aviation operations (Flight Safety Foundation 2005). Even short periods of sleep are better than no sleep at all in alleviating crew fatigue. Although in-flight bunk sleep can be difficult to properly schedule with reference to crew circadian rhythms and operational demands (Dinges 1984; Van Dongen and Dinges 2005), crews can estimate the times during a flight at which there is an increased risk of inadvertent sleepiness, and consider these the best times (from a circadian standpoint) to schedule in-flight bunk sleep opportunities. If flight demands permit, utilizing periods of increased physiological sleep propensity will help to increase both the quantity and quality of bunk sleep, reducing sleepiness across the flight (Carskadon 1989; Dinges 1984; Dijk and Franken 2005).

Cockpit naps

In-seat cockpit naps represent another in-flight sleep strategy that has been shown to benefit both alertness and performance. Rosekind et al. (1994) showed that 40-minute cockpit nap opportunities, resulting in an average of 26 minutes asleep, significantly improved pilot alertness and psychomotor performance on the flight deck. After napping, reaction times were faster and lapses (failures to respond) were fewer than when no nap was permitted, particularly during the last 90 minutes of flight. In fact, from top of descent to landing, the naps virtually eliminated inadvertent and uncontrollable drowsiness episodes. Many international airlines now utilize cockpit napping on long flights, and cockpit napping is sometimes authorized for US military flight operations as well (Goldsmith 1998; Caldwell et al. 2009). However, cockpit napping has not been approved for US commercial aviators.

Controlled rest breaks

Neri et al. (2002) found that simply offering a ten-minute hourly break during a six-hour simulated night flight significantly reduced slow eye movements, theta-band brain activity, unintended sleep episodes, and subjective sleepiness in pilots flying a 747-400 simulator. The benefits were transient (15–20 minutes), but noteworthy, particularly near the time of the circadian trough. Thus, it is clear that periodic breaks involving nothing more than simply leaving the flight deck and conversing with other crewmembers during long-duration flights can help to sustain alertness in the cockpit.

Caffeine

Numerous studies have shown that caffeine increases vigilance and improves performance in sleep-deprived individuals, especially those who normally do not consume high doses (Nehlig 1999). A caffeine dose of 100–200 mg noticeably affects the nervous system within 15 to 20 minutes after consumption, enhancing alertness for four to five hours. Personnel should consume caffeine sparingly and save the arousal effect until they really need it. For instance, pilots who are flying primarily daytime schedules and who are getting sufficient sleep every night should reduce or eliminate caffeine consumption until they find themselves on more fatiguing schedules. This will ensure that caffeine can exert the most beneficial effects when these effects are truly needed.

Hydration and nutrition

Although there is little scientific evidence to support the notion that adequate hydration exerts a significant effect on cognitive fatigue (Szinnai et al. 2005), aircrews should nonetheless maintain proper hydration if for no other reason than to maintain a higher level of motivation and a more positive self-appraisal of performance capabilities. As for the role of nutrition, it is important as well. Wells, Read, Uvnas-Moberg and Alster (1997) found that meals with high fat content increased sleepiness and fatigue. Thus it may be worthwhile to emphasize the ingestion of higher-protein meals immediately prior to flights and during flights. In addition, only light meals should be consumed since Lieberman (2003) has reported that while the precise macronutrient content of meals is apparently of little importance, large meals tend to be more soporific than smaller ones.

Pharmaceutical options

In some situations, prescription medication may be useful to aid with alertness when sleep opportunities are not adequate, or to optimize sleep when sleep opportunities come at times during which sleep is physiologically difficult to obtain. In US civil aviation operations, no alertness-enhancing medications are authorized with the exception of caffeine. However, in select military aviation operations the occasional use of alertness-enhancing medications is sometimes authorized to enhance the safety and effectiveness of personnel flying long missions when sleep deprivation is unavoidable. Currently, all three US services authorize at least one prescription alertness aid provided there is commander, flight surgeon, and pilot approval. Needless to say, these compounds should not be considered a replacement for adequate crew-rest planning, and they should never be considered a substitute for restorative sleep.

Sleep promoting compounds are sometimes useful for situations in which schedules allow time for sleep, but the opportunity occurs during a circadian time when sleep is difficult to obtain. In US civil aviation, the use of sleep aids is heavily discouraged, and the single aid that is occasionally permitted may not be taken within 24 hours of duty time, nor can it be used to aid in circadian rhythm sleep difficulties. However, all three US military services allow occasional use of certain prescription sleep aids to help with sleep difficulties associated with circadian disruptions, poor sleeping conditions, and poor sleep timing.

Use of prescription sleep and alertness aids during military operations is strictly controlled and implemented under a Performance Maintenance Plan. For a complete review of authorized alertness and sleep aids, see Caldwell et al. (2009).

Conclusions

Sleep and circadian factors are the primary underpinnings of human fatigue, and aviation schedules and duty demands exert a powerful effect on both. Unfortunately, the realities of modern aviation operations will continue to challenge human capabilities, but thanks to technological advances and scientifically-based sleep-optimization strategies combined with effective in-flight counter-fatigue strategies, pilots can significantly mitigate fatigue and improve operational safety in even the most demanding contexts.

References

Agnew, H.W., Webb, W.B. and Williomas, R.L. 1966. The first night effect: An EEG study of sleep. *Psychophysiology*, 2(3), 263–266.

Axelsson, J., Kecklund, G., Åkerstedt, T., Donofrio, P., Lekander, M. and Ingre, M. 2008. Sleepiness and performance in response to repeated sleep restriction and subsequent recovery during semi-laboratory conditions. *Chronobiology International*, 25(2), 297–308.

Balkin, T.J., Rupp, T., Picchioni, D. and Wesensten, N.J. 2008. Sleep loss and sleepiness Current issues. *Chest*, 134(3), 653–660.

Banks, S., Van Dongen, H.P.A., Maislin, G. and Dinges, D.F. 2010. Neurobehavioral dynamics following chronic sleep restriction: Dose-response effects of one night for recovery. *Sleep*, 33(8), 1013–1025.

Belenky, G., Wesensten, N.J., Thorne, D.R., Thomas, M.L., Sing, H.C., Redmond, D.P., Russo, M.B. and Balkin, T.J. 2003. Patterns of performance degradation and restoration during sleep restriction and subsequent recovery: A sleep dose-response study. *Journal of Sleep Research*, 12(1), 1–12.

Bliese, P.D., Wesensten, N.J. and Balkin, T.J. 2005. Age and individual variability in performance during sleep restriction. *Journal of Sleep Research*, 15(4), 376–385.

Caldwell, J.A., Caldwell, J.L., Brown, D.L. and Smith, J.K. 2004. The effects of 37 hours without sleep on the performance of F-117 pilots. *Military Psychology*, 16(3), 163–181.

Caldwell, J.A., Caldwell, J.L., Smythe, N.K. and Hall, K.K. 2000. A double-blind, placebo-controlled investigation of the efficacy of modafinil for

sustaining alertness and performance of aviators: a helicopter simulator study. *Psychopharmacology*,150(3), 272–282.
Caldwell, J.A., Mallis, M.M., Caldwell, J.L., Paul, M.A., Miller, J.C. and Neri, D.F. 2009. Fatigue countermeasures in aviation. *Aviation, Space, and Environmental Medicine*, 80(1), 29–59.
Caldwell, J.A., Mu, Q., Smith, J.K., Mishory, A., George, M., Caldwell, J.L., Peters, G. and Brown, D.L. 2005. Are individual differences in fatigue vulnerability related to baseline differences in cortical activation? *Behavioral Neuroscience*, 119(3), 694–707.
Caldwell, J.A., O'Hara, C., Caldwell, J.L., Stephens, R.L. and Krueger, G.P. 1993. Personality profiles of aviators screened for special operations duty. *Military Psychology*, 5(3), 187–199.
Caldwell, J.A., Smythe, N.K., LeDuc, P.A. and Caldwell, J.L. 2000. Efficacy of Dexedrine for maintaining aviator performance during 64 hours of sustained wakefulness. *Aviation, Space, and Environmental Medicine*, 71(1), 7–18.
Campbell, J.S., Moore, J.L., Poythress, N.G. and Kennedy, C.H. 2009. Personality traits in clinically referred aviators: Two clusters related to occupational suitability. *Aviation, Space, and Environmental Medicine*, 80(12), 1049–1054.
Carskadon, M.A. 1989. Ontogeny of human sleepiness as measured by sleep latency, in *Sleep and Alertness: Chronobiological, Behavioral, and Medical Aspects of Napping*, edited by D.F. Dinges and R.J. Broughton. New York: Raven Press, 53–84.
Chuah. Y.M.L., Venkatraman,V., Dinges, D.F. and Chee, M.W.L. 2006. The neural basis of interindividual variability in inhibitory efficiency after sleep deprivation. *Journal of Neuroscience*, 26(27), 7156–7162.
Civil Aviation Safety Authority Australia, 2010. Biomathematical fatigue modelling in civil aviation fatigue risk management – application guidance. Report prepared by Pulsar Informatics. www.casa.gov.au/wcmswr/_assets/.../fatigue/fatigue_modelling.pdf
Committee on Military Nutrition Research. 2001. *Caffeine for the Sustainment of Mental Task Performance: Formulations for Military Operations*. Washington, DC: National Academy Press.
Davenport, N.D. 2009. More fatigue! (Yawn). *Approach*, 54, 3–6.
Dijk, D.J. and Franken, P. 2005. Interaction of sleep homeostasis and circadian rhythmicity: Dependent or independent systems?, in *Principles and Practice of Sleep Medicine*, edited by M.A. Kryger, T. Roth, and W.C. Dement. Philadelphia: Elsevier Saunders, 418–434.
Dinges, D.F. 1984. The nature and timing of sleep. *Transportation Study, College of Physicians Philadelphia*, 6(3), 177–206.
Dinges, D.F., Graeber, R.C., Connell, L.J., Rosekind, M.R. and Powell, J.W. 1990. Fatigue-related reaction time performance in long-haul flight crews. *Sleep Research*, 19, 117.
Federal Register. 2012. Flightcrew member duty and rest requirements. *Federal Aviation Administration*, 77(2), 330–403.

Flight Safety Foundation. 2005. Lessons from the dawn of ultra-long-range flight. *Flight Safety Digest*, 24(8), 1–60.

Folkard, S. and Tucker, P. 2003. Shift work, safety and productivity. *Occupational Medicine*, 53(2), 95–101.

Goel, N., Van Dongen, H.P.A. and Dinges, D.F. 2011. Circadian rhythm in sleepiness, alertness and performance, in *Principles and Practice of Sleep Medicine*. 5th edition, edited by M.H. Kryger, T. Roth, and W.C. Dement. Philadelphia, PA: Elsevier, 445–455.

Goldsmith C. (1998). More carriers sanction their pilots' cockpit snoozes. *The Wall Street Journal* Jan 21, B1.

Hursh, S.R., Raslear, T.G., Kaye, A.S. and Fanzone, J.F. 2006. *Validation and Calibration of a Fatigue Assessment Tool for Railroad Work Schedules, Summary Report*, Technical report DOT/FRA/ORD-06/21, Washington, DC: US Department of Transportation, Federal Railroad Administration, Office of Research and Development.

Hursh, S.R., Redmond, D.P., Johnson, M.L., Thorne, D.R., Belenky, G., Balkin, T.J., Storm, W.F., Miller, J.C. and Eddy, D. 2004. Fatigue models for applied research in warfighting. *Aviation, Space, and Environmental Medicine*, 75, 3(Suppl), A44–A53.

International Civil Aviation Organization. 2011. *Operation of Aircraft Part I International Commercial Air Transport—Aeroplanes.*

Jacobson, B.H., Wallace, T., and Gemmell, H. 2006. Subjective rating of perceived back pain, stiffness and sleep quality following introduction of medium-firm bedding systems. *Journal of Chiropractic Medicine*, 5(4), 128–134.

Johnson, D.L., Wiebe, J.S., Gold, S.M., Andreasen, N.C., Hichwa, R.D., Watkins, G.L. and Boles Ponto, L.L. 1999. Cerebral blood flow and personality: A positron emission tomography study. *American Journal of Psychiatry*, 156(2), 252–257.

Killgore, W.S., Richards, J.M., Killgore, D.B., Kamimori, G.H. and Balkin, T.J. 2007. The trait of introversion-extraversion predicts vulnerability to sleep deprivation. *Journal of Sleep Research*, 16(4), 354–363.

Lack, L.C. and Wright, H.R. 2007. Treating chronobiological components of chronic insomnia. *Sleep Medicine*, 8(6), 637–644.

Lieberman, H.R. 2003. Nutrition, brain function, and cognitive performance. *Appetite*, 40(3), 245–254.

Lim, J., and Dinges, D.F. 2010. A meta-analysis of the impact of short-term sleep deprivation on cognitive variables. *Psychological Bulletin*, 136(3), 375–389.

Lindbergh, C. 1953. *The Spirit of St. Louis*. New York: Charles Scribner's Sons.

Moore-Ede M. 1993. Aviation safety and pilot error, in *The twenty-four-hour society: Understanding Human Limits in a World That Never Stops*. Reading, MA: Addison-Wesley Publishing Company, 81–96.

Morgenthaler, T., Alessi, C., Friedman, L., Owens, J., Kapur, V., Boehlecke, B., Brown, T., Chesson, A., Coleman, J., Lee-Chiong, T., Pancer, J. and Swick,

T.J. 2007. Practice parameters for the use of actigraphy in the assessment of sleep and sleep disorders: An update for 2007. *Sleep*, 30(4), 519–529.

Morgenthaler, T., Kramer, M., Alessi, C., Friedman, L., Boehlecke, B., Brown, T., Coleman, J., Kapur, V., Lee-Chiong, T., Owens, J., Pancer, J. and Swick, T. 2006. Practice parameters for the psychological and behavioral treatment of insomnia: An update. *Sleep*, 29(11), 1415–1419.

Mu, Q., Mishory, A., Johnson, K.A., Nahas, Z., Zozel, F.A., Yamanaka, K., Bohning, D.E. and George, M.S. 2005. Decreased brain activation druing a working memory task at rested baseline is associated with vulnerability to sleep deprivation. *Sleep*, 28(4), 433–446.

National Transportation Safety Board. 1986. *China Airline Boeing 747-SP, N4522V, 300 Nautical miles northwest of San Francisco, California, February 19, 1985*. NTSB/AAR-86-03. Washington, DC: National Transportation Safety Board.

National Transportation Safety Board. 1999. *Controlled flight into terrain, Korean Air Flight 801, Boeing 747-300, HL7468, Nimitz Hill, Guam, August 6, 1997.* NTSB/AAR-99-02. Washington, DC: National Transportation Safety Board.

National Transportation Safety Board. 2001. *Runway overrun during landing, American Airlines Flight 1420, McDonnell Douglas MD-82, N215AA, Little Rock, Arkansas, June 1, 1999*. NTSB/AAR-01-02Washington, DC: National Transportation Safety Board.

National Transportation Safety Board. 2006. *Collision with Trees and Crash Short of the Runway, Corporate Airlines Flight 5966 BAE Systems BAE-J3201, N875JX Kirksville, Missouri October 19, 2004*. NTSB/AAR-06/01. Washington, DC: National Transportation Safety Board.

National Transportation Safety Board. 2010. *Loss of Control on Approach, Colgan Air, Inc., Operating as Continental Connection Flight 3407, Bombardier DHC-8-400, N200WQ, Clarence Center, New York, February 12, 2009*. NTSB/AAR-10/01. Washington, DC: National Transportation Safety Board.

Nehlig, A. 1999. Are we dependent upon coffee and caffeine? A review on human and animal data. *Neuroscience Biobehavior Review*, 23(4), 563–576.

Neri, D.F., Oyung, R.L., Colletti, L.M., Mallis, M.M., Tam, P.Y. and Dinges, D.F. 2002. Controlled activity as a fatigue countermeasure on the flight deck. *Aviation, Space, and Environmental Medicine*, 73(7), 654–664.

Neville, H.J., Bisson, R.U., French, J., Boll, P.A., Storm, W.F. 1994. Subjective fatigue of C-141 aircrews during Operation Desert Storm. *Human Factors,* 36(2):339-49.

Onen, S.H., Onen, F., Bailly, D., and Parquet, P. 1994. Prevention and treatment of sleep disorders through regulation of sleeping habits. *Presse Med*, 23(10), 485–389.

Phillips, B.A., and Danner, F.J. 1995. Cigarette smoking and sleep disturbance. *Archives Internal Medicine,* 155(7), 734–737.

Ramakrishnan, K. and Scheid, D.C. 2007. Treatment options for insomnia. *American Family Physician*, 76(4), 517–527.

Reynolds, A.C. and Banks, S. 2010. Total sleep deprivation, chronic sleep restriction and sleep disruption. *Progress in Brain Research* 185, 91–103.

Rosekind, M.R., Graeber, R.C., Dinges, D.F., Connell, L.J., Rountree, M.S., Spinweber, C.L., Spinweber, C.L. and Gillen, K.A. 1994. *Crew Factors in Flight Operations IX: Effects of Planned Cockpit Rest on Crew Performance and Alertness in Long-haul Operations*. Report No: DOT/FAA/92/24, Moffett Field, CA: NASA Ames Research Center.

Rupp, T.L., Killgore, W.D.S. and Balkin, T.J. 2010. Socializing by day may affect performance by night: Vulnerability to sleep deprivation is differentially mediated by social exposure in extraverts vs introverts. *Sleep*, 33(11), 1475–1485.

Rupp, T.L., Wesensten, N.J., Bliese, P.D. and Balkin, T.J. 2009. Banking sleep: Realization of benefits during subsequent sleep restriction and recovery. *Sleep*, 32(3), 311–321.

Sadeh, A. and Acebo, C. 2002. The role of actigraphy in sleep medicine. *Sleep Medicine Reviews,* 6(2), 113–124.

Samel, A., Wegmann, H.M. and Vejvoda, M. 1995. Jet lag and sleepiness in aircrew. *Journal of Sleep Research*, 4(S2), 30–36.

Szinnai, G., Schachinger, H., Arnaud, M.J., Linder, L. and Keller, U. 2005. Effect of water deprivation on cognitive-motor performance in healthy men and women. *American Journal of Physiological Regulation Integrated Comparative Physiology*, 289(1), R275–R280.

Uchida, S., Shioda, K., Morita, Y., Kubota, C., Ganeko, M. and Takeda, N. 2012. Exercise effects on sleep physiology. *Frontiers in Neurology*, 3, 1–5.

Urponen, H., Vuori, I., Hasan, J. and Partinen, M. 1988. Self-evaluations of factors promoting and disturbing sleep: An epidemiological survey in Finland. *Social Science Medicine*, 26(4), 443–450.

Van Dongen, H.P.A. 2004. Comparison of mathematical model predictions to experimental data of fatigue and performance. *Aviation, Space, and Environmental Medicine*, 75(3 Suppl), A15–A36.

Van Dongen, H.P.A., Baynard, M.D., Maislin, G. and Dinges, D.F. 2004. Systematic interindividual differences in neurobehavioral impairment from sleep loss: Evidence of trait-like differential vulnerability. *Sleep*, 27(3), 423–433.

Van Dongen, H.P. and Dinges, D.F. 2005. Sleep, circadian rhythms, and psychomotor vigilance. *Clinical Sports Medicine*, 24(2), 237–249.

Van Dongen, H.P.A., Maislin, G., Mullington, J.M. and Dinges, D. F. 2003. The cumulative cost of additional wakefulness: Dose-response effects on neurobehavioral functions and sleep physiology from chronic sleep restriction and total sleep deprivation. *Sleep*, 26(2), 117–126.

Van Dongen, H.P., Mott, C.G., Huang, J.K., Mollicone, D.J., McKenzie, F.D. and Dinges, D.F. 2007. Optimization of biomathematical model predictions for cognitive performance impairment in individuals: accounting for unknown traits and uncertain states in homeostatic and circadian processes. *Sleep*, 30(9), 1129–1243.

Van Dongen, H.P.A., Vitellaro, K.M., and Dinges, D.F. 2005. Individual differences in adult human sleep and wakefulness: Leitmotif for a research agenda. *Sleep*, 28(4), 479–496.

Vandewalle, G., Maquet, P. and Dijk, D. 2011. Light as a modulator of cognitive brain function. *Trends in Cognitive Sciences*, 13(10), 429–438.

Warwick, G. 2011. Counting sheep won't help. *Aviation Week*. Available at: http://www.aviationweek.com/aw/blogs/commercial_aviation/ThingsWithWings/ [accessed 6 June 2012].

Wells, A.S., Read, N.W., Uvnas-Moberg, K. and Alster, P. 1997. Influences of fat and carbohydrate on postprandial sleepiness, mood, and hormones. *Physiology and Behavior*, 61(5), 679–686

Wesensten, N.J. and Balkin, T.J. 2010. Cognitive sequelae of sustained operations, in *Military Neuropsychology*, edited by C.H. Kennedy and J.L. Moore. New York: Springer Publishing, 297–320.

Williamson, A.M. and Feyer, A.M. 2000. Moderate sleep deprivation produces impairments in cognitive and motor performance equivalent to legally prescribed levels of alcohol intoxication. *Occupational and Environmental Medicine*. 57(10), 649–655.

World Health Organization, International Agency for Research on Cancer, 2010. Painting, firefighting, and shiftwork. *IARC Monographs on the Evaluation of Carcinogenic Risks to Humans*, 98, 561–764.

Wright, N. and McGown, A. 2001. Vigilance on the civil flight deck: Incidence of sleepiness and sleep during long-haul flights and associated changes in physiological parameters. *Ergonomics*, 44(1), 82–106.

Chapter 10
Aviation Neuropsychology

Gary G. Kay

Aviation neuropsychology refers to the field of Clinical Neuropsychology as it is applied in the context of civil and military aviation. According to the American Board of Professional Psychology (2012) a clinical neuropsychologist has:

> specialized knowledge and training in the applied science of brain-behavior relationships. Clinical neuropsychologists use this knowledge in the assessment, diagnosis, treatment, and rehabilitation of patients across the lifespan who have neurological, medical, developmental, or psychiatric conditions. The Clinical Neuropsychologist employs psychological, neurological, or physiological methods to evaluate patients' cognitive and emotional strengths and weaknesses and relates these findings to normal and abnormal central nervous system functioning. Clinical Neuropsychologists use this information, in conjunction with information provided by other medical/healthcare providers, to identify and diagnose neurobehavioral disorders, conduct research, counsel patients and their families, or plan and implement intervention strategies.

In the context of aviation, the aviation neuropsychologist evaluates the neurocognitive functioning of pilots, controllers, and other aircrew. These evaluations are generally conducted for the purpose of determining suitability for medical certification, assessing neurocognitive status of aviators with neurological trauma or illness, assessing fitness of aviators recovering from alcohol/substance abuse or dependence, and evaluating aviators and controllers with problems that arise in training, upgrading, transitioning aircraft, or in proficiency testing. This chapter will present a short history of aviation neuropsychology (see also Ryan, Zazeckis, French and Harvey 2006) and provide a description of the evaluation process and a discussion of current controversies in the field.

The advent of aviation neuropsychology appears to have occurred in the late 1970s. As the medical model of alcoholism became more accepted, the Federal Aviation Administration (FAA) developed a program whereby alcohol-dependent pilots who were committed to lifetime sobriety and to a recovery program characterized by intense monitoring, could be granted a Special Issuance medical certificate (see Chapter 5, this volume) allowing them to return to flying.

In 1981 a report was prepared for the FAA by the American Medical Association (AMA) in collaboration with the American Academy of Neurology and the American Association of Neurological Surgeons (Federal Aviation Administration

1981). This report documented the significance of various neurological conditions for aviator fitness. The FAA subsequently met with a small group of neuropsychologists in the early and mid-1980s to draft recommendations for a psychological testing protocol. In 1986 a review of the Aviation Medical Examiner (AME) standards and procedures was conducted by the AMA (Engelberg, Gibbons and Doege 1986) and stated that "the routine AME examination has never included tests that would detect a diminution of cognitive function, which if left unnoticed may result in poor pilot judgment or slow reaction time in critical operational situations." The report recommended the use of a brief cognitive mental status exam that could be administered at every examination. At present, AMEs do not routinely administer cognitive mental status testing as part of their examination.

Following publication of this report, the Chief Psychiatrist of the FAA, Dr Barton Pakull, convened an expert panel to develop a request for proposals for review of mental status examinations, neuropsychological screening tests, and neuropsychological test batteries, and to design a computer-administered neuropsychological screening test. This led to the development of CogScreen-Aeromedical Edition (CogScreen-AE; Kay 1995), a computer-administered neuropsychological screening test, and to the acquisition of normative data on conventional neuropsychological instruments in commercial aviators.

Over the last 20 years approximately 100 clinical neuropsychologists in the US have developed expertise in the evaluation of aviators, controllers, and other aircrew. The specific expertise required to function in the capacity of an aviation neuropsychologist is a comprehensive understanding of the flight environment and the cognitive and interpersonal demands of the job.

Task analyses have identified specific cognitive abilities required during different phases of flight (Imhoff and Levine 1981). Most task analyses and reports on the relationship between cognitive performance and flight performance are focused on the student pilot and originated in efforts to improve pilot selection procedures. Far less work has been directed at determining the relationship between cognitive abilities and performance of expert aviators or line pilots. Banich, Stokes and Elledge (1989) identified six cognitive skills critical for piloting performance: perceptual-motor activities, spatial abilities, working memory, attentional performance, processing flexibility, and planning and sequencing abilities. Existing task analyses, such as the Banich study are informative, yet they do not specify the level of each of the abilities required to operate different aircraft. How much psychomotor speed and coordination is needed to operate a modern glass cockpit airliner? How do the requirements change with years of experience flying the aircraft?

Aviation neuropsychologists are called upon to evaluate commercial, general aviation, and military pilots. Clearly, the cognitive, psychomotor, and interpersonal demands of flying are considerably different in these populations. The commercial aviation cockpit is operated by at least two fully qualified pilots flying under instrument conditions. In contrast, general aviation operations are almost always performed by a single pilot flying under visual flight rules. Commercial aircraft most often employ glass cockpit technology with advanced navigational and radar systems. By contrast, most general aviation aircraft are of the "stick and rudder" variety and have limited avionics. However, in spite of these differences, the medical requirements established by the FAA for commercial pilots and general aviation pilots differ mostly with respect to the frequency of required medical examinations. The military, which uses a wide array of high-performance aircraft that fly in hostile situations, has different standards and rules regarding flight status and is much more restrictive than both commercial and general aviation (see Chapter 6, this volume).

Neuropsychologists providing reports for the purpose of FAA medical certification are recommended to meet the following minimum qualifications requirement: neuropsychological evaluations must be conducted by a licensed psychologist who is either board certified or board eligible in Clinical Neuropsychology. In order to perform these assessments for the FAA the psychologist must be board certified as a clinical neuropsychologist or have the education, training, and clinical practice experience that would qualify him or her to sit for board certification with the American Board of Clinical Neuropsychology or the American Board of Professional Neuropsychology. Furthermore, it is recommended that the psychologist have familiarity with the Federal Aviation Regulations that pertain to medical certification and the Special Issuance program (Federal Aviation Administration 2010).

The FAA has developed testing protocols that specify the minimum requirements of evaluations to be conducted for a number of conditions including: HIV-seropositive pilots, pilots taking one of a specified list of selective serotonin reuptake inhibitor (SSRI) antidepressant medications (see Chapter 12, this volume), recovering alcohol-dependent pilots seeking Special Issuance (see Chapter 5, this volume), pilots recovering from head trauma and cerebrovascular conditions, and for medical certification applicants with a history of attention deficit hyperactivity disorder (ADHD) or treatment with medications used to treat ADHD. The protocols specify a core test battery to be administered for each of these conditions. The purpose in specifying a core test battery is to provide a standardized basis for the FAA's review of cases. The examiner is not in any way limited to administering only the tests specified in the core test battery. In fact the examiner is encouraged to use their professional judgment to decide what additional tests are needed to confirm or disconfirm the presence of potentially relevant deficits found in the course of the neuropsychological examination.

The following is a list of the tests included in the FAA core test battery for general neuropsychology cases:

Wechsler Adult Intelligence Scales (that is, WAIS-IV or WAIS-III).
CogScreen-Aeromedical Edition (CogScreen-AE).
Trail Making Test (Halstead Reitan version).
Paced Auditory Serial Addition Test (PASAT).

Continuous performance testing (for example, Test of Variables of Attention (TOVA).
Conners' Continuous Performance Test, or Integrated Visual and Auditory Continuous Performance Test (IVA).
Verbal memory testing (for example, Wechsler Memory Scale-IV (WMS-IV) subtests, Rey Auditory Verbal Learning Test or California Verbal Learning Test-II).
Visual memory testing (for example, WMS-IV subtests, Brief Visuospatial Memory Test-Revised, or Rey Complex Figure Test).
Language testing including Boston Naming Test and verbal fluency testing (Controlled Oral Word Association Test and semantic fluency task).
Psychomotor testing including Finger Tapping and Grooved Pegboard or Purdue
Pegboard.
Executive function tests including Halstead Category Test or Wisconsin Card Sorting Test (WCST), and Stroop Color-Word Test.
Personality testing including either the Minnesota Multiphasic Personality Inventory (MMPI-2) or Personality Assessment Inventory (PAI).

Specifying a core battery provides a fair and consistent basis for evaluating test findings by the regulatory authority. Examiners are informed that the use of an equivalent test is acceptable, though the burden of proof is on the examiner regarding the equivalence of the test with respect to sensitivity and the measurement of constructs. Tests were selected for the core battery according to the following criteria: standardization, suitability for repeat testing, and established construct validity as a measure of cognitive, perceptual and/or motor demands of flight operations. In addition, a preference was given to tests which have normative data for aviators and tests which have been demonstrated to be related to flight performance (that is, established criterion validity).

CogScreen-Aeromedical Edition

CogScreen-AE is a neuropsychological test specifically created for the purpose of aviation neuropsychology. It is a computer-administered cognitive screening test developed for the FAA as an automated cognitive screening instrument and was designed to be sensitive to mild brain dysfunction, resulting from conditions such as mild traumatic brain injury, Alcohol Dependence, early-stage dementia, toxic exposure, and other degenerative diseases.

CogScreen-AE includes measures of mental abilities required by flying, including measures of visual scanning and tracking, visual perceptual speed, divided attention, psychomotor tracking, math reasoning, and novel problem solving. In initial validation studies the test was found to be more sensitive to mild brain trauma (80 percent sensitivity) than conventional neuropsychological tests

(50 percent sensitivity), with no reduction in specificity (95 percent) compared to the conventional measures (Kay 1995). Normative data on commercial pilots was obtained from a large sample of aviators working for different size carriers and has subsequently been collected for the Russian, Spanish, and French editions.

The test consists of 13 subtests and is mostly focused on measures of attention, concentration, information processing, immediate memory span, working memory, multitasking and executive functions. It was designed for repeated administration and each test session presents different test items to minimize practice effects. The entire test battery requires approximately 45 minutes for administration, and all subject responses are input with a touchscreen except for the tracking component of one subtest, which requires the subject to use the arrow keys on a keyboard (Kane and Kay 1992).

Scores are provided for response speed, accuracy, thruput (number of correct responses per minute), and, on certain tasks, process measures (for example, impulsivity; perseverative errors, coordination, and tracking errors). Norm-based reports are generated by the program. The CogScreen-AE US aviator normative base includes 584 US pilots, screened for health status and alcohol and other substance abuse. The examiner can choose to compare the results according to age or by type of airline (from Major to Regional Carrier). A base rate analysis indicates the percentage of individuals in the comparison group who had an equal number of scores that fell at or below the 15th and 5th percentile. In addition, the report generates factor scores that are predictors of flight performance, and an index of the likelihood of brain dysfunction (that is, Logistic Regression Estimated Probability (LRPV) score).

Although CogScreen-AE was validated as a test of brain dysfunction, its links to the cognitive requirements of flying have led investigators to evaluate the test as a predictor of flight performance. A series of studies have shown that CogScreen-AE variables are predictors of actual and simulated flight performance in commercial and general aviation (Hoffmann, Hoffman and Kay 1998; Zuluaga et al. 2011; Taylor, O'Hara, Mumenthaler and Yesavage 2000; Yakimovich et al. 1994). As a result, CogScreen-AE is used internationally by airlines and military organizations in selection of pilots (see Chapters 2 and 3, this volume for more on selection of both military and commercial pilots).

CogScreen-AE has also been administered by airlines and air forces upon hiring or at the beginning of training to establish a baseline level of performance. Baseline neuropsychological data can be used to facilitate the return of aviators to flying status in the event of injuries or illness. The utility of CogScreen-AE in accelerating the return of military aviators who have sustained head trauma has been reported for Russian and US pilots (Yakimovich et al. 1995; Moore and Kay 1996).

The sensitivity of CogScreen-AE to HIV associated neurocognitive dysfunction led to the use of the test as a standard for the cognitive screening of HIV-seropositive pilots seeking Special Issuance Medical Certification. The test is also used by the pharmaceutical industry to evaluate the potential sedative effects of medications

(Kay et al. 1997). Medication studies conducted in pilots have helped in making a determination as to the safety of medication use by aviators (for example, Bower et al. 2003). The test has proven to be sensitive to changes in neurocognitive functioning. When an aviator performs well on CogScreen-AE it is unlikely that deficits will be found on conventional tests in the areas addressed by CogScreen-AE. The FAA recently reviewed neuropsychological test findings submitted by aviators seeking Special Issuance under the SSRI antidepressant program. There were no cases where a pilot who had performed normally on CogScreen-AE was found to be impaired on the conventional neuropsychological test battery. On the other hand, when an aviator performs poorly on CogScreen-AE it indicates to the practitioner the need to follow-up with additional conventional tests to confirm or disconfirm the presence of aeromedically significant cognitive deficits.

CogScreen-AE is mandated by the FAA for pilots seeking medical certification under several of the Special Issuance programs (see FAA Guide for Aviation Medical Examiners 2013), this includes: pilots taking specified SSRI antidepressant medications; pilots who are HIV-seropositive; pilots with a diagnosis of substance abuse or substance dependence; and for pilots with potential neurocognitive impairment (for example, secondary to traumatic brain injury, stroke, multiple sclerosis). In addition, the FAA has indicated that the core test battery should be administered when a psychological evaluation reveals findings suggesting cognitive deficits.

Interpretation of Neuropsychological Findings

The aviation neuropsychologist compares the examinee's results (that is, raw scores) to normative data in order to determine how the examinee's performance compares to relevant groups of individuals, such as healthy pilots of the same age as the examinee. Raw test scores are typically transformed into T-Scores (with a mean of 50 and standard deviation of 10) and reported in terms of percentiles. Scores are classified as falling within the normal, expected range, or as falling outside the normal, expected range. Performance that falls at or below the 15th percentile, for the examinee's normative comparison group, is below expectation and is considered to be evidence of impairment. Another criterion used to classify performance as impaired are scores that fall at or below the 5th percentile compared to aviator norms. Unfortunately, aviator norms are available for a limited number of tests. Normative data for pilots on the Wechsler Adult Intelligence Scale-Revised (WAIS-R; N=456) and for tests from the Halstead-Reitan Neuropsychological Test Battery (N=132) and other neuropsychological tests (N=132) were presented by Kay (2002) and are provided here in Tables 10.1, 10.2 and 10.3. Military aviator norms exist for the d2 Test of Attention (Hess, Kennedy, Hardin and Kupke 2010).

The norms for the intelligence test shown in Table 10.1 are derived from an earlier version of the Wechsler test (WAIS-R) which was superseded by the WAIS-III and current WAIS-IV. The Category Test score is for the 120-item abbreviated

Table 10.1 Pilot Norms on the Wechsler Adult Intelligence Scale-Revised (WAIS-R)

Subtest	INFO	DSPAN	VOCAB	ARITH	COMP	SIMIL	PCOM	PARR	BLDES	OBJAS	DSYM	VIQ	PIQ	FSIQ
Mean	13.1	11.6	12.3	13.1	12.1	11.5	11.3	11.1	12.3	10.7	10.5	116.2	117.0	118.5
SD	2.1	2.6	2.0	2.2	2.0	2.2	2.1	2.7	2.1	2.4	2.3	10.1	11.0	10.3
5th %ile	10.0	7.0	9.0	9.0	9.0	8.0	8.0	7.0	9.0	7.0	7.0	99.0	99.0	100.0

Note: $N = 456$

Table 10.2 Pilot Norms on the Halstead-Reitan Neuropsychological Test Battery

Subtest	Trails A	Trails B	TPT DH	TPT ND	TPT BH	TPT TOT	TPT MEM	TPT LOC	CATEG*	RHYTHM	SSPT A&B	TAP DH	TAP ND
Mean	21.8	47.5	4.6	3.7	2.2	10.5	7.9	4.5	34.4	27.7	2.5	53.3	48.2
SD	6.2	12.5	1.7	1.5	0.9	3.4	1.2	2.4	19.4	1.8	1.6	7.3	6.5
5th %ile	36.1	72.7	8.1	6.7	4.3	17.7	5.4	1.0	73.8	24.3	5.2	39.7	39.1

Note: $N = 132$

* 120-item abbreviation

Table 10.3 Pilot Norms on Additional Neuropsychological Tests: Rey Auditory Verbal Learning Test (RAVLT), California Verbal Learning Test (CVLT), Paced Auditory Serial Addition Test (PASAT), and Wisconsin Card Sorting Test (WCST)

Test	RAVLT			CVLT							PASAT		WCST
Subtest	**REY COPY**	**REY IMM**	**REY DELAY**	**CVLT 1**	**CVLT 5**	**CVLT 1-5**	**CVLT B**	**CVLT SD**	**CLVT LD**	**CVLT REC**	**PASAT 1**	**PASAT 2**	**WCST %PR**
Mean	35.7	25.4	24.9	8.3	13.8	59.6	8.1	12.5	13.1	15.3	46.0	42.1	7.4
SD	0.7	4.9	5.5	1.8	1.9	7.5	1.9	2.5	2.3	0.9	4.3	6.0	3.5
5th %ile	33.9	16.4	14.2	6.0	10.0	45.4	5.0	7.4	8.9	13.4	36.8	29.7	3.8

Note: $N = 132$

Source: Elsevier, Occupational Medicine State of the Art Reviews17, with permission

form of the test (Sherrill 1985). Also, these norms were collected in the early 1990s with commercial airline pilots, a high percentage of whom were former military pilots. It is suspected that the profession is no longer attracting individuals with the same level of education and general intelligence. Furthermore, relatively few of the current commercial pilots are former military pilots. For these reasons, the pilot norms presented in Tables 10.1–10.3, may be less representative of commercial pilots than they once were. Another concern with the use of these norms is that they may not apply to general aviation pilots. Because of this and consistent with findings presented by Kay et al. (1993), it is probably sufficient to use normative data that corrects for the effects of age, gender and education, such as those provided by the Heaton Comprehensive Norms (Heaton, Miller, Taylor and Grant 2004) when applicable aviator norms are not available. Heaton, Miller, Taylor and Grant classify scores as falling at or below the 15th percentile, as showing borderline impairment. This is an area of needed research.

In the general clinical practice of neuropsychology it is customary to use normative data, such as the Heaton Comprehensive Norms. This approach is well suited to detecting the presence of brain dysfunction and to assessing the severity of impairment. Age has a substantial effect on cognitive test performance, with the magnitude of the age-effect varying for different cognitive and psychomotor functions. For purposes of illustration, consider the situation where the same raw score which places a 31 year old patient at the 5th percentile would place a 57 year old (non-aviator) patient at the 30th percentile. It would be inappropriate to classify the score of this 57-year-old patient as indicating brain dysfunction. The patient's score clearly falls in the average range for his age. However, if the purpose of the examination had been to assess the examinee's absolute level of ability in relationship to individuals performing the same job, the use of age-corrected norms may be inappropriate. If, for example, on a measure known to predict flight performance, a pilot obtains a score that is in the average range compared to healthy 85 year olds, but which is below the level of performance seen in 99 percent of aviators across all ages, the pilot may not have sufficient functioning to perform the flight-related activity. Older pilots do not have the benefit of landing on "age-corrected runways" (Kay 1995). For more on aging aviators, please see Chapter 11, this volume.

The purpose of the neuropsychological examination is to resolve questions regarding the aviator's neurocognitive functioning as it relates to the requirements for certification/waiver. If the examination does not include sufficiently sensitive tests of neurocognitive functioning, an accurate determination of suitability for medical certification may not be reached.

The report of an aviation neuropsychological examination includes documentation of the findings from the psychosocial and medical history, clinical interview, flight performance history, behavioral observations/mental status examination, test results and interpretation, a summary/integration of findings, an explicit diagnostic statement, and the neuropsychologist's opinion/recommendation(s) regarding aeromedically significant findings and their potential impact on flight safety. The neuropsychologist makes inferences regarding the likelihood of brain dysfunction and the relationship between the test findings and the aviator's history. For example, for pilots seen for serial testing, such as HIV-seropositive pilots who are seen on an annual basis, the report describes significant changes in performance compared to the prior examination. The neuropsychologist provides an opinion as to whether any changes in test performance reflect significant improvement or deterioration in neurocognitive functioning.

Based upon their knowledge of neuropsychology and of the underlying ability requirements of aviation, the aviation neuropsychologist makes inferences as to the likelihood that the examinee's test findings reveal aeromedically significant neurocognitive deficits. Impaired performance on a single test may raise concerns, but needs confirmation with other tests of the same mental ability before being considered solid evidence of an impaired ability. In addition, the extent to which the ability is required for safe aviation performance is a critical question to be addressed. Mild symptoms of residual ADHD, for example, may be demonstrated in an adult previously treated with stimulants during their childhood. The aviation neuropsychologist is called upon to assess whether these residual mild signs would adversely impact aviation performance. In making this determination the neuropsychologist considers the examinee's flying history such as number of logged flight hours, types of aircraft flown, certifications and licenses held, and whether the pilot is a commercial, general aviation, or military pilot. In the context of general aviation mild residual ADHD may not be a disqualifying condition, however in the military it is contraindicated for flight status.

What magnitude of impairment of neurocognitive functions such as attention, memory, language, spatial abilities, or executive functions is aeromedically significant? The ability to answer this question, except in fairly obvious cases, is probably beyond the competence of the aviation neuropsychologist. The US Federal Aviation Regulations (FARs, Part 67.107, Part 67.207, and Part 67.307; National Archives and Records Administration 2012) indicate that to be eligible for a medical certificate the applicant cannot have a mental condition, based on the case history and appropriate, qualified medical judgment relating to the individual … "[which] makes the person unable to safely perform the duties or exercise the privileges of the airman certificate applied for or held." It is reasonable to expect that the aviation neuropsychologist possesses an understanding of (1) the required abilities, and (2) knowledge of the level of ability which would be considered impaired (relative to the examinee's age, gender and years of education). However, there are cases where it is less likely that the aviation neuropsychologist can accurately determine whether the impairment will interfere with the ability to safely perform flight operations.

Neuropsychological tests are designed to evaluate brain functioning. The ability of neuropsychological tests to predict skilled performance, such as flying, is generally not addressed by neuropsychologists. In fact, there has been relatively little work in this area of ecological validity of neuropsychological testing. For example, prediction of driving performance in patients with neurological conditions based upon their performance on conventional neuropsychological tests has shown only limited success (Dawson, Anderson, Johnson and Rizzo 2010). In contrast, CogScreen-AE, which is based on a task analysis of flight-required mental abilities, has been shown to be a predictor of simulated and actual flight performance.

Psychologists and neuropsychologists who are not accustomed to evaluating pilots commonly end their report with a recommendation, regardless of whether or not impairment was found on testing, that the airman would best be evaluated by a flight test. This is not particularly helpful. A flight test is not a gold standard for determining if the airman's condition has impacted their flight-related abilities. Automobile drivers with conditions that potentially impair driving are often examined using an 'over-the-road' driving test administered by a qualified driving instructor. This is not standardized and there exists no data validating the 'over-the-road' driving test as a predictor of driving safety. The driver is typically not driving under the conditions (that is, roadways, traffic and time of day) in effect when they operate their vehicle. Furthermore, the fact that they are being observed makes the test less ecologically valid. It is also a fact that each year qualified driving instructors are injured or killed in crashes during these tests (Martz 2011). An FAA medical certificate does not certify the pilot's competence or skill in flying. Physicians employed by the FAA's Office of the Federal Air Surgeon are not required to be pilots. While in the military, flight surgeons, who are responsible for monitoring and providing healthcare to pilots, are usually required to complete a period of flight training, neither civilian nor military flight doctors assess the flying competence of aviators.

The FAA employs flight standards pilots who conduct flight examinations and who are qualified to assess pilot proficiency. However, the criteria for a medical certificate as outlined above is not pilot proficiency but rather the absence of a statutory disqualifying medical or mental condition.

There are circumstances where the FAA does consider the flight proficiency of an applicant for a medical certificate. A person with a disqualifying condition can apply to the Federal Air Surgeon for a Special Issuance of a medical certificate "if the person shows to the satisfaction of the Federal Air Surgeon that the duties authorized by the class of medical certificate applied for can be performed without endangering public safety during the period in which the Authorization would be in force" (Federal Aviation Administration 2012, 10). The Federal Air Surgeon may authorize a special medical flight test for this purpose or, at the discretion of the Federal Air Surgeon, a Statement of Demonstrated Ability (SODA) may be granted to a pilot whose disqualifying condition is static or non-progressive. In granting one of these waivers the Federal Air Surgeon considers the person's operational experience and any medical facts that may affect the ability of the person to perform airman duties. In addition, when determining whether a waiver should be granted to an applicant for a general aviation (third-class) medical certificate, the Federal Air Surgeon considers "the freedom of an airman, exercising the privileges of a private pilot certificate, to accept reasonable risks to his or her person and property that are not acceptable in the exercise of commercial or airline transport pilot privileges, and, at the same time, considers the need to protect the safety of persons and property in other aircraft and on the ground" (FAA 2012).

Aviation neuropsychologists are well-qualified to evaluate the presence or absence of a disqualifying mental condition. In contrast, when making inferences as to whether an examinee can safely perform flight operations (consistent with a specific class of medical certificate) they have to make significant inferential leaps.

The FAA refers cases involving neurocognitive impairment to a panel of expert clinical neuropsychology consultants for independent review. The airman's entire medical record, including the report of the neuropsychologist and the underlying test materials, are reviewed. The FAA neuropsychology consultant is tasked with rendering an opinion regarding the airman's suitability for the class of medical certificate applied for, and indicating whether or not the certificate should be unrestricted or require a Special Issuance.

Non-Federal Aviation Authority (Military and International) Perspectives

Fiedler and Patterson (2001) contrasted the FAA and US Air Force (USAF) procedures for evaluating head-injured aircrew. While both agencies are responsible for oversight of aviation safety, in the military there is also the consideration of mission completion. In the FAA model the aviator selects the specialists who will be evaluating his/her head injury and making recommendations to the FAA. In contrast, in the military the aviator is assigned medical specialists. In the USAF model, in the absence of complications, there is a requirement of a five-year wait following a severe closed head injury. By comparison, the FAA requires only a one-year wait in the absence of complications. Fiedler and Patterson (2001) reported that a "working group involving representatives of several countries has developed a proposed standard for the evaluation of head injured aircrew that would be used by the militaries of these countries."

In the South African context, a practical flight test has, until recently, been considered the final arbiter where other findings are equivocal. However, it has been recognized that the practical flight test is not standardized. Notwithstanding the recognition of the significant shortcomings, practical flight tests are still preferred, although there is a movement to recognize that although the practical flight test is ecologically valid, it does not truly assess the airman's "quick and reasoned situation-appropriate performance shortcomings" (Trevor Reynolds, personal communication, 9 August 2012).

From an international perspective, there are considerable differences in the training of psychologists and neuropsychologists and in their reliance on standardized testing procedures. The American psychometric approach is not universally popular. A prominent non-US psychologist stated that "Americans are too obsessed with following instructions—we use tests more clinically."

It is also important to recognize that the formal education of the aviator varies widely in many parts of the world. Are US-based aviator norms applicable internationally? For CogScreen-AE few differences were found between US and Russian commercial airline pilots (Kay 1993). Similar studies are currently underway in India, Colombia, and in the Middle East. Preliminary findings from a Colombian Air Force study evaluating the validity of CogScreen-AE to predict simulated flight performance are consistent with results seen in US aviators (Zuluaga et al. 2011).

What are the appropriate tests and what are the appropriate norms for evaluating the pilot from a culture which differs dramatically with respect to formal education and even with respect to experience with standardized testing? Many of the neuropsychological tests used in the evaluation of the US aviator are likely culturally biased. On the other hand the commercial airline cockpit is certainly not culture-free. They are the product of Boeing and Airbus engineers and have become the de facto international standard. To quote a colleague, there are no "culture-specific runways" (Trevor Reynolds, personal communication, 7 August 2012). Therefore, assuming that a neuropsychological test provides a valid measure of aviation-related abilities, the test must be fairly universal.

Conditions Frequently Encountered in Aviation Neuropsychology

Head trauma

Traumatic brain injury is a common condition that is likely to be encountered by AMEs and other practitioners who provide evaluations of aviators. A head injury sufficient to cause a loss or alteration of consciousness generally requires evaluation before an aviator is permitted to resume flight activities.

Among the cognitive changes that accompany brain trauma the most common are in the areas of attention, concentration, information processing speed, mental flexibility, and executive functions. In addition, changes in mood and stress tolerance may accompany brain injuries. Neuropsychological testing is necessary to assess the changes in these abilities and to determine if functioning has returned to the level required for flying. The neuropsychological examination of the head-injured aviator is based on history, review of medical records, interviews with the aviator and family members, and the results of testing. The test battery needs to assess the abilities most vulnerable to the effects of brain trauma. Where possible it is preferable to compare the injured pilot's test scores with pre-morbid (baseline) levels of neuropsychological performance. However, for most pilots it is necessary to compare their post-injury neuropsychological performance with available norms. With resolution of symptoms and with the absence of disqualifying medications the head-injured aviator can be considered for medical certification. In pilots with severe brain injury a period of three years of observation is recommended due to concerns related to seizure risk. In less severe cases with subdural hematomas a minimum period of one year is generally required before issuance of a medical certificate. The FAA protocol for the evaluation of the head-injured aviator seeking medical recertification can be found in the *Guide for Aviation Medical Examiners* (FAA 2013). In addition to the core test battery the evaluation must include CogScreen-AE and the MMPI-2.

Internationally, regulations related to brain injury vary widely. By comparison, the South African protocols with regard to head injury are more stringent than those of the FAA. The South African regulations (CATS Part 67 MR; available at website www.CAA.co.za) indicate that a person who has suffered a severe brain injury is generally considered permanently unfit for certification, although the aviator may appeal after a period of ten years. The regulations are specific with respect to Mild, Moderate, Severe, and Very Severe closed head injury. Injuries with a loss of consciousness or post-traumatic amnesia greater than 30 minutes require neuropsychological evaluation (and a practical flight test).

Alcoholism (substance dependence)

The evaluation of pilots committed to abstinence and recovery from alcoholism and other substance dependence is reported in Chapter 5, this volume. The FAA recognizes the need to evaluate the cognitive functioning of these individuals and the initial inclusion of neuropsychology in the FAA medical program resulted from work done in the area of the neuropsychology of alcoholism. Current specifications for the neuropsychological evaluation of aviators seeking Special Issuance under this program can be found in the *Guide for Aviation Medical Examiners* (FAA, 2013). The evaluation must be conducted by a board certified (or board eligible) neuropsychologist who has completed the Human Interventional Motivation Study (HIMS) training. The evaluation must include CogScreen-AE, as well as the core test battery, and the MMPI-2.

Dementia and mild cognitive impairment

Mild neurocognitive impairment is a highly prevalent condition among adults over the age of 70. It has been estimated that 22.2 percent of the US population over the age of 70 suffer from cognitive impairment without dementia (Plassman et al. 2008). At 17 months follow-up 80.4 percent were again classified as having cognitive impairment without dementia. The annual rate of progression to dementia was reported as 12 percent.

Because mental status testing is not routinely conducted as part of an AME examination, it is very uncommon to have older aviators referred for neuropsychological assessment by AMEs. Older pilots at risk for mild neurocognitive impairment are more likely to be seen for neuropsychological testing due to a referral for other medical or neurological conditions, for alcohol-related issues, or for performance-related problems. It is not uncommon for older aviators to be referred after they have received notification from the FAA that other pilots or controllers have reported to the FAA that they suspect that the older aviator is no longer capable of safely performing the duties of their airman medical certificate.

The aviation neuropsychologist is well aware of the effects of normal aging on cognition, including the effects of age on flight-related abilities (see Chapter 11, this volume). In evaluating the older aviator the neuropsychologist must make every effort to accurately estimate the aviator's expected level of functioning based upon his/her occupational and educational history. Do the results indicate a decrement in performance? It is not sufficient to simply determine that the pilot is in the normal/expected range for their age. Does the pilot have the ability to perform essential aviation-related skills? The important caveat (worth repeating) is that there are no "age corrected runways." Neuropsychological testing can be used to quantify and monitor the rate and level of decline.

Human Immunodeficiency Virus

In 1992 it was the position of the Aerospace Medical Association (AsMA) that HIV seropositive (HIV+) pilots should not be in command of an aircraft (AsMA 1992). The basis for this recommendation was that there was no valid testing procedure for detecting subtle neuropsychiatric disease in the HIV+ pilot. Mapou, Kay, Rundell and Temoshook (1993) replied that there was no data to suggest that subtle neuropsychiatric dysfunction resulted in a decrement of aviation-related skills. Therefore, a blanket disqualification of all HIV+ aviators was considered unjustified. The AsMA position was also attacked by Miller and Selnes (1993) who recommended that pilots be examined on an individual, ongoing basis and only be disqualified if found to be impaired on testing. In 1996 the Board of the AsMA Aerospace Human Factors Committee decided to revisit the issue and requested that a review be performed of the neuropsychology of asymptomatic HIV infection and cognitive functioning. The findings of that review were that (1) the risk of cognitive impairment in asymptomatic HIV+ individuals is low; and (2) that there are available tests capable of detecting these HIV-related changes in neurocognitive functioning. Specifically, tests designed to measure subtle changes in speed of mental processing and with sensitivity to subcortical dysfunction were able to detect HIV-related changes. The Aerospace Human Factors Committee drafted a position paper which was adopted and published by AsMA (1997). The revised position of AsMA recommended medical certification of HIV+ aviators under a Special Issuance program. The FAA accepted this position and developed a Special Issuance (monitoring) program whereby HIV+ pilots are seen for annual testing with CogScreen-AE, which has been shown to be sensitive to HIV associated neurocognitive dysfunction in a study involving serial testing of HIV+ individuals (mean age=35.4±5.7; mean education 14.5±2.4) at six-month intervals (Kay 1991). The most sensitive CogScreen-AE measures to decline in functioning in HIV+ individuals were the Pathfinder subtest (coordination scores) and the

Symbol Digit subtest (memory accuracy scores). The HIV+ Special Issuance program has been in existence for 16 years and has allowed many commercial airline pilots to resume their career and reach retirement without any HIV-related career impediments. The current protocol (which can be found in the *Guide for Aviation Medical Examiners*, FAA, 2013) requires that only CogScreen-AE be administered if there is no evidence of cognitive impairment or a significant decline in functioning.

Interestingly, in South Africa, a country with a high incidence of HIV infection, the current medical certification protocol does not make provision for neurocognitive assessment. Based upon discussions with the Civil Aviation Authority it is clear that the Government recognizes the need for testing but a testing protocol has not been implemented due to the requirement that psychological tests have suitable South African normative data (comments made to author in meetings with the Civil Aviation Authority, August 2010).

Attention Deficit-Hyperactivity Disorder (ADHD)

Given that ADHD is currently the most prevalent psychiatric disorder found in children and adolescents it should not be surprising that aviation neuropsychologists are increasingly encountering referrals of aviators and individuals seeking initial aeromedical certification who have a history of treatment for ADHD. According to a recent report in the *New York Times* (Schwarz and Cohen 2013), data from the U.S. Centers for Disease Control and Prevention shows that "nearly one in five high school age boys and 11 percent of school-age children overall have received a medical diagnosis of attention deficit hyperactivity disorder." This disorder is characterized by distractibility, inattention, hyperactivity, impulsivity, disorganization, and restlessness. By definition, the condition first appears in childhood and often improves but does not disappear with maturation. The pharmacological treatment of ADHD typically involves the use of stimulant medications that generally preclude medical certification. Aviators with mild ADHD who do not require treatment with stimulant medications may be certified by the FAA. However, evaluation by a neuropsychologist is required. The evaluation focuses on the severity of attentional and executive function deficits in the absence of medication. The FAA's current specifications for evaluation of individuals with a history of ADHD or for those reporting use of medications used to treat ADHD can be found in the *Guide for Aviation Medical Examiners* (FAA 2013).

It is not uncommon when performing these examinations to find that the original diagnosis of ADHD was a misdiagnosis. In conducting these evaluations the examiner must be wary of learning disabilities, particularly problems with reading, spelling, and math, which had been misdiagnosed as ADHD. In these cases the neuropsychologist must carefully assess the individual to determine if their learning disabilities, particularly those involving reading or math, are severe enough to disqualify the individual for medical certification. The US military has a separate process as well, given the increasing numbers of service personnel entering the military with histories of ADHD and/or learning disorders. These requirements are described in Chapter 6, this volume.

Depression

A diagnosis of depression and use of psychotropic medication is considered medically disqualifying for applicants for FAA medical certification. In April 2010 a new FAA policy went into effect providing a Special Issuance certification program for airmen with a history of depression treated with (specified) selective serotonin reuptake inhibitor (SSRI) antidepressants. The FAA determined that aviators diagnosed with depression and taking one of four specified SSRI antidepressants (fluoxetine, sertraline, citalopram, or escitalopram) at a stable dose could be considered for Special Issuance of an airman medical certificate. Under the original terms of this program pilots were required to undergo a battery of conventional neuropsychological tests and CogScreen-AE. A revised protocol (January 2013) which can be found in the *Guide for Aviation Medical Examiners* (FAA 2013), only requires that CogScreen-AE be administered if testing shows no evidence of cognitive impairment. For an in-depth discussion of psychopharmacology and aviation, please see Chapter 12, this volume.

Aviation performance referrals

This type of evaluation is obviously unique to the practice of aviation neuropsychology and requires an understanding of the components of flight training (for example, survival training, ground school, line-oriented flight training, initial operating experience, simulator sessions), proficiency testing, military-specific issues if applicable, and of the management structure in airlines, as well as the role of airline Employee Assistance Program (EAP) providers and of the unions who represent pilots. Pilots are referred to aviation neuropsychologists for psychological fitness-for-duty evaluations as a result of aviation performance problems or training difficulties. The most common problem is repeated failure of proficiency checkrides. Pilots who fail transition training (that is, training to fly a different aircraft) and referrals for pilots who fail to complete upgrade training (that is, from first officer to captain) are also common. Please note that in the military the evaluation requirements also apply to air traffic controllers, flight crew, navigators/flight officers, and pilots of unmanned aerial vehicles (UAVs).

These cases raise important differential diagnostic issues. The information provided by the pilot's employer, or in the case of the military aviator the information provided by their command, and by the pilot during the clinical interview is especially critical in this type of evaluation. The evaluator needs to understand the context under which the performance problems occurred. It is also important to fully understand the pilot's perspective on how the problems occurred.

Performance difficulties may be the result of numerous factors: adjustment disorders, depressive disorders, anxiety disorders (especially performance anxiety), developmental learning disabilities, acquired neurocognitive deficits secondary to alcoholism, or other undiagnosed neurodegenerative disorders (for example, dementia or mild cognitive impairment).

Comprehensive psychological and neuropsychological examinations are performed for these evaluations. A detailed clinical interview and utilization of collateral sources is the key to performing this type of fitness-for-duty examination. In cases where a pilot has failed a component of training, it may be necessary to formally assess language and math skills with achievement testing. See Appendices for a case example. See also Chapter 4, this volume.

Summary

Aviation neuropsychology has played a key role in evaluating the medical certification status and fitness-for-duty of aviators with a wide range of neuropsychiatric conditions. Beginning with their role in the evaluation of pilots seeking Special Issuance medical certification for alcoholism in the 1970s, to monitoring asymptomatic HIV-seropositive pilots in the 1990s, and now with the Special Issuance program for pilots taking SSRI antidepressant medications which began in 2010, aviation neuropsychologists have been called upon to make recommendations that impact the aviator's career and public safety. The former chief psychiatrist of the FAA indicated that he engaged the services of neuropsychologists because of their "compulsive and thorough history taking" and their ability to assess the cognitive deficits that were associated with alcoholism (Barton Pakull, M.D., personal communication, 11 October 2011). Neuropsychology has continued to provide aviation medicine with reliable and valid testing methodologies for assessing the abilities of aviators.

In many instances neuropsychologists and neuropsychological tests have been instrumental in allowing aviators to more quickly return to flight status and to regain their medical certification. This has particularly been the case when pilots had baseline testing against which their post-injury or post-illness performance could be compared.

In spite of these accomplishments, aviation neuropsychology has many issues left to resolve. Some neuropsychologists are opposed to the FAA mandating tests to be administered. They feel that the decision about what tests to administer should be the prerogative of the professional conducting the examination. The FAA recognizes this sentiment and has established core test batteries that are meant to reflect a minimum test standard. The clinician is encouraged to administer any other tests that they feel are indicated.

Another problem that needs more attention and resolution is the tension that seems to be created by cases where there is a lack of overlap in the findings of different disciplines that contribute to aviation medical certification decision-making. For example, a pilot referred for evaluation of Posttraumatic Stress Disorder (PTSD) may be found, during the course of an aviation neuropsychology examination, to show evidence of significant neurocognitive deficits most likely the result of a head injury that might have occurred at the time of the trauma. The tension arises for a couple of reasons. First of all, neurological evaluation and neuroimaging may not show evidence of a traumatic brain injury. Secondly, the neurocognitive findings are considered to be an incidental finding. The purpose of the examination was to evaluate PTSD yet the exam uncovered aeromedically significant neurocognitive deficits. Other aviation medicine specialists have found this troubling, questioning whether the findings of neuropsychology "trump" the findings of other aeromedical experts.

This is not an unusual occurrence. Pilots seen by neuropsychologists for the SSRI Special Issuance Program or student pilots referred for evaluation due to their past reported use of medication for treatment of ADHD are not infrequently found to have unrelated, yet aeromedically significant neurocognitive impairments. For example, recently, a pilot referred for a substance abuse evaluation was denied certification due to the (incidental) finding of dyslexia. In another case a pilot referred due to a history of use of a stimulant medication was found not to have ADHD, but rather a severe visual memory deficit. Just as a routine medical examination may result in the discovery of an unexpected medical condition, regardless of the purpose of the neuropsychological examination, if the evaluation uncovers an aeromedically significant neurocognitive impairment the condition is medically disqualifying if in the view of the FAA it "makes the person unable to safely perform the duties or exercise the privileges of the airman medical certificate." While the aviator may have glowing reports from their primary care doctor, AME/flight surgeon, psychiatrist, neurologist, and their flight instructor, if there is evidence of an aeromedically significant neurocognitive impairment the aviator is unlikely to be certified/granted flight status.

Some military organizations and some civil aviation authorities have such cases presented before a panel of experts where they can be openly discussed and adjudicated. This allows for the input of different disciplines and from aviation experts. The FAA does not formally hold panels to review cases under consideration for medical certification or Special Issuance waivers. In the US the neuropsychologist presents his/her findings to either the Office of the Federal Air Surgeon or to the Medical Certification Branch.

The issue as to the need for aviator norms has also not been fully resolved. Is there a need for aviator norms? Most of the existing aviator norms were collected more than 20 years ago. Is there a need to update the norms with current populations of aviators including commercial, general aviation, and military pilots? There also remain the fundamental questions related to what deficits and what magnitude of deficits should be considered aeromedically significant. Should factors such as level of expertise in the cockpit be considered compensatory? In addition, from a global perspective, are there needs for specific country-based norms?

Finally, having shown the complexity of the practice of aviation neuropsychology it may appear obvious that there is a need for formal training and certification of aviation neuropsychologists. The FAA recently held a well-attended 2-day workshop on aviation psychology/neuropsychology. Work is already underway to develop an accredited, internationally-based curriculum for aviation psychologists which may be recognized by civil aviation authorities around the globe (see Chapter 15, this volume).

References

Aerospace Medical Association. 1992. HIV Positivity and aviation safety. *Aviation, Space, and Environmental Medicine*, 63(5), 375–377.

Aerospace Medical Association. 1997. New AsMA HIV policy. *Aviation, Space, and Environmental Medicine*, 68(3), 35.

American Board of Professional Psychology. 2012. Definition of a clinical neuropsychologist. Available at: http://www.abpp.org/i4a/pages/index.cfm?pageid=3304 [accessed 11 August 2012].

Banich M.T., Stokes, A. and Elledge V.C. 1989. Neuropsychological screening of aviators: A review. *Aviation, Space, and Environmental Medicine*, 60(4), 361–366.

Bower, E.A., Moore, J.L., Moss, M., Selby K.A., Austin, M. and Meeves, S. 2003. The effects of single-dose fexofenadine, diphenhydramine, and placebo on cognitive performance in flight personnel. *Aviation, Space, and Environmental Medicine*, 74(2), 145–52.

Dawson J.D., Uc, E.Y., Anderson, S.W., Johnson, A.M. and Rizzo, M. 2010. Neuropsychological predictors of driving errors in older adults. *Journal of the American Geriatrics Society*, 58(6),1090–1096.

Engelberg, A.L., Gibbons, H.L. and Doege, T.C. 1986. A review of the medical standards for civilian airmen: Synopsis of a two-year study. *Journal of the American Medical Association*, 255(12), 1589–1599.

Federal Aviation Administration. 1981. Neurological and neurosurgical conditions associated with aviation safety. *Federal Aviation Administration Report No. FAA-AM-81-3*. Washington, D.C.

Federal Aviation Administration, 2010. *Proposed Addendum to Federal Register Notification: Complete Neuropsychological Evaluation Specifications*. HIMS Conference, September 2010, Denver, Colorado.

Federal Aviation Administration. 2012. Guide for Aviation Medical Examiners. Available at: http://www.faa.gov/about/office_org/headquarters_offices/avs/offices/aam/ame/guide/app_process/general/appeals/authorization/index.cfm?print=go [accessed 11 August 2012].

Federal Aviation Administration. 2013. *Guide for Aviation Medical Examiners*. Available at: http://www.faa.gov/about/office_org/headquarters_offices/avs/offices/aam/ame/guide/dec_cons/disease_prot/ [accessed 7 April 2013].

Fiedler, E. and Patterson J.C. 2001. Assessment of head-injured aircrew: Comparison of FAA and USAF procedures. *DOT/FAA/AM-01/11*, Federal Aviation Administration, Department of Transportation, Washington, D.C.

Heaton, R.K, Miller, S.W., Taylor, M.J. and Grant, I. 2004. *Revised Comprehensive Norms for an Expanded Halstead-Reitan Battery: Demographically Adjusted Neuropsychological Norms for African American and Caucasian Adults Scoring Program.* Odessa, FL: Psychological Assessment Resources, Inc.

Hess, D.W., Kennedy, C.H., Hardin, R.A. and Kupke, T. (2010). Attention Deficit/Hyperactivity Disorders, in *Military Neuropsychology*, edited by C.H. Kennedy and J.L. Moore. New York: Springer, 199–226.

Hoffmann, C.C., Hoffmann, K.P. and Kay, G.G. 1998. *The Role that Cognitive Ability Plays in CRM.* Paper presented at the RTO Human Factors & Medicine Panel (HFM) Symposium on Collaborative Crew Performance in Complex Operational Systems, Edinburgh, Scotland, April 1998. North Atlantic Treaty Organization (NATO) Research and Technology Organization (RTO) Meeting Proceedings 4.

Imhoff, D.L. and Levine, J.M. 1981. *Perceptual Motor and Cognitive Performance Task Battery for Pilot Selection.* Brooks Air Force Base, TX, AFHRL-TR-10.

Kane, R.L. and Kay, G.G. 1992. Computerized assessment in neuropsychology: A review of tests and test batteries. *Neuropsychology Review*, 3(1), 1–117.

Kay, G.G. 1991. Application of performance assessment methodology in relating cognitive deficits in HIV and job performance. *The Clinical Neuropsychologist*, 5(3), 277.

Kay, G.G. 1995. *CogScreen: Professional Manual*, Odessa, FL: Psychological Assessment Resources, Inc.

Kay G.G. 2002. Guidelines for the psychological evaluation of air crew personnel. *Occupational Medicine*, 17(2), 227–245.

Kay, G.G., Berman, B., Mockoviak, S.H., Morris, C.E., Reeves, D., Starbuck, V., Sukenik, E. and Harris, A.G. 1997. Initial and steady-state effects of diphenhydramine and loratadine on sedation, cognition, mood and psychomotor performance. *Archives of Internal Medicine*, 157, 2350–2356.

Kay, G.G., Morris, S. and Starbuck, V. 1993. *Age and Education Based Norms Control for the Effects of Occupation on Pilot Test Performance.* Paper presented to the 13th Annual Conference of the National Academy of Neuropsychology, October, 1993, Phoenix, Arizona.

Kay, G., Strongin, G., Hordinsky, J. and Pakull, B. *Georgetown/Russian Collaborative Project: Development of Aviator Norms for CogScreen.* Paper presented at the 64th annual Scientific Meeting of the Aerospace Medical Association, May, 1993, Toronto, Canada.

Mapou, R.L., Kay, G.G., Rundell, J.R. and Temoshook, L. 1993. Measuring performance decrements in aviation personnel infected with the human immunodeficiency virus. *Aviation, Space, and Environmental Medicine*, 64(2), 158–164.

Martz, M. 2011. Examiners for DMV Face Dangers. *Richmond Times-Dispatch*, July 24 2011. Available at: http://www2.timesdispatch.com/news/2011/jul/24/4/tdmain01-examiners-for-dmv-face-dangers-ar-1192781/ [accessed 11 August 2012].

Miller, E.N. and Selnes, O.A. 1993. Aviation safety and asymptomatic HIV-1 infection. *Aviation, Space, and Environmental Medicine*, 64(11), 1059–1060.

Moore, J.L. and Kay, G.G. 1996. *CogScreen-Aeromedical Edition in the Assessment of the Head Injured Military Aviator*. Paper presented at the NATO Advisory Group for Aerospace Research & Development (AGARD) Conference, Köln, German. AGARD Conference Proceedings 579, April 1996.

National Archives and Records Administration. 2012. *Electronic Code of Federal Regulations (e-CFR). Title 14: Part 67*. Available at: http://ecfr.gpoaccess.gov/cgi/t/text/text-idx?c=ecfr&sid=eee7643bb2a79472381463dc569e99d9&rgn=div8&view=text&node=14:2.0.1.1.5.2.1.4&idno=14 [accessed 11 August 2012].

Plassman, B.L., Lang, K.M., Fisher, G.G., Heeringa, S.G., Weir, D.R., Ofstedal, M.B., Burke, J.R., Hurd, M.D., Potter, G.G., Rodgers, W.L., Steffens, D.C., McArdle, J.J., Willis, R.J. and Wallace R.B. 2008. Prevalence of cognitive impairment without dementia in the United States. *Annals of Internal Medicine*, 148(6), 427–434. Erratum in: *Annals of Internal Medicine*, 2009, 151(4), 291–292.

Ryan, L.M., Zazeckis, T.M., French, L.M. and Harvey, S. 2006. Neuropsychological practice in the military, in *Military Psychology: Clinical and Operational Applications*, edited by C.H. Kennedy and E.A. Zillmer. New York: Guilford, 105–129.

Schwarz, A. and Cohen S. A.D.G.D. seen in 11% of U.S. children as diagnoses rise. New Your Times, March 31, 2013, Accessed at: http://www.nytimes.com/2013/04/01/health/more-diagnoses-of-hyperactivity-causing-concern.html?pagewanted=all&_r=0 [accessed 8 April 2013]

Sherrill R.E. 1985. Comparison of three short forms of the category test. *Journal of Clinical and Experimental Neuropsychology*, 7(3), 231-238.

Taylor, J. L., O'Hara, R., Mumenthaler, M.S. and Yesavage, J.A. 2000. Relationship of CogScreen-AE to flight simulator performance and pilot age. *Aviation, Space, and Environmental Medicine*, 71(4), 373–380.

Yakimovich N.V., Strongin, G.L., Govorushenko, V.V., Schroeder, D.J., and Kay, G.G. 1994. *CogScreen as a Predictor of Flight Performance in Russian pilots*. Paper presented at Aerospace Medical Association, San Antonio, Texas, May, 1994.

Yakimovich N.V., Strongin, G.L., Govorushenko, V.V., Kay, G.G. and Hordinsky J. 1992. *CogScreen performance of Russian military aviators 6 to 18 months following closed head head injury.* Paper presented at Aerospace Medical Association, Anaheim, California, May, 1995.

Zuluaga, H., Gonzalez, L.A., Porges, C., De LaRosa, C.A., Serrano, G., & Sanchez, L.M. 2011. *Neuropsychological – Personality profiles and flight simulator performance correlation among Colombian Air Force pilots.* Paper presented at Aerospace Medical Association, Anchorage, Alaska, May 2011.

Appendix A: Case Example; First Examination

BACKGROUND: This is the case of a pilot referred by a psychiatrist who specializes in treatment of alcoholism and who sees many pilots in his practice. We will refer to the pilot as Mr Jones, a 53-year-old, Caucasian male.

REPORT OF NEUROPSYCHOLOGICAL ASSESSMENT
NAME: JONES, John
TEST DATE: August 2, 2011
AGE: 53

RELEVANT HISTORY:
Mr Jones has been employed as a first officer with a charter airline, for whom he has been flying B737 and Hawker jet aircraft for the past five years. Mr Jones reports more than 7,000 logged flight hours.

Mr Jones was raised by both parents in a major suburban center. His father was a successful business man. Mr Jones graduated from high school in 1974 with Bs and Cs. He reported that he had no learning disabilities and that he was never held back in school. Following graduation from high school he first enrolled in a community college and then transferred to a university where he majored in Sociology. He had considered going to law school but first went to work for a manufacturing company. During that time he decided that he wanted to become an Air Force aviator but was unable to pass the Air Force's math aptitude test. He re-enrolled in college and completed a degree in Math. He subsequently passed the Air Force aptitude test, received a commission, and entered Officer Candidate School. Mr Jones served on active duty with the Air Force as a decorated fighter pilot for more than nine years. Following his discharge he worked for a different charter company for six years. During his entire flying career Mr Jones never had a mishap and was never cited by the FAA.

According to Mr Jones there is no family history of alcoholism. He began drinking in high school but did not find it appealing. In college he reported that he drank "occasionally" but had no alcohol-related problems. After entering the Air Force he began drinking more heavily, explaining that in the Air Force alcohol was part of social activities. In recent years his family expressed concern to him that he was drinking a lot. His father visited him and confronted him about his excessive alcohol use. In the year prior to this examination he suffered a leg injury and was prescribed pain medication. He did not feel that the pain pills were working and he began drinking heavily. His girlfriend also expressed concern and orchestrated an intervention (primarily with Mr Jones' family members).

At the peak of his drinking Mr Jones was consuming alcohol every night after dinner. He reported that he was fully compliant with the FAA eight-hour cutoff rule. Ten years before this evaluation he was arrested for Driving While Intoxicated (DWI). He was found not guilty though his driving privileges were suspended until the case was resolved. Five years ago he was involved in a minor collision, refused to be tested with a breathalyzer, was arrested and had his driver's license suspended. He was not convicted. According to Mr Jones, during a typical week he would drink two to three martinis per night. On weekends he would drink an entire bottle of vodka. His physician expressed concern to him about elevated liver enzymes. He denied having any history of blackouts or withdrawal symptoms.

Following the intervention by his girlfriend, Mr Jones spoke to a counselor and was directed to an inpatient alcoholism treatment facility. He reported that there were no beds available at the facility except through the facility's detoxification unit. According to Mr Jones, in order to be admitted to the treatment facility, he drank a couple of bottles of wine, thus gaining admission to the detoxification unit. After a brief stay in the detoxification unit he was transferred to the inpatient unit. He was discharged to the outpatient treatment program after one week.

As of the time of this evaluation Mr Jones reported that he did not yet have an AA sponsor. He was attending AA meetings four or five days per week (a minimum of three meetings per week). He expressed concern about a lack of support from his employer but not from his fellow pilots. He reported that he is in good health. He stated that he was feeling the best that he has felt in years. He reported that he is being prescribed medication for hypertension. He reported no history of head trauma or neurological illness.

Mr Jones lives with his girlfriend. He reported that she is actively involved in his recovery program. Mr Jones was previously married for ten years and admits that his alcohol use probably had something to do with the failure of this marriage. He reported that he participated in marital counseling at the end of the marriage. He also received some counseling from a counselor following the death of his sister. He reports no other involvement with mental health professionals or past history of mental health problems.

TESTS ADMINISTERED:

CogScreen-Aeromedical Edition
Trail Making Test
Halstead Category Test
Wisconsin Card Sorting Test (WCST)
Conners' Continuous Performance Test
Paced Auditory Serial Addition Test (PASAT)
Personality Assessment Inventory

MENTAL STATUS EXAMINATION/TEST BEHAVIOR:
Mr Jones was fully alert, oriented, and cooperative. He was friendly and readily established good rapport with the examiner. He had no difficulty comprehending or following test instructions. He appeared to be motivated to perform at the best of his ability. His mood was neither anxious nor depressed. The following results are considered to accurately reflect his current level of functioning. Note that the examination was abbreviated when it became clear that his performance was going to require follow-up testing.

TEST FINDINGS (See also Appendix C):
Attention/Information Processing Speed. On a test of visual scanning and tracking, Part A of the Trail Making Test, Mr Jones' performance was within normal limits (27.5 seconds). He also performed well on a measure of sustained visual attention (vigilance). In contrast, on measures of working memory his performance was impaired. On the PASAT Mr Jones performed below the 5th percentile compared to aviator norms on the first two series. This test assesses both working memory and information processing speed. He also performed poorly on two CogScreen-AE measures of working memory; a test requiring him to sequence digits in reverse order and a test requiring him to follow written instructions following a brief delay. On the other hand, Mr Jones performed adequately on measures of multitasking and divided attention.
Memory. On memory screening with CogScreen-AE, which is based upon immediate and delayed recall of symbol-digit pairs, Mr Jones performed at the 66th percentile.
Executive Functions. Mr Jones' performance on measures of executive functioning was consistently abnormal. On CogScreen-AE he performed poorly on measures of math reasoning (2.5 percentile) and deductive reasoning (≤ 2.5 percentile). He made 60 errors (8th percentile) on the Halstead Category Test. On the WCST, 13 percent of his responses were scored as perseverative. This level of performance is mildly abnormal compared to aviator norms. On Part B of the Trail Making Test, a measure of mental flexibility, he performed below the 5th percentile compared to aviator norms.
Personality Testing. Mr Jones responded to test items in a consistent and reasonably open manner. The clinical profile was considered valid for interpretation. The configuration of the clinical scale is consistent with his history of substance abuse. The antisocial behaviors scale was elevated due to his admission of having been arrested. There were no other indications of antisocial personality traits. Mr Jones' substance abuse was considered to be resulting in severe disruptions in his social relationships and employment and was a major source of stress. He admitted that alcohol has had a negative impact on his life, including his relationships, health and work. Furthermore, personality testing revealed Mr Jones to be an adventurous, risk-taking, and somewhat impulsive individual. There was no evidence of depression, anxiety, or other psychopathology.

SUMMARY AND RECOMMENDATIONS:
Neuropsychological assessment revealed deficits in working memory, information processing speed, and executive functions. Results of the evaluation were considered to be consistent with a diagnosis of Alcohol Dependence (Alcoholism). The neurocognitive deficits seen on testing were seen as most likely resulting from Mr Jones' alcoholism.

It was noted that at the time of the evaluation he had been sober for less than three months. Further recovery of neurocognitive functioning was expected with continued sobriety. In addition to sobriety Mr Jones was believed to be in need of a more directed aftercare program, including counseling and monitoring. The referring psychiatrist further addressed these issues.

In addition it was recommended that Mr Jones return for follow-up neuropsychological assessment in approximately six months to further evaluate his recovery of function and fitness to resume flying. Based on these findings, Mr Jones's neurocognitive functioning was not considered to be adequate for an FAA Special Issuance 1st Class Medical Certificate.

Appendix B: Case Example; Second Examination

REPORT OF FOLLOW-UP NEUROPSYCHOLOGICAL ASSESSMENT
NAME: JONES, John
TEST DATE: February 5, 2012
AGE: 53

RELEVANT HISTORY:
Mr Jones has returned for follow-up neuropsychological assessment. He reported that he has not relapsed and that he has remained sober for nine months. He has an AA sponsor and attends three to four AA meetings per week. He is participating in an aftercare program. He joined a gym and exercises at least three days per week.

TESTS ADMINISTERED:
Wechsler Adult Intelligence Scale-IV (WAIS-IV)
CogScreen-Aeromedical Edition
California Verbal Learning Test-II (CVLT-II)
Trail Making Test
Halstead Category Test
Paced Auditory Serial Addition Test (PASAT)

MENTAL STATUS EXAMINATION/TEST BEHAVIOR:
Mr Jones indicated that he was anxious to return to duty and recognized the need to perform well on testing. He was fully alert, oriented, and cooperative. He readily re-established rapport with the examiner. He did not appear to be particularly

anxious or depressed. The following results are considered to accurately reflect his level of functioning.

TEST FINDINGS (see also Appendix C):

General Intelligence. Mr Jones was found to be functioning in the High Average range of intelligence (FSIQ=115; 84th percentile). He performed equally well on measures of Verbal Comprehension (110; 75th percentile) and Perceptual Reasoning (111; 77th percentile). On measures of verbal comprehension and problem solving his scores ranged from the 63rd to 84th percentile. On visuospatial tasks and non-verbal reasoning tasks his performance ranged from the 63rd to the 84th percentile.

Attention/Information Processing Speed. On the PASAT Mr Jones' score improved and now falls in the low end of the Normal range compared to aviator norms on the first two series (38 correct on both series). This reflects an improvement in working memory and information processing speed. On a measure of visual scanning and tracking, Part A of the Trail Making Test, his performance was relatively unchanged and remained well within the normal range (26 seconds). Performance was mixed on CogScreen-AE measures of working memory. He scored at the 12.5 percentile on a measure of his ability to sequence numbers in reverse order. He also had difficulty with mental arithmetic problems (5th percentile). In contrast, his ability to recall previously shown numbers on CogScreen-AE was at the 97.5 percentile when performing the test in a single task mode and his score showed very little deterioration under dual task conditions. He also showed improvement, compared to his prior evaluation, on a measure of his ability to follow written instructions following a delay. On the WAIS-IV Mr Jones scored at the 55th percentile on measures of Working Memory (Working Memory Index = 102). Similarly, on the WAIS-IV Processing Speed Index he performed well within normal limits (114; 82nd percentile). His speed of response on CogScreen-AE placed him at the 65th percentile compared to commercial aviators his age.

Memory. On memory screening with CogScreen-AE Mr Jones again performed at the 66th percentile. At this evaluation he was administered a test of word learning and memory (CVLT-II). Performance on this test was above average for his age and years of education. He was able to recall 13 of the 16 words after five presentations of the word list. After a 20-minute delay he was able to recall ten of the list words without prompts and 13 of the words with prompts. On delayed recognition testing he correctly recognized all 16 words and made only one false positive error.

Executive Functions. Mr Jones' performance on the CogScreen-AE measure of deductive reasoning improved to the Average range (T=49.7; 50th percentile). He also improved on the Pathfinder Combined test (35th percentile) and on Part B of the Trail Making Test (64 seconds). His performance on the Halstead Category Test remained mildly abnormal (64 errors; 10th percentile when compared to general population norms). He appears to be challenged in forming concepts and using trial and error approaches to novel problem solving.

SUMMARY AND RECOMMENDATIONS:
Follow-up neuropsychological assessment revealed improved functioning in nearly all areas of cognitive functioning and particularly in the areas of working memory and information processing speed. There was some improvement in executive functions, but he continues to struggle with novel problem solving. This was evident on measures of math reasoning and on a test of abstract concept formation and trial and error problem solving. On the other hand, he appears to be less perseverative and to be more capable of abstract reasoning than he was at the first evaluation. In addition he was able to attain Average scores on the WAIS-IV Arithmetic subtest and on the CogScreen-AE Deductive Reasoning factor.

Based upon these findings and the aviator's commitment to maintaining sobriety and remaining active in his alcoholism recovery and monitoring program, it is recommended that the FAA grant Mr Jones a Special Issuance 1st Class Medical Certificate.

Appendix C: Tabular Presentation of Test Results from Case Example

PATIENT: JONES, John
TEST DATES: August 2011, February 2012

TEST FINDINGS:

TEST	AUGUST 2011			FEBRUARY 2012		
	RAW	T-SCORE	%ILE	RAW	T-SCORE	%ILE
Wechsler Adult Intelligence Scale-IV				Scale Score		
Full Scale IQ				115	84	
Verbal Comprehension Index (VCI)				110	75	
Similarities				28	11	63
Information				19	13	84
Perceptual Reasoning Index (PRI)				111	77	
Block Design				47	12	75
Matrix Reasoning				21	13	84
Visual Puzzles				15	11	63
Working Memory Index (WMI)				102	55	
Digit Span				17	10	50
Arithmetic				15	10	50
Processing Speed Index (PSI)				111	77	
Symbol Search				35	12	75
Coding				78	12	75
Trail Making Test	(Aviator Norms)			(Aviator Norms)		
Part A	27.5	40	16	26	42	21
Part B	76	30		64	38	12

TEST	AUGUST 2011 RAW	T-SCORE	%ILE	FEBRUARY 2012 RAW	T-SCORE	%ILE
California Verbal Learning Test-II					z-Score	
Trial 1				8	1.0	84
Trial 5				13	0.5	70
Total Trials 1–5				63	1.3	91
List B				9	2.0	98
Short Delay Recall				12	1.0	84
Long Delay Recall				10	0.0	50
CogScreen-Aeromedical Edition	(Aviator Norms)			(Aviator Norms)		
LRPV	0.99			0.53		
# Speed Scores ≤ 5th percentile	3	37	10	1	48	45
≤ 15th percentile	7	35	7.5	1	54	65
# Accuracy Scores ≤ 5th percentile	3	34	5	1	46	35
≤ 15th percentile	4	42	20	2	48	45
# Process Scores ≤ 5th percentile	1	45	27.5	1	45	27.5
≤ 15th percentile	2	46	32.5	1	51	52.5
Taylor Aviation Factors						
Deductive Reasoning		34	5		49	45
Motor Coordination		78	99		70	98
Visual Association Memory		54	66		54	66
Speed/Working Memory		42	21		46	34
Tracking		56	73		60	84
Conners' Continuous Performance Test						
# Omissions	0	58	79			
# Commissions	8	58	79			
Hit RT	417	44	27			
Hit RT Std Error	4.8	53	60			
Variability	4.1	59	84			
Detectability	0.8	48	42			
Response Style	0.5	48	42			
Perseverations	0	55	70			
Hit RT Block Change	0	51	55			
Hit SE Block Change	0.1	30	2.3			
Hit RT ISI Change	0.0	52	58			
Hit SE Block Change	0.0	56	73			
Halstead Category Test						
Total Errors	60	36	8			
Wisconsin Card Sorting Test						
Categories Completed	6		>16			
Trials to Complete 1st Category	11		>16			
Failure to Maintain Set	3		6-10			
Perseverative Responses	14	45	32			

TEST	**AUGUST 2011**			**FEBRUARY 2012**		
	RAW	**T-SCORE**	**%ILE**	**RAW**	**T-SCORE**	**%ILE**
Paced Auditory Serial Addition Test (PASAT)	(Aviator Norms)			(Aviator Norms)		
Series 1	31	25	<5	38	37	10
Series 2	29	33	5	38	44	27
Personality Assessment Inventory						
ICN		37				
INF		51				
NIM		55				
PIM		50				
SOM		48				
ANX		42				
ARD		48				
DEP		47				
MAN		54				
PAR		49				
SCZ		53				
BOR		48				
ANT		60				
ALC		73				
DRG		72				
AGG		54				
SUI		49				
STR		53				
NON		39				
RXR		40				
DOM		61				
WRM		51				

Chapter 11
The Aging Aviator

Randy Georgemiller

Introduction

Several factors have converged in many parts of the world resulting in an aging workforce. Life span has increased, health care has improved, mandatory retirement age in many occupational sectors has been abandoned or increased, attempts at delaying retirement benefit age eligibility have been implemented to ease the societal cost of funding benefits, and more recently, due to the economic downturn experienced in many developed countries, investment funds have been decimated requiring workers to postpone retirement. Participation rates of older laborers in the workforce vary greatly around the world. Incentives to withdraw from the labor pool are dependent on a country's financial policies and culture. For example, recently in European Union countries, policies have been developed to purposely delay retirement to ease the social and economic burden of supporting workers who leave the work pool early (Barnes-Farrell 2005).

The 55 and over civilian workforce has been trending upward in the United States and about 37 percent of this group continues to participate, a 7 percent increase from only a decade before. Exceeding 60 percent, persons between the ages of 55 and 64 remain employed and of those between the ages of 65 and 74, 22 percent will continue to work (Barnes-Farrell 2005). Paralleling the graying of the US workforce, the US civil aviator community is advancing in years. A study commissioned by the Federal Aviation Administration (FAA) Civil Aerospace Medical Institute (Rogers et al. 2009) examined the aviator population frequencies from 1983 to 2005 and found that as a whole this group is shrinking and regardless of gender or medical certification by class, aviators are getting older. Over this 22-year period the median age for men and women aviators increased and the absolute number of pilots over the age of 56 (the highest age band in this study) increased by 55 percent when comparing 1983 to 2005 figures. With this demographic shift in the aviator community come challenges which have implications for federal regulatory rule making relative to ensuring public safety, the future of the commercial aviation industry, and medico/psycho/social factors.

In working with aging pilots the aviation psychologist will need to have an understanding of the age-related federal standards which regulate aviation privileges, the normal psychological age-related changes which may impact the aviator's competence to continue to fly, and abnormal conditions commonly associated with aging which may be aeromedically significant and therefore disqualifying for continued medical certification. Likewise, the aviation psychologist will need to be aware of developmental influences that promote competence among aging aviators and strategies for mitigating factors which may adversely impact performance. This chapter will discuss these aspects in the context of psychologically relevant dimensions of aging, provide guidance for the psychologist who assesses and/or treats aging aviators and provide an illustrative case example of an older adult aviator who was seen for neuropsychological evaluation, interacted with the federal regulatory system, had his medical certification to fly suspended, completed interventions designed to address aeromedically significant age-related changes, and secured continued medical certification necessary to return to active flying status.

Federal Aviation Regulations and Age

Federal legislation prohibits age discrimination in the workplace. The 1967 Age Discrimination in Employment Act (ADEA) (Pub. L. 90-202) prohibits persons over the age of 40 from being subject to workplace discrimination solely on the basis of age. One of the purposes of the ADEA is "to promote employment of older persons based on their ability rather than age" (Pub. L. 90-202). A 1986 amendment to the Act specifically extends this protection to all federal and non-federal employees but excludes three classes of workers: commercial airline pilots, air traffic controllers (ATC), and public safety officers (Popkin, Morrow, Di Domenico and Howarth 2008). From 1959 to 2007 Federal legislation set the age limit for commercial passenger pilots at 60 (24 Fed Reg 5247 1959) and what has commonly been called the "Age 60 Rule" was implemented in 1960 by Federal Aviation Regulations (FARs), Part 121 which prevented air carrier captains and first officers flying after their 60th birthday. The rationale for instituting this restriction by the FAA related to progressive and unpredictable deterioration associated with aging, lack of medical tests to confirm the presence of age-related functional loss, sudden incapacitation due to increased cardiovascular risks assumed to be tied to aging, and diminished cognitive capacity with advancing years (O'Connor 2009, Wilkening 2002, Aerospace Medical Association 2004). This rule did not apply to corporate, charter, or private pilots.

Given the demographic changes in US society relative to aging such as higher life expectancy, improved health in the aging population, extended years of work force participation, and resulting challenges and changes to age-related prejudice and stigma which are reflected in shifting social policy, the Age 60 Rule increasingly came under fire (Sirven and Morrow 2007). Over the course of almost five decades the Age 60 Rule withstood attempts to overturn or modify it with constitutional challenges, statutory arguments, and by administrative means (O'Conner 2009).

Mounting pressure in the international arena lent added pressure to change the US Age 60 Rule. Various countries such as Canada, Australia, and New

Zealand lifted mandatory retirement ages consistent with their countries' laws prohibiting age discrimination (International Civil Aviation Organization 2006). European members of the Joint Aviation Authorities (JAA) adopted rules allowing commercial air transport pilots between the ages of 60 and 64 to fly as members of multi-pilot crews as long as they are the only pilot at or above the age of 60 (JAR-FCL 3.060 2006). A similar rule was approved in November 2006 by the International Civil Aviation Organization (ICAO), a specialized agency of the United Nations (O'Conner 2009). The ICAO, an outgrowth of the first international aviation association, the International Commission of Air Navigation, went through various age-related policies for pilots beginning with an upper age limit of 45 until 1947. It went on to adopt the United States' 1959 Age 60 Rule in 1963. Cognizant of the JAA's rulings, but aware of the limited scientific data to support a more liberal age limit, the ICAO continued to poll its member countries and emerged with a consensus statement in 2005 allowing for an upper age limit of 65 for multi-crew commercial pilots which will be revisited in 2013 (Cornell, Baker and Li 2007). As a member country of the ICAO, the US withstood efforts to change and was as such, an outlier and contended "The 'Age 60 rule' has served well as a regulatory limit in the United States" (International Civil Aviation Organization 2006).

To address the explicit assumption contained in the Age 60 Rule that the aging aviator in the cockpit poses a potential risk to aviation safety, the FAA embarked on research to examine this assertion. In 1990 the agency sponsored research to study the connection between age, accident rates, and experience (Kay et al. 1994). In the third phase of the project, Kay et al. (1994) examined the relationship between age and accident rates for Classes I, II, and III medical certificates by accessing the FAA and National Transportation Safety Board (NTSB) databases from 1976 to 1988. The analysis was limited by the upper age band of 60 for Class I due to the FAA Age 60 Rule but for other classes in which pilots could hold medical certificates into later decades, aging pilot data was included. Regardless of medical certification class, the finding was consistent. The frequency of accidents dropped with advancing age and leveled off around age 60 and beyond. There was a "hint" of a slight increase in accident rates for Class III pilots older than age 63. However, the researchers found no evidence of increased safety risk based on accident rates for airline pilots as they approached the mandatory retirement age of 60.

In the fourth part of the FAA sponsored research, Hyland, Kay and Deimler (1994) set out to address the following objectives which are very relevant to the practicing aviation psychologist: (1) Quantitative, comprehensive, and objective measurement of pilot performance; (2) Relate performance measures to age; and (3) Determine performance factors which improved or declined with increasing age and experience. Two types of skills were measured, domain-independent cognitive skills which were unrelated to flight specific skills and domain-dependent flight skills. The former skills were laboratory measures of psychomotor, perceptual, and cognitive abilities. While unrelated to flying expertise, they were assumed to be influenced by and likely predict pilot performance. Of the test batteries utilized, CogScreen-Aeromedical Edition (CogScreen-AE; Kay 1995), a computer-administered set of tests which tap working memory, associative memory, selective attention, time sharing and visual spatial, verbal–sequential, and psychomotor functioning was found to be sensitive to the effects of age and discriminate simulator performance of pilots especially under "unusual, high workload emergency situations" (Hyland, Kay and Deimler 1994, 29). The researchers also sought to develop a more objective and quantifiable standard for evaluating domain-specific skill sets as opposed to traditional qualitative pass/fail check flight assessments in flight simulators. Rating criteria for simulator performance was determined with three scenarios that were "realistic, challenging, and had high face validity among pilots" (Hyland, Kay and Deimler 1994, 8).

The FAA contracted study results were limited by a small sample size (N=40), uneven age distribution of pilots (ages 41–71), and the fact that pilots over the age of 60 who participated were not actively flying. Despite the methodological limitations there were intercorrelations among CogScreen-AE scores, flight simulator performance, and pilot age.

The regulatory, political, and scientific debate surrounding the Age 60 Rule culminated in the passage of the Fair Treatment for Experienced Pilots Act (Pub L. No. 110-135 December 13, 2007) which was signed into law on December 13, 2007. This act amended the Age 60 Rule (Title 14 of the Code of Federal Regulations Part 121) pertaining to first class medical certification. Specifically, the legislation extends the federal age standards for commercial airline pilots from age 60 to 65. Amendments to the Act were implemented in 2009 (Federal Register Vol 74 No. 134 July 15, 2009). The amendment served "to harmonize FAA regulations with ICAO standards" (Federal Register Vol 74 No. 134 July 15, 2009). Interestingly, in the FAA's Final Rule and analysis of the estimated costs and benefits of the rule change extending the age limit it was found that, "The economic net benefits of the Act suggest that society is better off with the Act than without it" (Federal Register Vol 74 No. 134 July 15, 2009). It should be noted that the Act pertains solely to First Class Medical Certificates as defined in Part 121 and as such places a mandatory retirement age for aviators who pilot for commercial airline carriers.

To determine whether this Act endangered the flying public, the US General Accountability Office (GAO) was required to study the safety implications of the Fair Treatment for Experienced Pilots Act. In its report (United States General Accountability Office October 30, 2009) the GAO reviewed the FAA's and the National Transportation and Safety Board's (NTSB) accident and incident data from the inception of the Act through September 2009 and "showed that no accidents or incidents resulted from health conditions of pilots 60 years or older." A cautionary note in the report was that for a more definitive determination, a longer period of data collection should be undertaken to confirm or disconfirm the agency's initial finding.

Age-Related Changes in Aviation Competence

Given that there are many physiological changes which accompany aging that are relevant to the ability to safely function as an aviator, and given the space limitations and intended audience for this chapter, this section will restrict its discussion to the neurocognitive changes associated with aging and mitigating factors which have been demonstrated as relevant to aviation competence.

Tsang (2003) offered an explanation for the ecological validity of neurocognitive constructs underlying aviation performance by defining the following functional domains: "perceptual processing (e.g., instrument and out-the-window monitoring), memory (e.g., domain knowledge in long-term memory and air traffic controllers instructions in working memory), problem solving (e.g., fault diagnosis), decision making (e.g., whether to carry out a missed approach procedure), psychomotor coordination (e.g., flight control), and time-sharing or juggling multiple tasks (e.g., instrument monitoring while communicating with ATC; maintaining aircraft stability while navigating)" (Tsang 2003, 509). She also proposed that any understanding of domain-specific aviation skills should include executive skills/abstract novel problem solving, for example, synthesizing data, prioritizing, and responding to novel/emergency situations. Specifically, aspects of this "central cognitive control" should include "attention switching" and "attention sharing" (Tsang 2003, 510).

An ongoing study to understand the relationship between aging and flight competence has been undertaken by a research group from the Department of Veterans Affairs Health Care System in Palo Alto and Stanford University Department of Psychiatry and Behavioral Sciences (Kennedy, Taylor, Reade and Yesavage 2010; Taylor, O'Hara, Mumenthaler and Yesavage 2000; Taylor et al. 2001, 2005; Taylor, Kennedy, Noda and Yesavage 2007; Yesavage et al. 2010). Civilian aviators in varying age ranges have been studied cross-sectionally and longitudinally based on experience (that is, number of accumulated flight hours), expertise or "deliberate practice" (Tsang 2003, 517) (that is, FAA flight ratings from least to most experienced), domain specific flight competence (that is, flight simulator performance), and domain independent cognitive skills (that is, CogScreen-AE performance).

Of particular relevance to the aviation psychologist who conducts fitness-for-duty examinations with pilots is the Stanford/VA Aviation Study findings that CogScreen-AE provides the necessary specificity and sensitivity to be utilized in pilot medical certification determinations when underlying neurological and neuropsychiatric conditions are suspected, validly measures flight relevant cognitive skills, and by extension measures occupational performance given its relationship to flight simulator competence (Taylor, O'Hara, Mumenthaler and Yesavage 2000).

In terms of domain independent cognitive decrements due to aging, the most pronounced and consistent findings are that with advancing age, pilots are prone to reductions in cognitive processing speed, associative memory, dual tasking/divided attention, and concept formation and the least affected performance domains are motor speed, motor coordination, and tracking (Kennedy, Taylor, Reade and Yesavage 2010; Taylor, O'Hara, Mumenthaler and Yesavage 2000; Taylor et al. 2001). In particular, it was found that processing speed and executive function measures on CogScreen-AE could predict flight simulator performance which the authors took to mean that the test was validly tapping into the inherent complexity of aviation performance (Taylor, O'Hara, Mumenthaler and Yesavage 2000).

When it comes to determining the developmental effects of aging on actual flight performance, controlled laboratory tests utilizing flight simulators provide the most relevant domain-specific data. When testing skill level with various flight-related demands, research data has been rather inconsistent and most likely susceptible to numerous methodological limitations. Birth cohort effects, sample size, whether the design was cross-sectional or longitudinal, confounds among age, recency of flight experience, total flight hours, and FAA aviation flight rating, as well as the possibility that some older research participants may have been suffering from incipient clinically significant neurocognitive decline associated with Mild Cognitive Impairment (MCI) or dementia may help to explain the inconsistent research outcomes (Taylor et al. 2001; Taylor, Kennedy, Noda and Yesavage 2007; Tsang 2003).

In their longitudinal study of pilots Taylor et al. (2001) found no precipitous drop in flight simulator performance with advancing age for pilots with higher levels of flight hours and postulated that being a seasoned aviator had an insulating effect on performance decrement. Level of flight rating from least to most expert predicted baseline simulator performance in a longitudinal study of pilots but over a three year measurement of skills, the most expert pilots had less decline over time, confirming the positive impact of training on performance (Taylor, Kennedy, Noda and Yesavage 2007). The researchers hypothesized that expertise "builds an elaborate, integrated base of declarative and procedural knowledge" or "crystallized intelligence" (Taylor, Kennedy, Noda and Yesavage 2007, 652). Nonetheless, the study did not find that expertise moderated the effect of age when serial sampling was conducted.

To investigate the impact of age on two major contributors to air fatalities and accidents, weather conditions and pilot error, Kennedy, Taylor, Reade and Yesavage (2010) tested a series of pilots who varied by age and instrument rating, that is, expertise. They introduced flight simulation for executing holding patterns and landing in foggy conditions. While advancing age did lead to poorer execution of some tasks, for example, landing in too foggy conditions and instrument over-controlling, this was likely related to specific reductions in divided attention and processing speed. A higher level of expertise and faster processing speed did predict better simulator performance, leading the authors to recommend that focused training with older pilots in flight simulation coupled with processing speed training may improve flight safety for older aviators.

Another domain-specific skill which has been measured is the capacity for executing ATC communications (Morrow et al. 2003; Taylor et al. 2005). The underlying cognitive skills related to competently executing ATC communications have been related to online storage, information processing, processing speed, and sustained attention in the face of interference (Taylor et al. 2005). In various tasks related to read back, recall, and asking appropriate probe questions in response to ATC communication there was a decline with age and while Morrow et al. (2003) found no buffering effect of expertise, Taylor et al. (2005) found that greater experience as expressed as a higher flight rating as opposed to accumulated flight hours had a moderating effect on performance in the face of the deleterious impact of age. Various compensatory procedures to promote safety have been offered to address this age-related decline such as shortening ATC messages, redundant visual and auditory presentation of ATC communications, reliance on notes (Morrow et al. 2003), standardizing and organizing the format of communications, and use of "cue words and intonation" to minimize confusion (King 1999).

Much has been made of the impact of accumulated flight experience and graded levels of expertise in moderating the impact of age-related flight competence. While Worthy et al. (2011) did not conduct their research with an aviator sample, their work advances the concept of expertise in their examination of "insight and wisdom." The researchers differentiated two types of decision-making, one with discrete choices that are not contingent on building a model of the environment and the other which requires a "cognitive map of the environment that describes how different options and their associated rewards are connected to one another" (Worthy et al. 2011, 2). The latter strategy is likened to chess playing in which one decision needs to be considered in the context of future choices. Their laboratory findings confirmed that younger adults were more efficient in maximizing immediate rewards whereas older adults were more focused on developing hypotheses with greater emphasis on the interconnection of responding to reward within the context of an environment. They concluded, "While aging may lead to some cognitive declines, it may also lead to gains in the insight and wisdom used to make the best decisions" (Worthy et al. 2011, 6).

Abnormal Aging and Aviation Competence

There are aeromedically relevant conditions associated with all the major organs of the body which may develop with advancing age. A complete listing and discussion of these disorders is beyond the scope of this chapter, but highly relevant to aviation psychology are the progressive neurological disorders which impact the cognitive skills and functional capacity that are more prevalent in older adults, the dementias.

Dementia is defined as a disorder involving "multiple cognitive deficits that include memory impairment and at least one of the following cognitive disturbances: aphasia, apraxia, agnosia, or a disturbance in executive functioning.

The cognitive deficits must be sufficiently severe to cause impairment in occupational or social functioning and must represent a decline from a previously higher level of functioning" (American Psychiatric Association 2000, 134). Other diagnostic criteria which may be included are "preserved awareness of the environment" and "a decline in emotional control or motivation" characterized by features such as "irritability, apathy, and coarsening of social behavior" (World Health Organization 1993, 45–47). A discussion of dementia has added significance for aviation psychology since inherent in the definition of dementia is diminution of functional capabilities expressed in the occupational setting as well as a reduction of cognitive skills necessary for safe operation of aircraft.

Epidemiological studies estimate that the incidence of people with dementia worldwide exceeds 24 million as of 2001 and will reach 80 million by 2040 (Ballard et al. 2011), the majority of whom will suffer from Alzheimer's disease. A global prevalence study (Ballard et al. 2011) found that the US is a region in the world which has one of the highest rates along with our counterparts in Western Europe, Latin America, the developed countries within the Western Pacific and China. For people over the age of 60 in North America 6.4 percent suffer from some form of dementia and the absolute numbers will balloon from 3.4 million in 2000 to 9.2 million in 2040 given the doubling of the aging population as baby boomers advance in years (Ballard et al. 2011). Other estimates (Alzheimer's Association 2012) place the number of Americans with Alzheimer's disease at 5.4 million in 2012 with the majority of these cases occurring in those 65 years and above. Again confirming that dementia is predominately a disease of the elderly, frequency by age band estimates that 4 percent are under the age of 65 and 13 percent of those over the age of 65 suffer from the disorder (Alzheimer's Association 2012). Population estimates differ due to variation in sampling and the definition of dementia utilized in diverse studies.

But of major concern for the aviation psychologist who evaluates and/or treats aging pilots is the fact that within the general population in the US it is estimated that one-half of the cases of Alzheimer's disease go undetected (Alzheimer's Association 2012). Fortunately, FAA standards which require frequently recurring aviation medical examinations may reduce the likelihood of missed cases in the aviator population. Also, one author (Murphy 2011) has noted that dementia-related cognitive deficits in the cockpit may be manifested earlier than in other functional arenas given the high skill level expected and strenuous demands placed on the aviator. The assumption is that given the greater likelihood that dementia will be manifested earlier, it may also be detected earlier. Given the hallmark features of dementia, inefficiencies and difficulties may first be noted in negotiating ATC communication, mastering new or upgraded equipment, missing items on a checklist, and responding to an in-flight emergency (Murphy 2011). Nonetheless, if the prevalence estimate of approximately 6.4 percent of persons over the age of 65 suffer from the most common form of dementia, Alzheimer's disease (Alzheimer's Association 2012), is applied to US aviators there may be more than 10,000 possible cases of dementia within this group (Rogers et al.

2009). This figure is a very imprecise estimate since the latest age-related figures for aviators dates back to 2005 and the highest age band was 56 years and above. Further research is required to determine if this estimate is relevant to the aviator population, especially since prevalence of the disorder varies within the population by gender, education, and race (Alzheimer's Association 2012).

The dementias have diverse classifications and suspected etiologies but the detection of this progressive disorder by the aviation psychologist may be critical to the determination of the aging aviator's fitness-for-duty. Neuropsychological evaluation has been included as a component of the diagnostic workup for dementia for 25 years (National Institutes of Health 1987) and even with advances in the development of biological markers for dementia, cognitive testing is still an integral part of a dementia evaluation (Knopman, Boeve and Petersen 2003) and has demonstrated good sensitivity and specificity (greater than 80 percent) for differentiating patients with and without dementia (Ballard et al. 2011). Besides the administration of a neuropsychological test battery, as part of a thorough neuropsychological evaluation the aviation neuropsychologist should include in the history taking evidence of dementia risk factors for medical conditions, that is, midlife hypertension and hypercholesterolemia, stroke, and diabetes and lifestyle features, such as obesity, smoking, physical activity level, components related to "cognitive reserve" (intelligence, occupation, and education), and alcohol consumption (Ballard et al. 2011, 1027). Given the insidious nature of the onset of these disorders and the deleterious implications of this set of cognitive deficits for aviation safety, early detection is critical.

Assuming the progressive nature of dementia, various formats for staging the disease have been proposed to identify the continuum based on early, middle, or late-stages or mild, moderate, or severe categories. More recently, in 2011, guidelines for Alzheimer's disease diagnostic criteria include the staging from preclinical Alzheimer's disease, to MCI, to full diagnostic criteria for the presence of Alzheimer's disease (Clifford et al. 2011). For the aviation neuropsychologist the pre-clinical phase has limited relevance since there are no established criteria for diagnosis, there is no behavioral or cognitive manifestation, and its presence is based on preliminary findings of biomarkers which may be present for decades before the symptoms express themselves.

The second phase of dementia, MCI, should be of particular interest to the aviation neuropsychologist who undertakes to determine whether there are aeromedically significant cognitive deficits which could jeopardize aviation safety when evaluating the senior adult aviator. From its inception as a relevant clinical entity, MCI has been variously characterized as a "boundary or transitional state between normal aging and dementia," "incipient dementia," and "isolated memory impairment" (Petersen et al. 1999, 303). As mentioned above, it is now seen as a progressive stage which eventuates in a diagnosis of full-blown dementia for a sizeable portion of patients with MCI. It is estimated that between 10 and 20 percent of persons over the age of 65 have MCI (Alzheimer's Association 2012). Based on variations in sampling, diagnostic criteria, and research method, conversion rates for MCI to dementia range significantly but the recurrent finding is that MCI patients go on to develop dementia at a much higher rate than their same aged peers without MCI. For example, it is estimated that 15 percent of patients who consult with their physician with MCI symptoms go on to develop dementia each year. In three to four years almost half of this group will meet criteria for dementia (Clifford et al. 2011). When community sampling of persons with MCI symptoms is done the rate of conversion to dementia is 10 percent per year (Alzheimer's Association 2012). For example, Schinka et al. (2010) determined that by applying various stringent levels in terms of the number of domain-specific cognitive tests impaired, and the cutoff scores used to define impairment, (1.5 SD or 2.0 SD) the relative rates of MCI diagnosis was affected. Left unanswered is the fact that not all MCI cases convert to an eventual diagnosis of dementia. It is estimated that 40 percent of persons with MCI do not convert to a diagnosis of dementia or may return to a normal baseline (Schinka et al. 2010). Nonetheless, the MCI conversion rates are in stark contrast with the typical incidence rate of dementia for the general population. For example, meta-analysis finds that the annual incidence rate for dementia for persons age 60–64 is only 0.11 percent, ages 65–69 is 0.33 percent, and ages 70–74 is 0.84 percent (Petersen et al. 2001). As expected, the percentage of persons diagnosed with dementia by age band continues to increase precipitously until the ninth decade where it levels off at approximately 7 to 8 percent. However, even at these upper age ranges, the incidence rate does not come anywhere close to the number of persons previously diagnosed with MCI that are subsequently found to have a dementia.

The generally accepted criteria for MCI are as follows: cognitive complaint with preferred collateral confirmation, more often involving memory; cognitive impairment on standardized tests in the context of overall intact cognitive functioning; preserved daily functioning; and not meeting criteria for a dementing disorder (Petersen et al. 2001; Kelley and Petersen 2007).

While the complaint of memory loss and documentation of mild memory impairment on standardized neurocognitive measures has been the hallmark of MCI, there is now greater elaboration on the subtypes (Kelley and Petersen 2007). Amnestic versus non-amnestic MCI is characterized by the presence or absence of documented memory loss. Secondly, MCI may include a single domain or multiple domains of cognitive impairment. The typical non-memory cognitive functions measured are speech/language, attention/executive skills, and visuospatial abilities (Kelley and Petersen 2007). Neuropsychological testing's utility for establishing a baseline for future comparison and disease staging and domain-specific standardized measurement of relevant cognitive functions makes it a critical aspect of any workup for MCI (Kelley and Petersen 2007; Schinka et al. 2010; Petersen et al. 1999; Petersen et al. 2001).

Recent research indicates that MCI diagnostic criteria may need to be reconsidered. Aretouli and Brandt (2010) found that regardless of MCI subtype (amnestic versus non-amnestic and single domain versus multi-domain) there was a decrement in instrumental aids to daily living when compared to a normal elderly sample extending to difficulty with keeping appointments, losing belongings, and recalling current events. Of greatest relevance to the evaluating aviation psychologist was that 20 percent of the MCI sample reported impaired complex skills including driving and using transportation. This finding would not bode well for the aviator with MCI who is expected to meet even greater mental rigors in the air than on the road.

For the purposes of the aviation neuropsychologist who is evaluating the aging pilot for the presence of MCI and the potential safety risk posed by this disorder, the current FAA protocol for psychological examination (Federal Aviation Specifications for Psychiatric and Psychological Evaluations Rev 1-22-1996) extends the necessary clinical latitude and judgment to the examining professional to create a battery of standardized measures to tap the cognitive domains of memory, speech and language, attention and executive skills, and visuospatial ability.

Case Example: Aviation Neuropsychological Examination of an Aging Aviator

The following case study is offered as an example of the practical implications of assessing for aeromedical competence when examining an aging aviator. The case example is illustrative and in no way a complete report of findings which typically would include a thorough psychological/social/medical history, review of academic, occupational and medical records, behavioral observations, interviews with the examinee and pertinent collateral sources, and administration of standardized clinical psychological and neuropsychological measures (see Chapter 10, this volume for an example aeromedical neuropsychological report).

Time 1

RB is a 71-year-old man who has been a captain for a charter carrier for seven years. He was referred for neuropsychological evaluation after his employer determined he had "cognitive diminution." The opinion was the result of a corporate investigation after the first officer who flew with RB noted that he was confused about a flight departure time despite repeated reminders, carelessly drove an automobile with flight crew as passengers en route to the airport, miscalculated fuel in preparation for a flight, provided incorrect call signs while taxiing and did not realize the error until it was pointed out by ATC, and implemented some other flight procedures which were judged by the first officer as non-standard and potentially dangerous.

As part of the data-gathering phase, the airman was requested to reflect in writing on the events pertaining to the complaint. He demonstrated some awareness of his deficiencies while not fully agreeing with the first officer's complaints. In particular, he was aware of the adverse impact of fatigue after flying

12-hour duty days and various stresses in his personal life. Additional collateral data was gathered via structured telephone interviews with three crewmembers who had recently flown with RB to ascertain their perception of his performance. Questions focused on the length of the relationship, extent of on the job and off the job contact with RB, knowledge of him in other work-related settings such as committee work, behavior changes (for example, assessment of mood, anxiety, frustration tolerance, stress level), and cognitive and functional skill changes (for example, memory, situational awareness, communication, attention, visual spatial, planning, decision-making, and judgment). None of the informants admitted to being aware of the complaints lodged against RB and their reports did not support the allegations of the first officer.

The assessment battery and all subsequent assessments conformed to FAA standards for neuropsychological testing (Federal Aviation Administration 1996) and utilized aviator norms (Kay 1995; Kay 2002) for interpreting test results where applicable.

The examination found standardized test results which were compatible with the complaints lodged against the airman with demonstration of diminished memory skills relative to intellect, differentially-impaired visual memory relative to auditory memory, visual choice reaction time, mental processing speed, divided attention, and cognitive set shifting. These mild cognitive deficits were noted in conjunction with contributory characterological features, defensiveness, and externalization that could lead to interpersonal strain. The final opinion was that the deficits were attributed to multiple factors such as age-related changes, sensory loss, characterological features, acute psychological stress, and increased susceptibility to fatigue. A recommendation for short-term individual psychotherapy with behavioral and cognitive behavioral modalities was offered to enhance his resiliency, reaction to stress, self-monitoring skills and improve his adaptation to fatigue, and learn focused strategies such as relaxation training and other psychophysiological desensitization techniques. After successful completion of a course of treatment it was recommended that he undergo a focused neuropsychological re-evaluation to determine if he was eligible for a recommendation for return to active flight status.

Time 2

Approximately four months after the initial contact with RB he was seen for re-evaluation. In the interim he had completed a course of short-term individual psychotherapy with the goals of developing a monitoring system to detect stress and fatigue and he successfully identified external and internal markers for these conditions. RB also became proficient at implementing relaxation procedures. Selected measures which are sensitive to the presence of aeromedically significant cognitive deficits were administered and for which aviator norms were available (Kay 1995; Kay 2002). Based on his behavioral presentation and test performance the opinion was rendered that he demonstrated adequate skills for return to active duty flight status.

Time 3

Data from the two prior neuropsychological examinations with RB were independently reviewed by an FAA neuropsychologist consultant with a determination by the FAA Civil Aerospace Medical Institute that the aviator suffered from aeromedically disqualifying "cognitive deficits" (Title 14 of the Code of Federal Regulations, revised Part 67, paragraphs 107 (c), 207 (c), 307 (c), and 113(c), 213 (c), 313 (c)) and was therefore ineligible for medical certification. For reconsideration RB was to submit to a follow-up neuropsychological examination which would be reviewed by the FAA to determine eligibility for a Special Issuance medical certification. The examining neuropsychologist recommended that prior to retesting RB complete a course of psychologist administered cognitive rehabilitation.

To date, there are no published studies addressing amelioration of cognitive deficits or age-related changes with aviators by means of cognitive rehabilitation programs. Case study has indicated that alcohol-dependent pilots suffering from MCI as measured by standardized neuropsychological measures have profited by completing a course of computer-assisted, psychologist-administered cognitive rehabilitation (Georgemiller, Bracy and Altman 2009). Nonetheless, there is extensive research documenting short-term and long-term improvements of up to two to five years in cognitive skills such as memory, reasoning, and processing speed and visual search for normal seniors when structured training exercises have been employed (Ball et al. 2002; Rebok, Carlson and Langbaum 2007; Willis et al. 2006). Hertzog, Kramer, Wilson and Lindenberger (2009) introduced the concept of "enrichment" and its impact on cognitive performance during adult development, especially in old age. "The cognitive-enrichment hypothesis states that the behaviors of an individual (including cognitive activity, social engagement, exercise, and other behaviors) have a meaningful positive impact on the level of effective cognitive functioning of old age" (Hertzog, Kramer, Wilson and Lindenberger 2006, 3). This concept subsumes and expands upon the commonly held "use it or lose it" belief regarding mitigation of cognitive decline with aging.

RB completed a three-month course of computer-assisted cognitive rehabilitation (PSSCogRehab 2012; Assis et al. 2010) and was released by the treating neuropsychologist as successfully completing. He also demonstrated ongoing personal enrichment strategies such as engaging in cognitively challenging activities, brisk exercise, and social interaction in his day-to-day life.

Upon completion of an extensive battery of neuropsychological measures, it was the examiner's opinion that RB had the requisite cognitive skills to maintain his aviation medical certification. Most striking was his serial improvement across three examinations on a standardized measure with alternate forms that is quite sensitive to aviation performance, CogScreen-AE (Kay 1995).

Time 4/Time 5

Subsequent to the third examination, RB was granted a Special Issuance medical certification and was returned to active duty status flying for his employer. There was mandated yearly screening with CogScreen-AE and he was able to maintain his medical certification upon two yearly screenings. After two screenings and no in-flight incidents or further complaints of reduced competence by flight crew, he petitioned to be released from yearly monitoring. His request was granted and he was reissued an unrestricted medical certificate. RB eventually retired from the charter company at the age of 73 and at age 74 petitioned for a Third Class medical certificate to pilot a personally-owned single-engine Cessna airplane. Therefore, there was no further need to seek a First Class Medical Certificate. To date, he has successfully completed annual training and has maintained Instrument Flight Rating, that is, proficiency which allows an aviator to fly during conditions of poor visibility with reliance on navigational instruments.

Conclusions

With changes in the age demographics in many countries around the world paralleling the graying of the aviator population, as well as the inherent tension between advancing fair and equal treatment for seniors and governments' charge to ensure aviation safety, the discussion surrounding the competence of aging aviators will continue to be a necessary but emotionally charged issue. Given the education and training of the aviation psychologist, this discipline is uniquely qualified to contribute to the clinical, research, and policy fronts regarding aviation certification determinations for aging pilots. With evidence that laboratory-based neurocognitive procedures are specific and sensitive for diagnosing aeromedically significant age-related changes, aviation neuropsychologists should seek additional credentialing to supply such services (American Psychological Association 2010) and advance further development of measurement tools. Aviation psychologists in their role as behavioral scientists and building upon performance enhancement psychology research findings should be engaged in further development and promotion of enrichment regimes for aging aviators which include social engagement, physical exercise, promotion of positive emotional states, and specific cognitive activities to ameliorate age decrements. Working in collaboration with other human factors scientists, aviation psychologists should also contribute to the development of prosthetic aviation environments which could promote optimal functioning and accommodate normal age-related changes while safeguarding aviation safety.

References

24 Federal Regulation 5247 1959, NRPM maximum age limitations for pilots.

Aerospace Medical Association. 2004, January. *Position Paper: The Age 60 rule*. Available at: http://www.age60rule.com/docs/2004_asma_position.pdf [accessed: 15 May 2012].

Alzheimer's Association. 2012. 2012 Alzheimer's disease facts and figures. *Alzheimer's and Dementia*, 8(2), 1–66.

American Psychiatric Association. 2000. *Diagnostic and Statistical Manual of Mental Disorders, Fourth Edition, Text Revision*. Washington, DC: American Psychiatric Association.

American Psychological Association. 2010. Amendments to the 2002 "Ethical principles of psychologists and code of conduct". *American Psychologist*, 65(5), 493.

Aretouli, E. and Brandt, J. 2010. Everyday functioning in mild cognitive impairment and its relationship with executive cognition. *International Journal of Geriatric Psychiatry*, 25, 224–233.

Assis, L., Tirado, M., de Melo Pertence, A., Pereira, L. and Mancini, M. 2010. Evaluation of cognitive technologies in geriatric rehabilitation: A case study pilot project. *Occupational Therapy International*, 17(2), 53–63.

Ball, K., Berch, D., Helmers, K., Jobe, J., Leveck, M., Marsiske, M., Morris, J., Rebok, G., Smith, D., Tennstedt, S., Unverzagt, F. and Willis, S. 2002. Effects of cognitive training interventions with older adults. *Journal of the American Medical Association*, 288(18), 2271–2281.

Ballard, C., Gauthier, S., Brayne, C., Aarsland, D. and Jones, E. 2011. Alzheimer's disease. *Lancet*, 377(9770), 1019–1031.

Barnes-Farrell, J. 2005. Older workers, in *Handbook of Work Stress*, edited by Barling, J., Kelloway, E. and Frone, M. Thousand Oaks, CA: Sage Publication, 431–454.

Clifford, J., Albert, M., Knopman, D., McKhann, G., Sperling, R., Carrillo, M., Thies, B. and Phelps, C. 2011. Introduction to the recommendations from the National Institute on Aging-Alzheimer's Association workgroups on diagnostic guidelines for Alzheimer's disease. *Alzheimer's and Dementia: The Journal of the Alzheimer's Association*, 7(3), 257–262.

Cornell, A., Baker, S. and Li, G. 2007. Age-60 rule: The end is in sight. *Aviation, Space, and Environmental Medicine*, 78(6), 624–626.

Fair Treatment for Experienced Pilots Act of 2007, Pub. L. No. 110-135.

Federal Aviation Administration. 1996, January 22. Federal Aviation Specifications for Psychiatric and Psychological Evaluations.

Federal Register Vol 74 no. 134 July 15, 2009. Part 121 Pilot Age Limit. Doc. No. FAA-2006–26139.

General Accountability Office. 2009, October. Aviation Safety: Information on the Safety Effects of Modifying the Age Standard for Commercial Pilots (GAO-10-107R).

Georgemiller, R., Bracy, O. and Altman, D. 2009, September. *Psychiatric and Psychological Evaluation and Cognitive Rehabilitation*. Presentation at the Human Intervention and Motivation Study Conference, Denver, Colorado.

Hertzog, C., Kramer, A., Wilson, R. and Lindenberger, U. 2009. Enrichment effects on adult cognitive development. *Psychological Science in the Public Interest*, 9(1), 1–65.

Hyland, D., Kay, E. and Deimler, J. 1994, October. Age 60 study, part IV: Experimental Evaluation of Pilot Performance (Contract No. DTFA02-90-C-90125), Civil Aeromedical Institute, Federal Aviation Administration.

International Civil Aviation Organization 2006. Comments of States and International Organizations in Response to State Letter AN 5/16.1-05/17. [Online: ICAO]. Available at: http://www.icao.int/cgi/goto_m.pl?/icao/en/new.htm [accessed: 12 May 2012].

Joint Aviation Requirements 2006, JAR-FCL 3, Joint Aviation Authorities.Kay, G. 1995. *CogScreen Aeromedical Edition: Professional Manual*. Odessa, FL: Psychological Assessment Resources, Inc.

Kay, G. 2002. Guidelines for the psychological evaluation of air crew personnel. *Occupational Medicine*, 17(2), 227–245.

Kay, E., Harris, R., Voros, R., Hillman, D. Hyland, D. and Deimler, J. 1994, October. Age 60 study, part III: Consolidated Database Experiments Final Report (Contract No. DTFA02-90-C-90125), Civil Aeromedical Institute, Federal Aviation Administration.

Kelley, B. and Petersen, R. 2007. Alzheimer's disease and mild cognitive impairment. *Neurologic Clinics*, 25(3), 1–34.

Kennedy, Q., Taylor, J., Reade, G. and Yesavage, J. 2010. Age and expertise effects in aviation decision making and flight control in a flight simulator. *Aviation Space and Environmental Medicine*, 81(5), 489–497.

King, R. 1999. *The cost/benefit of aging on safety and mission completion in aviation professions*. Paper presented at the NATO Research and Technology Organization Human Factors Medicine Symposium on Operational Issues of Aging Crewmembers, Toulon, France. Available at: http://www.dtic.mil/cgi-bin/gettrdoc?location=42&doc=gettrdoc.pdf&ad=adp010569 [accessed: 9 April 2012].

Knopman, D., Boeve, B. and Petersen, R. 2003. Essentials of the proper diagnosis of mild cognitive impairment, dementia, and major subtypes of dementia. *Mayo Clinic Proceedings*, 78, 1290–1308.

Morrow, D., Menard, W., Ridolfo, H., Stine-Morrow, E., Teller, T. and Bryant, D. 2003. Expertise, cognitive ability, and age effects on pilot communication. *The International Journal of Aviation Psychology*, 13(4), 345–371.

Murphy, R. 2011. The aging brain, cognition, and aeromedical concerns. *Federal Air Surgeon's Medical Bulletin*, 49(4), 1–6.

National Institutes of Health. 1987, July 6–8. Differential diagnosis of dementing diseases. *Consensus Development Conference Statement*, 6(11), 1–27.

O'Connor, N. 2009. Too experienced for the flight deck? Why the age 65 rule is not enough. *The Elder Law Journal*, 17(2), 375–399.

Part 121 Pilot Age Limit2009. 14 CFR Parts 61 and 121. 74 Federal Register No. 134.

Petersen, R., Smith, G., Waring, S., Ivnik, R., Tangalos, E. and Kokmen, E. 1999. Mild cognitive impairment: Clinical characterization and outcome. *Archives of Neurology*, 56(6), 303–308.

Petersen, R., Stevens, J., Ganguli, M., Tangalos, E., Cummings, J. and DeKosky, S. 2001. Report of the Quality Standards Subcommittee of the American Academy of Neurology. Practice parameter: Early detection of dementia: Mild cognitive impairment (an evidence-based review). *Neurology*, 56(9), 1133–1142.

Popkin, S., Morrow, S., Di Domenico, T. and Howarth, H. 2008. Age is more than just a number: Implications for an aging workforce in the US transportation sector. *Applied Ergonomics*, 39(5), 542-549.

PSSCogRehab Website. 2012. Available at: http//:www.neuroscience.cnter.com/pss/psscogrehab.html [accessed: 15 April 2012].

Rebok, G., Carlson, M. and Langbaum, J. 2007. Training and maintaining memory abilities in health older adults: Traditional and novel approaches. *Journals of Gerontology: Series B*, 62B (Special Issue), 53–61.

Rogers, P., Veronneau, S., Peterman, C., Whinnery, J. and Forster, E. 2009 May. *An analysis of the US pilot population from 1983–2005: Evaluating the effects of regulatory change* (Report No. DOT/FAA/AM-09/9). Civil Aerospace Medical Institute, Federal Aviation Administration.

Schinka, J., Lowenstein, D., Ashok, R., Schoenberg, M., Banko, J., Potter, H. and Duara, R. 2010. Defining mild cognitive impairment: Impact of varying decision criteria on neuropsychological diagnostic frequencies and correlates. *American Journal of Geriatric Psychiatry*, 18(8), 684–691.

Sirven, J. and Morrow, D. 2007. Fly the graying skies. *Neurology*, 68(9), 630–631.

Taylor, J., Kennedy, Q., Noda, A. and Yesavage, J. 2007. Pilot age and expertise predict flight simulator performance. *Neurology*, 68(9), 648–654.

Taylor, J., Mumenthaler, M., Kraemer, H., Noda, A., O'Hara, R. and Yesavage, J. 2001. Longitudinal study of older small-aircraft pilots: Changes in CogScreen-AE performance. [Online: Federal Aviation Administration]. Available at: http://www.faa.gov/library/online_libraries/aerospace_medicine/sd/media/taylor.pdf [accessed: 15 April 2012].

Taylor, J., O'Hara, R., Mumenthaler, M., Rosen, A. and Yesavage, J. 2005. Cognitive ability, expertise, and age differences in following air-traffic control instructions. *Psychology and Aging*, 20(1), 117–133.

Taylor, J., O'Hara, R., Mumenthaler, M. and Yesavage, J. 2000. Relationship of CogScreen-AE to flight simulator performance and pilot age. *Aviation, Space, and Environmental Medicine*, 71(4), 373–380.Tsang, P. 2003. Assessing cognitive aging in piloting, in *Principles and Practice of Aviation*

Psychology, edited by Tsang, P. and Vidulich, M. Mahwah, NJ: Lawrence Erlbaum, 507–546.

Wilkening, R. 2002. The age 60 rule: Age discrimination in commercial aviation. *Aviation, Space, and Environmental Medicine*, 73(3), 194–202.

Willis, S., Tennstedt, S., Marsiske, M., Ball, K., Elias, J., Koepke, K., Morris, J., Rebok, G., Unverzagt, F., Stoddard, A. and Wright, E. 2006. Long-term effects of cognitive training in everyday functional outcomes in older adults. *Journal of the American Medical Association*, 296(23), 2805–2814.

World Health Organization, 1993. *The ICD-10 Classification of Mental and Behavioral Disorders: Diagnostic Criteria for Research*. Geneva: World Health Organization.

Worthy, D., Gorlick, M., Pacheco, J., Schnyer, D. and Maddox, W. 2011. With age comes wisdom: Decision-making in younger and older adults. *Psychological Science*, 22(11), 1375–1380.

Yesavage, J., Jo, B., Adamson, M., Kennedy, Q., Noda, A., Hernandez, B., Zeitzer, J., Friedman, L., Farichild, K., Scanlon, B., Murphy, G. and Taylor, J. 2010. Initial cognitive performance predicts longitudinal aviator performance. *The Journals of Gerontology, Series B: Psychological and Social Sciences*, 66(4), 444–453.

Chapter 12
Psychopharmacology in Aviation

Bradford C. Ashley and Gary G. Kay

Introduction

This chapter will discuss various pharmacological treatments and their current use in the management of mental health disorders in aviators. This is a dynamic and controversial area of modern aviation medicine. As newer medications are being developed aeromedical regulatory agencies are rethinking age old prohibitions against treating pilots with psychiatric medications. However, these changes in policy have not been universally supported by those who oppose the use of medication in the management of mental illness (Breggin 2011). A decade ago this chapter would have been less than a couple of pages in length and limited to a few scattered case reports and human psychopharmacology studies looking at the potential use of selective serotonin reuptake inhibitor (SSRI) medications. A decade from now this chapter may expand into its own textbook covering a multitude of different pharmacological treatment strategies considered appropriate for use in aviation and aerospace environments.

This psychopharmacological discussion will focus on medication use by pilots and other individuals in direct control of aircraft. With the global expansion of civil aviation and the dawning of space tourism it is clear that individuals being treated with a wide range of psychopharmacological agents will be exposed to the aerospace environment.

One might ask why the use of psychiatric medication by pilots is being considered. Shouldn't they just be grounded or replaced with pilots free of mental health conditions? This was the approach that dominated aeromedical regulatory policy for decades. However, as these medications became readily available, resourceful aviators were able to obtain the medications and use them without adequate supervision (Sen, Akin, Canfield and Chaturvedi 2007). Additionally, the military and civil aviation regulatory authorities found that replacing pilots who were prescribed psychopharmacological treatments was not quick, easy, or cost effective (McKeon, Persson, McGhee and Quattlebaum 2009). Eventually it was realized that a change in the policy related to the use of these medications in aviation medicine was needed.

After all, in the non-aviation civilian world these medications were revolutionizing the management of psychiatric illness and returning many patients to full pre-morbid levels of function. Rigid thinking that banning the use of these medications would somehow ensure aviation safety or prevent pilots from using the medications without oversight is rather naive and perpetuates the stigma of mental illness. Some experts assert that psychiatric medication in general is too risky (for example, potentially turning people into dangerous individuals; Breggin 2008). This viewpoint ignores the reality that untreated mental illness may represent a much more serious risk to aviation safety than the use of medications.

The prohibition against the use of medications for treatment of mental health conditions has encouraged pilots to deny symptoms of psychiatric illness. In the past, a pilot admitting to mental health symptoms believed that this would most certainly lead to a loss of their medical certificate. As a result, aviators simply denied symptoms and found ways to surreptitiously obtain and use prohibited medications. There is unfortunately no method for detecting psychiatric illness in aviators who deny mental health symptoms, unless symptoms are severe. Consequently, it is better to accept that pilots are a unique population and understanding their attitudes and behaviors is critical in helping them access mental health services when needed (Jones 2002).

The treatment of mental health disorders in pilots ultimately comes down to a basic question. Which is of greater concern: the untreated mental health disorder, or the potential side-effects of the psychiatric medication? The answer is that neither is acceptable. Untreated mental illness can dramatically impair a pilot's ability to safely operate an aircraft. Some forms of psychiatric medication can also produce significant cognitive side-effects. Therefore, it is important to limit our discussion of psychopharmacology to only those medications that, when used to treat relatively mild forms of mental illness, allow the pilot to perform at a normal level of functioning and which are unlikely to produce sudden incapacitation during flight.

Aeromedical regulatory agencies are justifiably conservative and have developed lists of specific psychiatric conditions that are permanently disqualifying. These are conditions that either render the pilot too incapacitated to safely operate an aircraft or have a chronic progressive course. It should be noted that any serious or untreated psychiatric illness is always grounds for medical disqualification. For example in the US the Federal Aviation Administration (FAA) has established the following four psychiatric conditions (or a history of the condition) as permanently disqualifying for current pilots and those applying for new medical certification (psychosis, bipolar illness, substance dependency, and severe personality disorder; FAR part 67.107; please also see Chapter 4, this volume for more on mental health and aviation). This effectively limits our discussion on pilots to a few select classes of medication that are both effective at treating relatively mild mental health conditions and for which evidence exists that the medication does not impair aviation-critical cognitive or psychomotor abilities (for example, use of certain SSRI medications; Paul, Gray, Love and Lange 2007).

It is critical to state that all potential treatment modalities be initially considered in addressing a mental health disorder. Medication should never be considered the only treatment nor should it necessarily be the initial treatment when other non-pharmacological treatment options are available. There is now robust evidence suggesting that psychotherapy, such as cognitive behavioral therapy (CBT) can be just as effective as medication in treating many common forms of psychiatric illness (Roshanaei-Moghaddam et al. 2011). Another example is a recent study showing that interventions such as exercise can be as effective as the SSRI antidepressant medication sertraline in the management of major depressive disorder (MDD) (Hoffman et al. 2011). Therefore, it is recommended that prescription of psychotropic medication to aircrew be made only after it has been established that medication is the most effective (and/or the only effective) treatment available and that other (non-pharmacological) modalities have either failed or are not indicated for treatment of the condition.

Overview of Psychiatric Medications in Aviation

It is helpful when beginning a discussion on the use of psychiatric medication in aviation to look at analogue populations that share similarities to pilots. The best analogue available presently would be driving and drivers since they both require much of the same skill set. In 2000 the US National Transportation Safety Board issued a recommendation to the Department of Transportation for the development of a list of "safe medications" for use by drivers. In response to this recommendation, the National Highway Transportation Safety Administration convened an international panel of scientists including experts in toxicology, epidemiology, pharmacology, and behavioral scientists (that is, driving assessment experts) to consider development of the list (Kay and Logan 2011). There was agreement among the panel that a limited number of drugs generally pose a low risk to driving when taken according to approved prescribing information and when used with appropriate medical oversight. However, the panel also recognized that even among drugs generally considered safe for driving, adverse reactions may occur, and interactions may occur between these medications and other drugs or alcohol that could impair driving performance. The panel was of the opinion that some specific drugs and drug classes are clearly impairing, including sedatives, hypnotics, sedating antihistamines, narcotic analgesics, hallucinogens, antipsychotics, and muscle relaxants, even at therapeutic or sub-therapeutic doses. However, for many drug classes, the question of impairment falls in a grey area, either because the potential adverse effects are highly dose-dependent or because there is much heterogeneity within the class. For some of the impairing drugs in this group, it is likely that drug tolerance alone determines the extent of impairment. Examples of drugs that fall in this category include many SSRI antidepressants, anticonvulsants, and antihypertensives. There was consensus among the panel that no definitive list of non-impairing, "safe," substances could be prepared, because of varying effects from dose, tolerance, drug combinations, latency between dosing and driving, and metabolic differences. However, the panel did propose a structured, standardized protocol for assessing the driving impairment risk of drugs. Specifically, they designed a "tiered, parallel process, consisting of pharmacological/toxicological, and epidemiological reviews, and a standardized behavioral assessment." Applied to aviation, this would provide a more transparent and defensible approach to evaluating the suitability of medications (of all types) for use in aviation.

In addition to considering the medications themselves, there are a limited number of psychiatric conditions for which an aviator would be considered suitable for medical certification (with or without an aeromedical waiver). This allows for the elimination of a great number of potential psychiatric medications from this discussion. Medications that are normally used to treat serious psychiatric illness (for example, psychosis) can be eliminated from this discussion. Among the medications to be excluded from discussion would be antipsychotic medications (including newer generation atypical antipsychotics), mood stabilizers, older-generation antidepressant medications, and dementia medications. All of these medications are used to treat conditions deemed far too unsafe for aeromedical waiver. This raises another common concern regarding the broad use of modern psychiatric medication for non-psychiatric illness. For example, Bupropion, a common antidepressant medication is known to be useful in helping patients stop smoking (Culbertson et al. 2011), carries a label indicating its use for smoking cessation. This opened the door for anyone who wanted to take this medication to obtain the drug by asking their primary care provider for something to help them quit smoking. This "backdoor" to the medication made it possible to be prescribed an antidepressant medication for a non-psychiatric condition. Similarly, SSRI medications are prescribed for a number of health conditions including irritable bowel syndrome (IBS; Tack et al. 2006) and premenstrual syndrome (Shah et al. 2008) further opening the door to the use of psychiatric medications for non-psychiatric indications. This ability to be prescribed psychiatric medication for non-psychiatric indications can make it difficult to know the true reason a person is being treated with antidepressant medication. Aeromedical regulatory agencies have disqualified aviators taking these prohibited medications even if they had been prescribed the medication for a non-psychiatric condition. In summary, any medication used in aviation medicine or psychiatry must not have significant effects on a pilot's ability to safely operate their aircraft, regardless of the condition for which the medication is being prescribed.

Another major area of concern involves a tenet of basic pharmacology: medications can have idiosyncratic effects. A drug which is highly effective in one patient may be quite problematic in another. As the science of pharmacology advances we now appreciate that there are both environmental as well as genetic factors that shape a patient's individualized response to a medication. Concerns that need to be addressed include different responses based on ethnic background, gender, age, and state of physical condition. All of these variables come into play when trying to understand whether a particular medication would be safe to treat aviators. Unfortunately, data obtained from general psychiatric populations participating in clinical research trials may not be suitable for determining the safety of use of the drug for aviators in general or for a particular aviator. Therefore, it is generally well accepted that, even for drugs which may be allowed, it is essential to first assess the aviator's subjective and objective response to the drug before issuing a waiver. Pilots are warned against the use of any medication during flight operations until they have determined that they personally tolerate the drug under non-flight conditions.

In looking at various medications suitable for aviation a final consideration is the flight mission. Commercial aviation operations by definition are for-profit undertakings aimed at earning income for both employees and owners. These activities are essential in daily commerce and without aviation the economy would quickly grind to a halt. There are also non-profit aviation activities that are quite critical including weather monitoring, traffic reporting, and law enforcement; all of which are important public safety activities. However, it is in military operations where aviation is pushed to the limit. Lives (and national defense) depend on pilots being able to complete missions safely and effectively. Without aviation assets most military conflicts such as the first Persian Gulf War (also known as the 100-hour war) would have been protracted. Military aviation is one aspect of modern aviation that places extraordinary demands on pilots. War fighters must be able to perform at peak levels for sustained periods of time and take risks that would be deemed too dangerous for civilian operations. Therefore, considering the exceptional demands placed on the military aviator and the need to maintain flight crew readiness, the determination of the suitability of psychiatric medications for use by these aviators should be viewed as distinct from civil aviation (for more information on military aviation and psychopharmacology see Chapter 6, this volume).

Basic Psychopharmacology

Neurons are the target of most psychiatric medications. They can be controlled to fire with less stimulation or more stimulation depending on their state of polarization. This is conceptually important since it helps providers to understand how certain drugs can speed things up in the brain while others can slow things down by influencing how much of a signal is needed to fire the neuron. Nerve cells, like all cells in the body, are surrounded by a cellular membrane. This membrane protects the cell and also contains various receptors that are specific for different chemical agents (including drugs). Once activated by a specific substance these receptors can cause immediate changes in the firing of cells or by changing how the cell functions overall. Most psychiatric medications are believed to act through each of these pathways depending on their structure and receptor binding affinities. This explains why some psychiatric drugs produce immediate effects while others take weeks to show a response. Understanding these basic cellular principals can help one to better understand why medications have their differing effects on mood and behavior.

Classic models assert that mental illness is due to structural abnormalities in the brain, neuronal dysfunction, and/or neurochemical imbalances. However, recent research, for example into the genetics of mental illness, has shown that these models are overly simplistic and fail to take into account the complexity of mental illness (Kohli et al. 2011). Regardless, current treatment of mental illness is based on restoring balance to several key neurotransmitters believed to be central in the pathogenesis of psychiatric illness. Most of this work dates back to the monoamine hypothesis of depression and the dopamine hypothesis of schizophrenia (Howes and Kapur 2009). These key transmitters include norepinephrine (NE), serotonin (5HT), and dopamine (DA). It was thought that psychiatric medications worked by simply changing the synaptic concentrations of these neurotransmitters in key areas of the brain. This is a helpful model but again overly simplistic. The model does explain some of the side-effects that are seen when these substances activate receptors outside the central nervous system (CNS). For example gastrointestinal (GI) distress can be caused by activating serotonin receptors found in the GI tract. It is now believed that the key neurotransmitters merely act as intermediate steps in more complex changes at the cellular level. The precise mechanism of action of most psychotropic medications has yet to be fully determined.

The principles of both pharmacokinetics (that is, what the body does to a drug) and pharmacodynamics (that is, what a drug does to the body) must be considered when reviewing a particular medication and its application to aviation. Medications can be delivered to the human body through several different routes. This includes injection into muscle (IM) or blood stream (IV), oral ingestion (PO), or finally through direct contact with a mucous membrane or by diffusion through skin (TD). The mechanism of entry is important since it can control how quickly the medication is absorbed and transported to its intended site of action as well as having an impact on how the drug is metabolized. The time it takes from administration of a medication until the desired effect occurs is termed time of onset. Drugs are designed with this in mind so that they can have varying speed of onset. Psychiatric medication needs to enter the blood stream and make its way to the brain in order to exert an effect on the CNS. The chemical structure of the medication can directly influence its ability to produce effects since it must pass through the blood–brain barrier in order to reach the targeted brain cells. Once a medication enters the body it is treated like any other substance and it starts to be eliminated. This can be as simple as a one-step process or may involve a very complex multi-step process that can take weeks. The time it takes the body to remove one half of the original concentration is termed the drug's half-life and is important in discussing how long a particular drug will likely remain in the body and exert its effects. It is generally accepted that a drug no longer has a meaningful effect on the body after a period of five half-lives has elapsed. Again this is important since drugs with long half-lives may remain in the body for months. The other important consideration is that once a drug starts to be metabolized (that is, processed for elimination) it may form a number of metabolites before it is completely eliminated. These metabolites can be inactive or may be even more active than the parent drug itself. Therefore, a drug with a long half-life and many active metabolites can have a long duration of ac'ion, perhaps days or even weeks after it is administered.

Adding to this complexity is the impact of drug interactions. Particularly in older adults polypharmacy is the rule, not the exception. Drug interactions can have an enormous impact upon metabolism and can be responsible for dangerous side-effects (for example, cardiac arrhythmias). This is the primary reason for trying to limit aviators to monotherapy (that is, use of a single agent for a particular indication). Multi-drug cocktails, although commonly used in clinical psychiatry, are avoided in aviation psychopharmacology. Unfortunately, patients generally fail to recognize that over-the-counter (OTC) medications are also drugs, and these medications, when combined with prescribed psychiatric medication, can result in serious and unanticipated side-effects which could impair aviation operations. This can occur because both the prescribed drug and the OTC drug may be competing for the same exit pathway from the body. As a result toxic metabolites from either of the two drugs can build up and cause aeromedically significant side-effects. Another reason for limiting pilots to monotherapy is to minimize the possibility of providing a waiver for a more serious psychiatric illness that requires polydrug therapy.

Another consideration in aviation medicine relates to the use of older drugs versus newer drugs. A case can be made for using the newest, state-of-the-art, psychopharmacological treatments. However, it is also important to understand that most new medications, when first approved, will have limited time on the market to reveal side-effects that were not seen in the limited pivotal clinical trials. After several years and millions of patient treatment days it is easier to determine if a drug is likely to generate side-effects incompatible with aviation safety. This helps explain why most aeromedical regulatory agencies limit psychopharmacological treatment to older drugs with a long track record or at least require a minimum number of years on the market before considering a drug suitable for a waiver request.

Go/No Go Pills

In beginning the discussion of specific classes of psychiatric medication in aviation it is helpful to start with some historical background. The use of performance-enhancing drugs dates back centuries. In warfare having a soldier fight longer and more aggressively could easily sway the battle and win the conflict. Amphetamines were first synthesized in the later part of the nineteenth century and human experimentation began shortly thereafter. It was soon discovered that this class of medication could reduce fatigue, prolong wakefulness, and enhance performance. World War II (WWII) saw the first large-scale use of these medications in military operations with pilots frequently using these medications to complete various long-duration missions (Rasmussen 2009). The US military has continued the use of these medications commonly referred to as Go pills by the pilots who are given them. In military aviation there are many high-value missions that require prolonged wakefulness for successful completion. In these high-stakes circumstances pilots may be given stimulants when the benefits outweigh the risks. Upon their return, if the pilots have difficulty falling asleep, they may be given sedative hypnotics or No Go pills to induce immediate sleep. This practice went essentially unnoticed by the general public until a tragic event occurred during the first Persian Gulf conflict. The Tarnak Farms incident which occurred on 17 April 2002 resulted in the friendly fire bombing deaths of four Canadian service members and the injury of eight others (Friscolanti 2005). The lawyers defending the pilots asserted that the sedating medication taken by the two airmen had resulted in impaired judgment.

This tragic occurrence illustrates the possible hazards associated with using psychoactive medication in aviators. The increased public awareness from this event and the traditional conservative stance by aeromedical regulatory agencies makes it clear that any use of psychiatric medication must be undertaken with caution.

Psychostimulants

Psychostimulants can be thought of as medications which increase brain neuronal activity through a variety of CNS mechanisms. This class of medication generally has a relatively quick onset of action and a short half-life. The most commonly used psychostimulant in the world is generally not even considered to be a drug but rather a food additive. Caffeine is found commonly in coffee, carbonated beverages, energy drinks, and in OTC tablets. Since the introduction of Red Bull in 1987 the use of energy drinks has skyrocketed (Persad 2011). This multibillion dollar industry has promoted the use of these caffeinated beverages worldwide. Given this universal access it is not uncommon to see aviators consuming these beverages sometimes to excess while engaged in flight operations. Studies have shown that caffeine increases alertness, improves self-reported energy and mood, and enhances reaction time and response consistency in fatigued or sleep-deprived individuals (Childs and de Wit 2008). Caffeine is used ubiquitously throughout the aviation world. Unless taken in excess caffeine has a wide therapeutic window and only becomes toxic at extremely high doses. The US Food and Drug Administration (FDA) has reported that consuming up to 300mg of caffeine a day can be considered safe (Temple 2009). Most energy drinks contain at least as much caffeine as a cup of coffee (roughly 100mgs per serving). However, it is important to look closely at the product label since caffeine content is often stated for the 'per serving' amount and there may be multiple servings in a single beverage container. Some energy drinks exceed the FDA recommendations and contain more than 300mg of caffeine per single serving. Even though caffeine is quite socially accepted it should not be viewed as a replacement for sleep. If used, caffeine should be taken only when needed, preferably early in the pilot's work day when sustained wakefulness is required for mission completion. The amount of caffeine taken in any one dose should be the minimum necessary to improve wakefulness with minimal side-effects. Prolonged, elevated use of caffeine such as continuous daily consumption of energy drinks can result in tolerance and significant withdrawal side-effects when stopped suddenly. The insomnia which can result from excess use of caffeine can lead to sleep deprivation which may result in serious psychiatric conditions. Although it is considered relatively safe and socially acceptable, caffeine use needs to be factored into any mental health evaluation. A pilot may be using caffeine to self-medicate a more serious underlying disorder (for example, excessive daytime sleepiness) or recklessly abusing caffeine for perceived performance enhancement.

Amphetamines have been available for decades and have been used for many different psychiatric indications. Today they are commonly used to manage Attention Deficit Hyperactivity Disorder (ADHD) and they are also widely prescribed for weight loss (particularly phentermine). This class of psychostimulants produces immediate effects through the release of large amounts of excitatory neurotransmitters primarily norepinephrine and dopamine from brain neurons. This in turn results in increased blood pressure, faster (and more consistent) reaction times and increased vigilance. Although operationally helpful in the short run for fatigue management, the side-effects associated with long-term abuse are horrific as exemplified by the current methamphetamine epidemic in the US. The military recognized the value of amphetamines for use in sustained operations (SUSOPS) where personnel were required to maintain long periods of wakefulness. Under these conditions, the benefits of the use of amphetamines or Go pills may outweigh their risks (Caldwell, Caldwell, Crowley and Jones 1995). Newer wakefulness-promoting agents (for example, modafinil) appear to be particularly promising in aviation medicine and may eventually replace amphetamines.

There has been some aeromedical discussion regarding the use of amphetamines in pilots who suffer from ADHD. In Germany aviators can receive a waiver for the use of stimulants to treat ADHD. However, in the US and elsewhere there is significant concern that ADHD may not be compatible with aviation operations due to the impact on aviation-related neurocognitive functions (that is, vigilance, impulsivity, inattentiveness). Although medication has been shown to be highly effective, for example in improving (or normalizing) the driving performance of young adults with ADHD (Kay et al. 2009), there is concern regarding treating the disorder on an ongoing basis with medication that has a short half-life (and therefore is dependent upon daily compliance) and which may cause insomnia when taken too close to bedtime.

Newer wakefulness-promoting agents represent possibly the most helpful class of psychostimulant medication. They are designed to increase wakefulness and are believed to act by stimulating central alpha 1 adrenergic brain neuron receptors (Minzenberg and Carter 2008). Modafinil has been used operationally in the military and has been shown to be effective at improving wakefulness and to be no more dangerous than other more traditional psychostimulants (Estrada et al. 2012). It is important to point out that excessive daytime sleepiness (EDS) may be a symptom of a more serious underlying medical condition (for example, obstructive sleep apnea). Modafinil and armodafinil are not currently permitted for use by aviators in the US. If these medications are permitted for use by pilots it will be critical that a detailed medical history, preferably by a sleep specialist, be obtained before using either of these agents to ensure that it is not being prescribed to treat a more serious medical or psychiatric condition.

Psychostimulants have obvious value in aviation for the short-term management of fatigue and sleepiness from prolonged operational activities. However, they have potential for aeromedically significant CNS and cardiovascular side-effects. They should never be used as a substitute for rest and sleep. In military operations they may be suitable for specific high-value missions, however, in civilian aviation there should be no need for these psychoactive drugs with proper flight crew scheduling and fatigue management (for more on fatigue and aviation, see Chapter 9, this volume). Regarding the specific use of amphetamines in the chronic management of ADHD in pilots it appears that further research is necessary before a sound aeromedical decision can be made regarding their suitability.

Nicotine deserves special mention in this section on psychostimulants. Even with antismoking efforts tobacco use is still quite wide spread among pilots. Whether smoked, chewed, or snorted tobacco introduces nicotine into the body and produces a wide variety of CNS effects. Nicotine has been shown clinically to produce different effects on the cognitive performance of users and non-users (Newhouse, Potter, Dumas and Thiel 2011). One interesting and relevant finding is that nicotine use in active smokers enhances arousal while simultaneously impairing working memory (Ashor 2011). When smokers are not permitted to smoke, such as when they are flying, they can suffer from degraded cognitive performance secondary to nicotine withdrawal (Atzori et al. 2008). This decline occurs in as little as 12 hours, or in less than the length of many international flights (Giannakoulas et al. 2003). Furthermore, the use of nicotine in pilots can compound the use of other psychoactive agents and represents yet another source of psychopharmacological concern. At this point further study is needed on how to manage nicotine-dependent aviators safely, since restrictions on access to nicotine could result in a significant cognitive side-effect and degradation of aviation safety.

Central Nervous System Depressants

The class of drugs referred to as CNS depressants includes tranquillizers (for example, barbiturates), anxiolytics (for example, benzodiazepines), older generation antihistamines (for example, diphenhydramine), and sedative-hypnotics (for example, zolpidem). These medications decrease, or slow down, brain neuronal activity. When this happens the usual result is sedation and cognitive impairment. While some of these medications have a fairly clear mechanism of action (for example, antihistamines and benzodiazepines) others work through multiple neurotransmitter systems (for example, ethanol; Valenzuela 1997).

Ethanol is undoubtedly the most widely-used CNS depressant by aviators. Early fliers described use of alcohol to calm their nerves. Early flight surgeons prescribed alcohol to help their aviators cope with the loss of fellow pilots, manage sleep problems, and for treating "jitters" (Guly 2011). Even without the "assistance" of a flight surgeon, pilots have long used alcohol to self-medicate for a variety of psychological conditions; particularly since the use of other CNS depressants was prohibited for use by pilots. Aeromedical regulatory agencies have established strict "bottle to throttle" rules (for example, the eight-hour rule, FAR 91.17) in an attempt to keep alcohol out of the cockpit.

Ethanol is eliminated from the body through zero order kinetics which means that for most people it is metabolized at the rate of 1 ounce (of 100 proof) per hour (that is, 15ml of 100 percent ETOH per hour). A pilot can abide by the eight-hour rule yet still have a Blood Alcohol Concentration (BAC) that would land them in jail. The diagnosis, treatment, and the waiver program for pilots recovering from alcohol dependence (alcoholism) is the topic of Chapter 5 in this volume.

The next two popular groups of CNS depressants include the benzodiazepines and barbiturates. These medications are incompatible with flying due to their adverse effects on wakefulness, alertness, and cognitive functioning. Barbiturates are rarely used today except in the treatment of certain types of seizure disorders. Benzodiazepines are a newer class of medications than the barbiturates. Benzodiazepines have proven useful in the short-term management of anxiety and anxiety-related disorders (for example, panic attacks). However, patients treated with benzodiazepines quickly develop tolerance to the medications. To produce the same result higher doses are required over time. These drugs are commonly prescribed in urgent care settings by non-aeromedically trained providers. Therefore aviators need to be educated about the danger associated with these medications. This class of drugs is detected by most drug surveillance programs. A pilot found to be using benzodiazepines or barbiturates is likely to lose his/her license.

The newer class of short-acting "sedative-hypnotics" appears to have a potential application in aviation medicine. These ultra-short half-life medications have minimal residual CNS sedation and have been used in the military setting for sleep induction following mission completion. Also known as No Go pills they have been deemed safe enough that NASA has allowed their use by orbiting astronauts not tasked with flight duties. There appear to be differing opinions with regard to the use of these medications in aviation (for example, Sicard et al. 1993; Wesensten, Balkin and Belenky 1996). Zolpidem (brand name Ambien) was one of the first of these new sedative-hypnotic medications. Its mean elimination half-life is 2.6 hours, which indicates that most of it has left the body in 12 hours. The FAA Aeromedical Certification Division has previously allowed the limited use of this sedative under the following conditions: "The Federal Air Surgeon's Medical Bulletin states that Ambien [zolpidem] may be used if no more than twice a week and not within 24 hours of flight duties" (Aviation Medical Advisory Service, 2013).

In summary, CNS depressant use is not compatible with aircraft operations. The one exception may be the use of short-acting sedative-hypnotics for occasional treatment of acute insomnia not due to a more serious underlying psychiatric or medical disorder.

Over-the-Counter Medications and Nutraceuticals

Over-the-counter (OTC) drugs and nutritional supplements (nutraceuticals) may represent a generally unrecognized risk in aviation medicine. Consumers are promised cures to afflictions that challenge Western medicine. To avoid losing their medical certificate due to use of prohibited drugs or documentation of a disqualifying condition, pilots have attempted to treat their condition with less tested (and potentially unsafe or ineffective) forms of treatment. We have already discussed caffeine and alcohol which are frequently used for management of mood symptoms and sleeping disorders. Nutraceuticals can have detrimental effects when taken in combination with prescription medications, most often by altering drug metabolism (for example, St John's Wart). Of greater concern is the lack of demonstrated efficacy and safety for some of these alternative medicines. Pilots should be cautioned when combining their prescribed medication with alternative medications, treatments, or nutritional supplements.

One of the more popular alternative medicine treatments is melatonin, a natural substance produced in the pineal gland that can be very helpful in regulating circadian rhythms (Dillie 1996). Melatonin has been studied quite extensively and has been shown to produce a variety of different beneficial CNS effects in addition to sleep regulation (Carpentieri et al. 2012). However, it should be treated as any other psychoactive substance by pilots and only used under the strict guidance of an aviation medical specialist when allowed by aeromedical regulations. A once common nutraceutical, L-tryptophan, was also shown to be useful in promoting sleep onset and for use as an antidepressant. Prior to 1989, l-tryptophan had become a popular OTC nutritional supplement. However, its popularity disappeared following an outbreak of what was believed to be l-tryptophan-related eosinophilia myalgia syndrome in 1989 which caused 1,500 cases of permanent disability and at least 37 deaths. Another dietary supplement, Vitamin D, has shown some promise in helping those who have low vitamin D levels who also suffer from depression (Hoang et al. 2011).

In aviation medicine there is concern when alternative medicines are used in order to avoid receiving proper treatment for a mental health disorder that would be otherwise disqualifying for a medical certificate. This is reflected in a case discussed in the FAA 2003 Quarterly Bulletin. AMEs were instructed that, "the FAA does not prohibit the use of most herbal remedies; however, since St. John's Wart can be used to relieve the symptoms of 'depression,' prior to issuing a medical certificate you should question the airman about the signs/symptoms of this condition. Should the airman give a positive history, I would not grant issuance and suggest an evaluation for depression" (Silberman 2003). The policy of the US Air Force (USAF) is to restrict the use of any OTC supplement in safety-sensitive personnel. For obvious reasons, the military has a much lower tolerance for experimentation by aviators with untested remedies. In the aviation world it is very important to appreciate the widespread use of these interventions. It is recommended that clinicians take detailed histories when working with aviators and exercise caution when recommending treatments that have limited scientific research to back their claims.

One of the most commonly used OTC products is sedating antihistamines. These medications are prohibited for use by pilots. However, aviators are often unaware that diphenhydramine, the active ingredient found in Benadryl, common cold and allergy medications, nighttime pain relievers, and most OTC sleep aides, is a sedating drug prohibited for use by pilots. Sadly, OTC sedating antihistamines are among the most common medications found by the FAA's Civil Aviation Medical Institute (CAMI) when performing post-morbid toxicology testing following fatal aviation crashes. A 50mg dose of diphenhydramine, which is 50 percent less than the amount of diphenhydramine present in an adult dose of Excedrin-PM, impairs driving performance to the same extent as alcohol at a 0.07 BAC level (O'Hanlon and Ramaekers 1995). This is a BAC a level which is beyond the legal limit in Europe, Canada, and Japan. Furthermore, roughly two-thirds of individuals receiving a 50mg dose of diphenhydramine report not feeling sleepy, even though they are impaired on aeromedically relevant cognitive and psychomotor tasks (Kay et al. 1997).

Antidepressant Medications

It is unclear if the prevalence of MDD (*American Psychiatric Association 2000 DSM-IV-TR*) in aviators is similar to that found in the general population due to under-reporting of symptoms by pilots. Current statistics assert that MDD has a lifetime prevalence of approximately 16 percent and a one-year (point) prevalence of roughly 6 percent (Ebmeier, Donaghey and Steele 2006). If these numbers are applied to the 620,000 holders of FAA pilot certificates (FAA Civil Aerospace Medical Institute), it is clear there should be quite a few depressed pilots. However, pilots tend to not report anything that will put their medical certificate in jeopardy. In the case of MDD, for which there is no lab test or scan, there is significant concern. Patients who suffer from MDD usually have other affective symptoms such as anxiety or panic attacks that co-occur along with the depression which only worsen the person's disease state (Kessler et al. 2003).

Historically MDD has been managed with a combination of psychotherapy and medication. The first medications developed were those based upon the monoamine hypothesis of depression. These early medications include monoamine oxidase inhibitors (MAOIs) and tricyclic antidepressants (TCAs). These are effective treatments for MDD, however, their side-effects render them unsuitable as maintenance therapy for depressed aviators. In the past, pilots would be diagnosed with depression, prescribed one of these medications, and treated for a fixed period of time. Once it was determined that the MDD episode was in full remission, the medication was discontinued, and then, after several months the pilot was eventually re-evaluated and if found to be free of depression was issued a medical certificate. This process often took years. During this time the pilot was not able to perform their occupation or enjoy general aviation flying. As a result, pilots often flew while depressed, self-medicated with non-efficacious "remedies," received surreptitious treatment (off the books), and failed to disclose their condition to their AME since the diagnosis of MDD could potentially ruin their flying careers.

In their landmark paper, Jones and Ireland (2004) tackled complex questions related to the use of the new class of antidepressant medications, SSRIs, in the aeromedical management of psychiatric illness. The paper was cited by the FAA as strong supporting evidence in establishing the policy announced in 2010 (Federal Air Surgeon's Medical Bulletin 2010) regarding the maintenance use of SSRI medications by civil aviation pilots with all classes of medical certification in the US. The Jones and Ireland paper proposed a process by which SSRI medications could be introduced in a controlled fashion to manage pilots who needed them for maintenance treatment of MDD.

An entire chapter could be written about these newer antidepressant medications. When considering a specific antidepressant medication for use in the maintenance treatment of pilots it is helpful to use the basic conceptual template presented in this chapter. Given the rapid development of SSRI and related medications it is clear that each one will need to be considered individually with regard to its suitability for use in the aviation environment. Key points include low potential for sedation, minimal side-effects when used in maintenance therapy, minimal withdrawal side-effects if doses are missed, minimal or no drug–drug interactions, no long-term medical dangers with prolonged use, no masking of underlying medical conditions, and finally a demonstrated track record of safety, tolerability, and effectiveness. This may seem like an impossible set of requirements. However, several medications meet most of these criteria. These include sertraline, citalopram, escitalopram, and fluoxetine which are now listed as being eligible for aeromedical waiver (for example, FAA Special Issuance) by aeromedical regulatory agencies.

Current Aeromedical Policy on Psychiatric Medication

Aeromedical regulatory agencies must carefully consider many diverse variables when approaching the use of psychiatric medication by aviators. There are a number of factors that go into formulating new aeromedical regulations. These policies are driven by advances in medical (and psychiatric) care including the introduction of new medications, legal issues (for example, court decisions regarding the use of drugs such as Lithium; Bullwinkle Case 1993), pressure from pilot advocacy groups, and recommendations from aviation medicine organizations (for example, Aerospace Medical Association and Civil Aviation Medical Association). As one can imagine, this is a difficult process since no one wants to take the lead in allowing a medication into the cockpit that is later found to be linked to a tragic outcome. The public's lack of awareness and knowledge of mental illness is another factor. When the FAA announced the SSRI Special Issuance program there was intense criticism regarding permitting depressed pilots back in the cockpit. Given these factors, it is understandable that regulatory agencies have found it very difficult to proceed with development of programs supporting maintenance use of psychiatric medication by pilots.

There are two primary driving forces behind the change in policy regarding the maintenance use of antidepressant medication. The first driving force is the need to retain trained aviation assets (that is, pilots) and the second force is the evidence demonstrating the safety and efficacy of certain SSRI antidepressants. Australia can be given credit for launching the first large-scale study of the use of newer antidepressant medications in commercial aviators in 1987. Based on their findings the Australians formulated what may be considered a "best practices" model for treating MDD in flight crew and air traffic control (ATC) personnel. During a 10-year follow-up study (1993–2004) of 481 pilots who had been permitted to use specified SSRI medications, there was no increase in accidents or mishaps that could be attributed to the use of the antidepressant medications (Ross, Griffiths, Emonson and Lambeth 2007). Subsequently, the US Army began to consider waiver requests for aviation rated personnel on active flight status who were prescribed certain antidepressant medications for maintenance therapy of depression (McKeon, Persson, McGhee and Quattlebaum 2009).

Table 12.1 summarizes the current policies regarding use of antidepressant medications for four regulatory agencies. The agencies differ with respect to their lists of allowed medications, personnel to whom the regulation applies, duration of grounding, requirements for testing, and qualifications of examiners.

The FAA regulates the greatest number of aviators for whom Table 12.1 would apply. There is considerable potential for serious political fallout if something should go wrong and therefore it is understandable that the FAA has formulated a conservative policy. The most recent revision of the policy, available in the FAA's *Guide for Aviation Medical Examiners* (2013), reduces the burden and cost to the aviator. The pilot is required to have been clinically stable on a stable dose of medication for 12 months prior to applying for a Special Issuance medical certificate. The aviator provides a written statement describing his/her history of antidepressant use and mental health status. The aviator is required to provide medical/treatment records related to his/her history of antidepressant use. An evaluation report is to be submitted by their treating provider. If the treating provider is not a psychiatrist then a report of a board certified psychiatrist must

Table 12.1 Summary of Current Aeromedical Regulatory Policies

Country Agency	**Australia Civil Aviation Safety Authority**	**Canada Civil Aviation Authority & Military**	**USA US Army**	**USA Federal Aviation Administration**
Psychiatric Diagnosis	Major Depressive Disorder	Major Depressive Disorder	Major Depressive Disorder, Generalized Anxiety Disorder, Post Traumatic Stress Disorder	Mild-Moderate Major Depressive Disorder, Dysthymic Disorder, Adjustment Disorder with Depressed Mood, Any non-depression related condition for which SSRIs are indicated
Approved Medication	Sertraline Citalopram Venlafaxine	Setraline Bupropion Citalopram	Newer medication for depression FDA approved for at least three years, dual therapy with bupropion	Citalopram Escitalopram Sertraline Fluoxetine
Minimum Required Grounding	Four weeks stable on medication	Six months stable on medication	Four months stable on medication	12 months stable on medication
Personnel Covered by Policy	Pilot with or as Co-pilot, Air Traffic Control Operator	Civilian: Pilot with or as Co-pilot Military: (non tactical) Pilot with or as Co-pilot, Aircrew, Air Traffic Control Operator	Pilot, Flight Surgeon, Aviation Physician Assistant, Non-rated Aircrew, Air Traffic Control Operator	1st, 2nd, 3rd Class medical certificate holders
Required neuropsychological testing	None Required	CogScreen-AE	CogScreen-AE, Simulator, In flight test	CogScreen-AE
Required Clinical Monitoring	Expert in treating major depressive disorder	Psychiatrist	Aviation psychiatrist/ other qualified provider	Human Intervention Motivation Study Aviation Medical Examiner
Follow-up Interval	Every month for one year	Every six months	Every six months	Per Special Issuance requirements for certificate class

be submitted. The aviator must also submit a report from a neuropsychologist on the results of CogScreen-AE. If the aviator's performance on CogScreen-AE reveals possible abnormalities the neuropsychologist conducts additional testing to determine if there are aeromedically significant neurocognitive findings. The application is then submitted by a HIMS-certified AME. The FAA's SSRI Special Issuance Program has been viewed by aviators as costly and burdensome. The number of pilots seeking Special Issuance under the program is believed to be a marked under-representation of those who take SSRI medications and those who would benefit from taking these medications.

Summary and Future Considerations

This chapter has been designed to provide a basic conceptual framework that can be applied to aeromedical waivers for psychoactive medications. Drug regulatory agencies (FDA, European Medicine Agency) employ very specific guidelines and policies for determining the efficacy and safety of drugs. While a drug may be safe and effective for a particular psychiatric indication, it may be completely inappropriate for use by aviators. The underlying condition (for example, psychosis) may be disqualifying regardless of whether or not it is controlled by medication. However, there are also conditions where treatment is effective, where the aviator can be evaluated to determine if there are residual effects of the disorder, and for which medication can be taken to maintain the aviator's mental health. For these medications aeromedical regulatory authorities need to apply procedures to determine if the medication is safe for use in aviation.

To be considered safe for use in aviation a medication must be free of CNS depressant effects. It cannot cause sleepiness, drowsiness, decreased alertness, or impact aeromedically significant neurocognitive functions. It cannot increase the risk of incapacitation or diminish aviation performance due to its effects on other bodily systems (for example, increased cardiac risk, effects on vision). There are also considerations unique to aviation. Are there interactions between the drug and hypoxia? Psychoactive drugs, by definition, cross the blood–brain barrier. Therefore, they act upon various neurotransmitter systems and affect various networks. Is there a range of doses that are more or less safe for use in aviation? As stated previously, what are the risks of poor compliance (for example, missing a dose)? Are there dangers for the pilot who discontinues treatment? What is the existing epidemiology and toxicology data regarding the use of the drug in drivers? Is the drug associated with increased risk of injuries and accidents?

This chapter has been written to provide a broad overview of the key topics relevant to a meaningful discussion of aviation psychopharmacology. The chapter has highlighted the general classes of medication that are used for treatment of psychiatric conditions in aviation medicine, reviewed basic psychopharmacological principals and summarized the current aeromedical regulatory policies regarding the use of psychiatric medication. In examining the subject it is clear that there is limited data specific to aviation however, analogue populations such as drivers provide evidence that with careful monitoring pilots can benefit greatly from treatment of mental illness with modern psychiatric medication. Up until recently the lack of specific aeromedical policy has driven pilots to use both prescription and non-prescription medication with limited supervision, sometimes causing dire results. Neurobiological research continues to provide us with a better understanding of how psychiatric medications exert their effect on the CNS. Armed with this information there have been great advances in the development of psychiatric medication that is compatible with aviation activities and could safely be used to treat pilots who suffer from minor types of mental illness not responsive to non-pharmacological forms of treatment.

It is also important that all clinicians involved in the care of pilots take detailed medical and psychiatric histories. Questions such as, "Tell me everything you put in your body during the week and weekend?" should always be asked. This information is critical given the wide spread use of OTC and nutraceutical products in our modern world. Once discovered, further follow-up on these products is critical to determine what they contain and how often they are used. As has been shown, these products can contain dangerous amounts of caffeine, sedating antihistamines, or other potentially cognitively impairing substances. Thus it is critical that aeromedical providers involved in the care of pilots understand the importance of monitoring for psychoactive medication and screen for them accordingly.

In summary, aviation psychopharmacology is a complex and promising area of aviation medical care. The recent changes in aeromedical regulatory policy have opened the door to provide pilots with very effective means of treatment for minor forms of mental illness. In making these changes aeromedical regulatory agencies are now expecting pilots to be more open with providers in disclosing symptoms. Aeromedical regulatory agencies have taken a much-needed first step by allowing pilots to use newer effective means of treatment for their mental health issues and still retain the ability to fly. Clearly, progress is being made in psychopharmacology, aviation medicine, and regulatory policies aimed at identifying safe and effective psychopharmacological treatments for aviators.

References

American Psychiatric Association. 2000. *Diagnostic and Statistical Manual of Mental Disorders—Text Revision* (4th Edition). Washington DC: American Psychiatric Association.

Ashor, A. 2011. Degree of dependence influences the effects of smoking on psychomotor performance and working memory capacity. *Neurosciences (Riyadh)*, 16(4), 353–357.

Atzori, G., Lemmonds, C.A., Kotler, M.L., Durcan, M.J. and Boyle, J. 2008. Efficacy of a nicotine (4mg)-containing lozenge on the cognitive impairment of nicotine withdrawl. *Journal of Clinical Psychopharmacology*, 28(6), 667–674.

Aviation Medicine Advisory Service 2013. Medication Classs- Sleep-inducing Medications. Available at http://aviationmedicine.com/medications/index.cfm?fuseaction=medicationDetail&medicationID=32 [accessed 7 April 2013].

Berry, M.A. 2010. New SSRI Certification Guidelines In Place apositive step forward that will be a benefit to aviation safety. *Federal Air Surgeon's Medical Bulletin,* 48(2), 1-4.

Breggin, P. 2008. *Medication Madness: The Role of Psychiatric Drugs in Cases of Violence, Suicide and Murder.* New York, St. Martin's Press.

Breggin, P.R. 2011. Psychiatric drug-induced Chronic Brain Impairment (CBI): implications for long-term treatment with psychiatric medication. *International Journal of Risk and Safety in Medicine*, 23(4), 193–200.

Caldwell, J.A., Caldwell, J.L., Crowley, J.S. and Jones, H.D. 1995. Sustaining helicopter pilot performance with Dexedrine during periods of sleep deprivation. *Aviation, Space, and Environmental Medicine*, 66(10), 930–937.

Carpentieri, A., Diaz de Barboza, G., Areco, V., Peralta Lopez, M. and Tolosa de Talamoni, N. 2012. New perspectives in melatonin uses. *Pharmacological Research*, 65(4), 437–444.

Childs, E. and de Wit, H. 2008. Enhanced mood and psychomotor performance by a caffeine-containing energy capsule in fatigued individuals. *Experimental Clinical Psychopharmacology*, 16(1), 13–21.

Culbertson, C.S., Bramen, J., Cohen, M.S., London, E.D., Olmstead, R.E., Gan, J.J., Costello, M.R., Shulenberger, S., Mandelkern, M.A., Brody, A.L. 2011. Effect of bupropion treatment on brain activation induced by cigarette-related cues in smokers. *Archives of General Psychiatry*, 68(5), 505–515.

Dillie, J.R. 1996. Melatonin: A wonder drug for jet lag, daytime sleep, better sex, and longer life? *Aviation, Space, and Environmental Medicine*, 67(8), 792.

Ebmeier, K.P., Donaghey, C. and Steele, J.D. 2006. Recent developments and current controversies in depression. *Lancet*, 367(9505), 153–167.

Estrada, A., Kelley, A.M., Webb, C.M., Athy, J.R. and Crowley, J.S. 2012. Modafinil as a replacement for Dextroamphetamine for sustaining alertness in military helicopter pilots. *Aviation, Space, and Environmental Medicine*, 83(6), 556–567.

Friscolanti, M. 2005. *Friendly Fire: The Untold Story of the US Bombing that Killed Four Canadian Soldiers in Afghanistan.* Mississauga, Ontario: John Wiley & Sons.

Guide for Aviation Medical Examiners. 2013. Available at: http://www.faa.gov/about/office_org/headquarters_offices/avs/offices/aam/ame/guide/4April2013

Giannakoulas, G., Katramados, A., Melas, N., Diamantopoulos, I., Chimonas, E. 2003. Acute effects of nicotine withdrawl syndrome in pilots during flight. *Aviation, Space, and Environmental Medicine*, 74(3), 247-251.

Guly, H. 2011. Medicinal brandy. *Resuscitation*, 82(7), 951–954.

Hoang, M.T., DeFina, L.F., Willis, B.L., Leonard, D.S., Weiner, M.F. and Brown, E.S. 2011. Association between low serum 25-Hydroxyvitamin D and depression in a large sample of healthy adults: The Cooper Center Longitudinal Study. *Mayo Clinical Proceedings*, 86(11), 1050–1055.

Hoffman, B.M., Babyak, M.A., Craighead, W.E., Sherwood, A., Doraiswamy, P.M., Coons, M.J., Blumenthal, J.A. 2011. Exercise and pharmacotherapy in patients with major depression: one-year follow-up of the SMILE study. *Psychosomatic Medicine*, 73(2), 127–133.

Howes, O.D. and Kapur, S. 2009. The dopamine hypothesis of schizophrenia: version III—the final common pathway. *Schizophrenia Bulletin*, 35(3), 549–562.

Jones, D.R. (2002). Aerospace psychiatry, in *Fundamentals of Aerospace Medicine*, 3rd Edition, edited by R.L. DeHart and J.R. Davis. Philadelphia, PA: Lippincott Williams & Wilkins, 403–419.

Jones, D.R. and Ireland, R.R. 2004. Aeromedical regulation of aviators using selective serotonin reuptake inhibitors for depressive disorders. *Aviation, Space, and Environmental Medicine*, 75(5), 461–470.

Kay, G.C., Berman, B., Mockoviak, S.H., Morris, C.E., Reeves, D.S., Starbuck, V., Sukenik, E., Harris, A.G. 1997. Initial and steady-state effects of diphenhydramine and loratadine on sedation, cognition, mood, and psychomotor performance. *Archives of Internal Medicine*, 157(20), 2350–2356.

Kay, G.C. and Logan, B.K. 2011. *Drugged Driving Expert Panel: A Consensus Protocol for Assessing the Potential of Drugs to Impair Driving (DOT HS 811 438).* Washington DC: National Highway Traffic Safety Administration.

Kay, G.C., Michaels, M.A., Pakull, B. Simulated driving changes in young adults with ADHD receiving mixed amphetamine salts extended release and atomoxetine. *Journal of Attention Disorders,* 12(4), 316-329.

Kessler, R.C., Berglund, P., Demler, O., Jin, R., Koretz, D., Merikangas, K.R., Rush, A. J., Walters, E.E., Wang, P.S. 2003. The epidemiology of major depressive disorder: results from the National Comorbidity Survey Replication (NCS-R). JAMA, 289(23), 3095–3105.

Kohli, M. A., Lucae, S., Saemann, P.G., Schmidt, M.V., Demirkan, A., Hek, K., Czamara, D., Alexander, M., Salyakina, D., Ripke, S., Hoehn, D., Specht, M., Menke, A., Hennings, J., Heck, A., Wolf, C., Ising, M., Schreiber, S., Czisch, M., Müller, M.B., Uhr, M., Bettecken, T., Becker, A., Schramm, J., Rietschel, M., Maier, W., Bradley, B., Ressler, K.J., Nothen, M.M., Cichon, S., Craig, I.W., Breen, G., Lewis, C.M., Hofman, A., Tiemeier, H., van Duijn, C.M., Holsboer, F., Müller-Myhsok, B., Binder, E.B. 2011. The neuronal transporter gene SLC6A15 confers risk to major depression. *Neuron*, 70(2), 252–265.

McKeon, J.F., Persson, J.L., McGhee, J. and Quattlebaum, M. 2009. A review of the US Army experience using selective serotonin reuptake inhibitors in aircrew. *RTO-MP-HFM-181—Human Performance Enhancement for NATO Military Operations (Science, Technology and Ethics).* Sofia: NATO Research and Technology Organisation.

Medical Standards and Certification part 67. 2012. *CFR 14, Chapter I—Federal Aviation Administration, Department of Transportation, Subchapter D–Airmen.*

Minzenberg, M.J. and Carter, C.S. 2008. Modafinil: A review of neurochemical actions and effects on cognition. *Neuropsychopharmacology*, 33(7), 1477–1502.

Newhouse, P., Potter, A., Dumas, J. and Thiel, C. 2011. Functional brain imaging of nicotinic effects on higher cognitive processes. *Biochemical Pharmacology*, 82(8), 943–951.

O'Hanlon, J.F. and Ramaekers, J.G. 1995. Antihistamine effects on actual driving performance in a standard test: a summary of Dutch experience, 1989–94. *Allergy*, 50(3), 234–242.

Paul, M.A., Gray, G.W., Love, R.J., Lange, M. 2007. SSRI effects on psychomotor performance: assessment of citalopram and escitalopram on normal subjects. *Aviation, Space, and Environmental Medicine*, 78(7), 693-697.

Persad, L.A. 2011. Energy drinks and the neurophysiological impact of caffeine. *Frontiers in Neuroscience*, 5:116, 1–8.

Petition of Benton W. Bullwinkel Docket SM-3938, NTSB Order No. EA-3823 (United States of America National Transportation Safety Board March 22, 1993).

Rasmussen, N. 2009. *On Speed: The Many Lives of Amphetamine*. New York: New York University Press.

Roshanaei-Moghaddam, B., Pauly, M.C., Atkins, D.C., Baldwin, S.A., Stein, M.B. and Roy-Byrne, P. 2011. Relative effects of CBT and pharmacotherapy in depression versus anxiety: is medication somewhat better for depression, and CBT somewhat better for anxiety? *Depression and Anxiety*, 28(7), 560–567.

Ross, J., Griffiths, K., Emonson, D. and Lambeth, L. 2007. Antidepressant use and safety in civil aviation: a case-control study of 10 years of Australian data. *Aviation, Space, and Environmental Medicine*, 78(8), 749–755.

Sen, A., Akin, A., Canfield, D.V. and Chaturvedi, A.K. 2007. Medical histories of 61 aviation accident pilots with postmortem SSRI antidepressant residues. *Aviation, Space, and Environmental Medicine*, 78(11), 1055–1059.

Shah, N.R., Jones, J.B., Aperi, J., Shemtov, R., Karne, A. and Borenstein, J. 2008. Selective serotonin reuptake inhibitors for premenstrual syndrome and premenstrual dysphoric disorder: a meta-analysis. *Obstetrics and Gynecology*, 111(5), 1175–1182.

Sicard, B.A., Trocherie, S., Moreau, J., Viellefond, H. and Court, L.A. 1993. Evaluation of zolpidem on alertness and psychomotor abilities among aviation ground personnel and pilots. *Aviation, Space, and Environmental Medicine*, 64(5), 371–375.

Silberman, W.S. 2003. Certification Issues and Answers. *The Federal Air Surgeon's Medical Bulletin*, 41(3), 3–4.

Tack, J., Broekaert, D., Fischler, B., Van Oudenhove, L., Gevers, A.M. and Janssens, J. 2006. A controlled crossover study of the selective serotonin reuptake inhibitor citalopram in irritable bowel syndrome. *Gut*, 55(8), 1095–1103.

Temple, J.L. 2009. Caffeine use in children: What we know, what we have to learn and why we should worry. *Neuroscience & Behavioral Review*, 33, 793–806.

Guide for Aviation Medical Examiners. Federal Aviation Administration. 2013, March. Available at http://www.faa.gov/about/office_org/headquarters_offices/avs/offices/aam/ame/guide/ [accessed 6 April 2013].

Valenzuela, C.F. 1997. Alcohol and Neurotransmitter Interactions. *Alcohol Health & Research World*, 21(2), 144–148.

Wesensten, N.J., Balkin, T.J. and Belenky, G.L. 1996. Effects of daytime adminstration of zolpidem and triazolam on performance. *Aviation, Space, and Environmental Medicine*, 67(2), 115–120.

Chapter 13
Aviation Disaster Crisis Management: Multidimensional Psychological Intervention

Idit Oz and Orit Lurie

Flying has always been considered a dangerous activity. The professional life of an aircrew member exposes him/her on a regular basis to a variety of stressors. At times, exceptional stressors, such as aircraft accidents, are additionally woven into that life. Such mishaps have many potential negative ramifications, not only for the aircrew member but also for the emergency personnel who respond to the mishap, the bystanders on the ground, and additional populations such as colleagues and family members.

In an effort to facilitate and moderate these expected consequences, it is necessary to create and provide relevant professional intervention strategies, tailored to match the needs and resources of the target populations. The purpose of this chapter is to present intervention options for the aviation population. Studies show that early preventative psychological intervention may prevent the worsening of legitimate and expected symptoms and shorten their duration, thus facilitating the rehabilitation process and the return to functioning (Barnea and Lurie 1989).

Definition of the Problem

An aircraft mishap is not an acute event but rather a prolonged one with negative consequences on several levels. An aviation mishap can significantly change crew dynamics, impact an individual's motivation for flying, lead to the development of a fear of flying (see Chapter 7, this volume for an in-depth discussion of motivation to fly and fear of flying), and diminish the aviator's self-confidence. This can result in a period of adjustment (or maladjustment). The intensity of the struggle to regain balance and confidence can result in problems with physical, emotional, cognitive, and behavioral functioning, any of which may make it difficult for the aviator to return to flight status. These effects may last for a significant period of time and have an impact on multiple populations. Even the mental health providers working with the survivors of the mishap may develop reactions which require attention.

Aviators are a particularly challenging population when it comes to providing psychological education and intervention. They are often resistant to seeking help due to stigma and concerns about placing their flight status at risk by admitting to psychological problems. In this chapter, we will describe a military mishap intervention that addresses the needs of a number of impacted populations and enhances aircrew resources. Following presentation of this case example, civil aviation responses will be discussed.

Military intervention conceptualization

The intervention is designed to keep aircrew from becoming dependent on external resources, to prevent prolongation of symptoms and relies upon internal resources and self-help techniques applied by survivors. This approach avoids the stigma that would otherwise be associated with receiving mental health treatment. The intervention described here is based on four principles/treatment methods: Stress Inoculation, The Continuity Principle, The Salmon Principle, and Debriefing.

Stress Inoculation Training (SIT)

In routine situations there is a balance between the demands placed upon the person and the resources at the individual's disposal. The classic theory of stress holds that certain events which constitute significant changes may affect this balance and require an adaptation period during which the system fights to regain equilibrium. Stress is defined as the relationship between the environmental demands placed on the individual and the resources that he/she has to meet these demands (Lazarus and Folkman 1984).

Stress Inoculation Training (SIT) is a treatment that was designed to enhance and strengthen the individual's resources so that he/she becomes less vulnerable or "immune" to the stresses of a variety of circumstances. It has been applied to a broad range of situations, such as preparing patients for painful medical procedures, as a treatment to improve emotional control, as a therapy for generalized anxiety and specific phobias, to help individuals deal with interpersonal problems and addictions and to enhance the ability of professionals to deal with the stress they are exposed to as part of their daily work routine.

SIT is a flexible cognitive–behavioral intervention technique which was created in the 1980s as a means to facilitate an individual's ability to cope with significant stress (for additional discussion of cognitive behavioral therapy see Au, Marino-Carper, Dickstein and Litz 2012). It has been successfully applied to members of occupations who face routine danger and critical events, such as police, military, aviation, and medical personnel (Doran, Hoyt and Morgan 2006; Meichenbaum 2009).

SIT can be applied preventatively and/or post-mishap. Initial inoculation efforts enhance the individual's coping ability and resources prior to a mishap. Secondary inoculation efforts enhance the individual's coping ability in order to moderate or shorten maladaptive reactions. According to Meichenbaum (2009), "the central notion is that bolstering an individual's repertoire of coping responses to milder stressors can serve to build skills and confidence in handling more demanding stressors" (628).

SIT is implemented in three phases: (1) conceptual and educational (for more on psychoeducation see Au, Marino-Carper, Dickstein and Litz 2012); (2) skills acquisition, consolidation, and rehearsal; and (3) application and follow-through. The intervention is tailored to the individual, and provides a set of coping guidelines rather than a specific and rigid formula. The intent is to strengthen and enhance the individual's coping repertoire in order to moderate the gap between the demands of the stressor and the individual's coping resources (Meichenbaum 1993).

It should be recognized that flight personnel commonly engage in mandatory activities geared toward stress inoculation. For example, they routinely participate in life-saving/survival-training exercises. Through this training they are exposed to frightening and potentially dangerous activities, such as practicing bailing out of an aircraft, parachuting, simulating helicopter crashes into water, surviving ocean rescue by helicopter, as well as experiencing simulated fires, engine loss, stalls, and other flight emergencies. These exercises are geared to increase the effectiveness and confidence of the flight crew to respond in a real emergency. In this way, "challenging and realistic training develops trainees' ability to perform … and exposure to realistic levels of stress is intended to inoculate them from the negative effects" (Doran, Hoyt and Morgan 2006, 250; see also Lurie 2007). As a result of their training experiences, aircrew are likely to be particularly receptive to SIT-based intervention, to include the addition of (secondary) SIT to simulation and realistic training exercises. Secondary SIT helps the individual to be more able to react appropriately to their immediate environment when encountering a crisis as he/she grows to believe they have an adequate response in their repertoire making effective response and coping possible (Lurie 2007; Kushnir and Lurie 2002). The aim of any inoculation program is to increase the individual's self-efficacy pertaining to their ability to cope with the demands imposed by the situation. SIT helps the individual have a better understanding of the situation, helps to normalize their emotional response, and provides them with a stress coping "tool set."

When SIT is conducted following an incident the purpose is to provide the aircrew that has directly experienced the disaster, as well as other affected populations, with information pertaining to the incident, likely psychological ramifications, and a relevant set of tools, with the purpose of shortening the post-incident period and promoting the rehabilitation process. One of the important outcomes of the intervention is promotion of self-efficacy. Self-efficacy is the term used by Bandura (1977, 1986) to describe the individual's belief in their ability to successfully cope with difficulties (Ferren 1999; Kwok and Wong 2000). According to Bandura, the individual's self-efficacy dictates what action will be taken and the amount of effort and time that will be invested in coping when encountering problems.

The Continuity Principle

The Continuity Principle is primarily an organizational principle that refers to the fact that a disaster is experienced as a deep fracture within the daily routine and violates the balance and continuity of different aspects of life. The aviation mishap has effects at multiple levels of the organization and at the level of the individual. According to Omer and Alon (1994), following a disaster, "management and treatment should aim at preserving and restoring functional, historical, and interpersonal continuities, at the individual, family, organization, and community levels." Decision errors result from underestimating individual's abilities to cope with disaster but also from underestimating the probability or extent of expected disruption. The Continuity Principle suggests that it is critical that organizations maintain communication between the individuals who provide crisis services in order to enhance synchronization, reduce the likelihood that organizations work at cross-purposes, and increase the likelihood that the needs of potentially traumatized individuals are met without unnecessary distractions.

The Salmon Principle

Salmon (1919), a psychiatrist during World War I (WWI), proposed three principles for intervention in the case of combat trauma victims, known more widely as PIE (that is, proximity, immediacy, and expectancy; see also Campise, Geller and Campise 2006). Proximity refers to the notion that intervention should be provided in close proximity to the incident site and at the location of the impacted organizational unit in order to maintain the benefits of unit cohesion. In the military context, this principle recognizes the conflict that exists between the service member and the unit. Removal of the service member from the unit causes a disruption in supportive relationships and can lead to feelings of guilt and loneliness which make the normal recovery process more difficult. As a result, separation from the organizational unit following a mishap is likely to exacerbate potential maladaptive reactions and make the prognosis less favorable.

Immediacy refers to the fact that interventions should be implemented soon after the mishap in order to improve outcomes by preventing the formation of anxiety and promoting adaptive coping patterns. Expectancy refers to setting the expectation that this is a temporary situation, that the response is normal and that there will be a return to full functioning. The expectation of recovery is considered to be the most central and meaningful message. These principles have been proven to be effective in military interventions (Solomon 1993). Evidence for their efficacy in response to aviation disasters is more anecdotal than empirical but they and their evolved iterations (Campise, Geller and Campise 2006) remain a fundamental building block of many crisis intervention techniques.

Psychological Debriefing

Debriefing principles were implemented during World War II (WWII). Brigadier General Marshall (Shalev 1994) developed a group intervention method based on debriefing which was designed to collect factual data on the unit's activities in the battlefield. Debriefing was conducted within the unit, while seated in a circle and following the removal of ranks by the various officials, in order to create a feeling of comfort and partnership. Marshall conducted the debriefing close to the time of the occurrence of the incident and in close proximity to the battlefield, in order to receive information based on fresh memories of the incident. Beyond information gathering, Marshall found that the command structure was strengthened and that the level of unit cohesion increased following these historical debriefings.

Psychological group debriefing as later described by Mitchell (1983, 1986) was developed from these principles with the primary objective being prevention of the development of posttraumatic disorders and reduction of stress and anxiety. This specific form of debriefing is termed Critical Incident Stress Debriefing (CISD). Research on debriefing does not present a clear and consistent viewpoint pertaining to the effectiveness of this intervention. On the one hand, there are reports of the efficacy of this intervention and authors who advocate its use (for example, Everly, Flannery and Eyler 2002, Flannery and Everly 2004). Other researchers suggest that the intervention is ineffective (Neria and Solomon 1999) and potentially harmful (Devilly and Cotton 2003; for a review of psychological debriefing following disasters see Au, Marino-Carper, Dickstein and Litz 2012). Because of the disparate views of psychological debriefing today, it is recommended that it only be used by trained providers in order to avoid negative reactions related to potential re-traumatization via the intervention. For a review of the battle mind debriefing model which addresses some of these concerns, see Adler et al. 2009.

Identifying Target Populations

Intervention is designed to meet the needs of individuals involved in the accident both directly and indirectly and it should be offered to each of the relevant sub-populations. One way to conceptualize the various sub-populations is in the context of vulnerability circles (Ayalon 1989) which suggests that the degree of emotional and adaptive vulnerability of an individual is influenced by their proximity to the disastrous incident. Just as a stone thrown into water creates ripples, the traumatic incident creates vulnerability circles. Proximity can be physical, that is, direct involvement in the incident, or it can be a psychological proximity, derived from a subjective internal experience (Ayalon 1989). For example, friends, family members, search-and-rescue and emergency medical providers will all, to some extent, be affected by a mishap. (It is important to note that though all will be impacted, certainly not all will experience maladjustment or long term consequences.)

Because there are varying degrees of stress-related consequences following an accident, the potential casualties need to be prioritized by identifying those considered to be at high risk for the development of problems, that is, those who are in the initial vulnerability circle, those considered at risk for other reasons due to their involvement in the incident or because of predisposing factors known to contribute to the development of problems (see Bensimon 2012; Taylor et al. 2009). Understanding vulnerability emphasizes the need to address the multiple populations impacted by the mishap and expands the field of intervention. Applying this model to an aviation mishap makes it clear that the adverse effects can be prolonged and simultaneously affect a number of populations. Therefore, an appropriate intervention should address all of these variables and strive to meet the needs of all relevant sub-populations.

Case Example

Implementation of this multidimensional model will be demonstrated by presenting the intervention conducted following a military helicopter accident which occurred in the summer of 2010. In this accident a team of six military aircrew flying a CH-53 helicopter were killed while performing training exercises in a foreign country. The accident occurred in an unfamiliar environment, far from resources that would otherwise be available. This necessitated intervention in a number of areas.

The intervention plan took a multidimensional perspective focusing on the prevention of post-traumatic symptoms by mapping stressors, identifying target populations, coordinating with squadron leadership, and providing relevant stress coping tools. The intervention operations were conducted in three locations: the disaster site, and at the two home-bases in the country of origin of the aircrew.

Intervention at the Disaster Site

An aeromedical psychologist joined the forces at the disaster site abroad. The forces in the disaster location included the aircrew, technical teams, and the administrators who were joined by search-and-rescue units. In the first stage, the psychologist joined the commanding officer in the field to obtain a situation report of the psychological needs of the commanding officer, the unit, and the rescue teams.

Following consultation with the psychologist, the commanding officer identified the specific populations requiring intervention. The messages conveyed by the psychologist included the need to normalize stress reactions under such circumstances, the importance of having interventions conducted by unit commanders and the need to keep individuals in contact with their normal support systems (that is, other members of the command, family and so on). The commanders were instructed to provide background on the mishap and to provide appropriate expectations for their subordinates. They were trained to emphasize that these are normal reactions. Once forces returned to their home-base in the country of origin, debriefing principles were applied in sessions conducted with the units jointly by their officers and psychologists.

Intervention at the Home-Bases

Psychological intervention was conducted simultaneously at the two bases where the aircrew had been previously stationed. Psychologists mapped the intervention points through observation and discussion with the field commanders that were directly or indirectly involved in the incident. The majority of the interventions were jointly conducted by an officer and a psychologist. These sessions were focused upon legitimizing grief responses, normalizing, and ventilation of feelings. Later, guidance was given to enlist coping resources among the officers and service members that were in the initial vulnerability circle or who had reacted at a strong intensity. Additionally, group discussions were held for relevant populations such as aviation mechanics teams and other technical teams, in order to increase the team's strength and cohesion and to associate the group members with the overall sequence of the mission.

In addition, both of the bases have family housing in which the aircrew and the support crew reside with their families. The population residing in these houses is usually a young population of individuals in their late 20s and 30s, and includes many young children and adolescents. The family housing area intervention was conducted independently and in parallel to the military unit.

Families of the casualties The psychologist and/or social worker accompanied the unit commander and other officers and prepared them to support the families. The needs of families were determined and the family was connected with appropriate support resources. This intervention included providing support to commanders and to service members in dealing with bereavement such that they were able to support family members.

Housing residents The disaster naturally caused anxiety and emotional distress among the residents of the family housing area. Psychological teams identified the sub-populations in distress. They escorted the unit commander when assembling the housing residents for a meeting, and provided specific response to sensitive groups. For example, a support group was formed for pilot spouses who had reacted with anxiety.

Intervention with education staff in the preschools The preschool staff is part of routine base life and is in itself an important aircrew support resource. The preschool staff provides support for both the children directly impacted as well as to the other children in the preschool. Work with the education staff was provided in collaboration with the preschool coordinator and the psychologist. The teachers were markedly distressed. They felt responsible for the children's mental health and for relieving the mothers so that they could mourn. The teachers themselves felt a need to mourn the loss and became concerned about their ability to cope with the children's response to the incident.

The goal of the intervention was to help the children and the teachers cope with feelings of anxiety, uncertainty, lack of control, and concern following the accident. Some of the parents remained abroad for several days following the accident until the search was complete and the bodies had been recovered. This increased anxiety among the children and the spouses who had remained in the country of origin.

An additional goal of the intervention was to increase the sense of security and protection among the children and their parents. The teachers provided a model for supportive coping behavior, knowing and recognizing that the children's responses are greatly affected by the responses of the adults in their surroundings. The teachers were asked to identify children who were continuing to show signs of distress and refer them to the psychological team. The teachers received training on the perception of death among children of various ages and were given an opportunity to discuss individual cases and possible responses among the children. Recognition was given to the group of educators and child care personnel as a support factor and they played a key role in helping the children return to a healthy routine. This experience greatly empowered the education staff.

Following the intervention an evaluation questionnaire was distributed to this population. They reported that they perceived the accident to be a stressor with various negative ramifications. They felt that these ramifications had mostly disappeared, especially in their work and with their families. The impact had diminished to a lesser extent in the emotional sphere, where the process was longer and lasted several months. In the respondents view, interventions to help them return to routine were effective.

Interventions among the therapists Fourteen therapists took part in the incident, including enlisted and reserve clinical psychologists and social workers. Recognizing the dangers of compassion fatigue, burnout, and potential vicarious trauma (see Kennedy 2012), the therapeutic team was supported and drew strength and guidance from the psychology central headquarters, whose main involvement was to establish boundaries (where to go and where not to), proper resource allocation, and intervening, as needed, to prevent team members from developing their own maladaptive reactions. Two of the psychologists were housing residents themselves, and as such, were also in a vulnerability circle. The headquarters psychologists provided counseling to these therapists, allowing them to grieve and to create a buffer between them and the individuals they had treated. When possible, those therapists were replaced with others in the various interventions.

Intervention with surviving aircrew Aircrew members who had survived the crash were accompanied as they went through the process and in their return to their squadron and their home. The intervention was based on the self-help form of stress inoculation training described by Lazarus (1984, 1994), Michenbaum (1993), and Lurie (2007). The goal of the intervention was to equip the aircrew directly involved in the accident with tools to enable them to cope with its consequences and formulate an appropriate response to the situation despite potential initial feelings of helplessness, depression, and anxiety (Barnea and Lurie 1989). Research indicates that people can be taught to apply techniques, which have previously been applied by professionals, on their own, allowing them to take a more active role in their coping.

A self-help kit was created, based on the content analysis of interviews with prior "accident graduates." The kit includes four components geared by population, that is, individual crewmember involved in the accident, squadron commander, technical division commander (that is, the individual responsible for the technical condition of the aircraft), and the families of the crew involved in the accident.

The self-help component intended for the survivors assists them in learning to self-assess their reactions (prevalence and intensity), so that they can monitor their progress and understand what to do in order to return to their prior level of functioning. Common responses are detailed following an aviation mishap at three different time points: during the incident, up to 48 hours following the incident, and during the first six months following the incident. The kit contains evaluation criteria specifying when an individual should seek outside assistance.

There are several topics presented in the component intended for the squadron commander. The squadron commander has been found to be the most relevant monitoring agent and is provided information about common difficulties faced by the crewmembers involved in the accident and their ramifications, tools for assisting with professional intervention, possible responses of other squadron personnel, and a description of commander responses and behaviors which may be disruptive to the process.

The components of the kit intended for the technical division commander and for the family members provide details regarding the processes that surviving crew experience during their return to routine. They are provided information on how their responses can affect the process (positively or negatively) and information that will aide them in being additional sensors for the crewmembers and the commander. In addition, the component intended for the family also refers to their specific needs within the process.

The self-help kit was designed with the aviation population in mind. In other words, it was developed with the recognition of the strong need for control, the population's adversity to showing weakness, the need to avoid the stigma associated with accepting/needing external assistance, and a busy and rigid schedule that does not allow time for traditional treatment. However, the psychoeducational kit is only one part of a comprehensive action plan that forms the intervention for surviving aircrew.

In addition, several individuals receive instruction on how to observe and monitor recovery and when to seek outside help. The commander is enlisted to help with reference to emotional and behavioral reactions. Attention is also directed toward other members of the squadron who were not directly involved in the mishap for the identification and resolution of problems. The commander takes on an almost case manager role following the mishap, and has a role in debriefing, discussions, and monitoring. The goal is to create a local intervention, within the squadron, and consistent with the flight crew's professional jargon. The strategy is tailored to the goal of returning the crewmember to flying; a "return to Olympus."

Problem Definition—"The Descent from Olympus"

Studies based on integrated work of squadron medical staff and psychologists have shown that return to flying after an accident may be problematic (Fowlie and Aveline 1985; Barnea and Lurie 1989). Despite the fact that many flight crew will eventually be touched by an air mishap either directly or indirectly, it is counterproductive to refer to accidents as a routine part of the professional life of aircrew and demand a return to regular operations as though nothing out of the ordinary has happened. This disregards the short- and long-term consequences of accidents and the resulting emotional responses. Ignoring these ramifications or investing too much energy in order to conceal them may actually intensify them, prolong their duration, and even make them irreversible. Despite the passage of time, unresolved issues may continue to trouble surviving crewmembers during flight operations as well as on the ground, resulting in further casualties.

In order to cope with the constant danger of flying most crewmembers use coping skills such as denial, rationalization, intellectualization, and humor (Jones 1982). Aviators are characterized by their high levels of self-confidence and a strong belief in their ability to rely on themselves. It is precisely these coping skills and personality traits (effective when used in the face of danger), which can have negative consequences following an aircraft accident. These characteristics may contribute to the emotional and professional problems seen subsequent to the accident (Aitken 1972). An air accident is an event which can cause the individual to reconsider the daily risk involved in flying and it is likely to evoke negative and unfamiliar emotions.

Medical and behavioral science professionals now recognize that aviation accidents may result in outcomes such as phobias, development of anxiety due to awareness of personal vulnerability, and channeling of tension into somatic symptoms. These adverse outcomes may expand into various flight situations and lead to deterioration in performance, motivation, and commitment (Aitken 1972; Jones 1986). Popplow (1984) labeled this phenomenon Post Accident Anxiety Syndrome, which is reported to occur in 30–40 percent of survivors (Fowlie and Aveline 1985; Zeller 1973). (For a comprehensive discussion of the development and treatment of fear of flying, see Chapter 7, this volume.)

A study conducted in the Israeli Air Force by Lurie and Barnea (1989) found that following an aircraft accident 24 percent of the survivors had thoughts about not returning to flight. It was reported by 51 percent that there had been a change in the way they felt about flying; expressed as feelings of insecurity, stress, fear, and anxiety during similar flight situations. Forty-four percent reported a change for the worse in their flying abilities for periods ranging from several days to several months following the accident. Ninety-one percent felt the need to talk about the accident for emotional ventilation purposes and for strengthening their professional abilities.

The destruction of the myth of pilot invulnerability, "this will not happen to me," impacts the aviator as well as other individuals in the squadron. They become aware of the risks of flying—hence the feeling of "the descent from Olympus." For this reason, an intervention is required to facilitate a return to functioning and to prevent secondary casualties.

Findings Six Months Following the Incident

Approximately six months after the incident it was found that there had been a quick return to flight routine among the aircrew, technical, and administrative teams. Individuals had fully resumed their prior work routine. Among the search-and-rescue forces there was one individual who was being treated for Post Traumatic Stress Disorder (PTSD). Within the family housing area, no additional requests for support had been received on behalf of residents, preschool staff, or children following the first week after the accident occurred. Among therapists, one experienced difficulties for approximately three months following the incident.

Civil Aviation

Although the frequency of aviation mishaps is far lower in commercial aviation than in military operations, air disasters also occur in airline operations. Airlines are required in the US to have a formal disaster plan describing their response to survivors and to family members. This includes providing psychological support for the survivors and for family members of crash victims. In 1996 the Aviation Disaster Family Assistance Act was enacted in the US. This law requires airlines to have a plan for aircraft accidents involving major loss of life. The plan provides for notification of families of the passengers, handling of personal effects, assistance with traveling to the site of the accident, physical care of the family while staying at the location, and most importantly, an assurance that the air carrier will work with a designated organization on an ongoing basis to ensure that families of passengers receive an appropriate level of services and assistance following the accident. An airline's organizational response to an air disaster can have a marked impact on family members. TWA's handling of the 1996 Flight 800 crash off Long

Island has been described as "the manual on how not to do things" (Airliners.net 2000). Family members reported receiving no information, counseling services, or assistance in traveling to New York to the wreckage site.

In the US, the American Red Cross is the agency responsible for coordination of Family Care and Mental Health. Following an air disaster the Red Cross is notified and activates a national response system. They coordinate resources for debriefing of support personnel prior to departure as well as volunteer counseling and support services. They also coordinate child care and spiritual support services (including a suitable interfaith memorial service). They are the agency responsible for coordinating, together with the airline, the delivery of crisis health support to family members, including those who do not travel to the incident site.

Butcher and Hatcher (1988) described the factors that differentiate an air crash from other disasters. This includes the sudden and unexpected nature of these disasters. Furthermore, these disasters tend to create extensive destruction and the accident site is often a "scene of incredible carnage" (725). The crashes typically occur with the victims and survivors located far away from their normal and familiar settings. In addition, the airline crash has a long-ranging effect impacting many lives, including nearly all of the employees of the air carrier and families of survivors and friends distributed across a wide geographic area. Butcher and Hatcher (1988) described the impact of crashes on airline employees and the four common cognitive processes used by employees to regain a sense of mastery over the incident; search for information about family and friends possibly on board the aircraft; search for causation; fantasy replays of the events leading to the crash; and decisions about continuing to work in the industry.

The Air Line Pilots Association (ALPA) advises pilots on post-crash behavior. Due to the focus on protecting the pilots' legal rights it may be difficult for the pilot to receive full access to psychological care. Flight attendant unions provide information, legal counsel, and referrals for the psychological care of flight attendants.

Organizations responding to the air crash site include the Red Cross and National Organization for Victim Assistance. Lessons learned from crashes such as the 1978 San Diego airline crash and the 1989 Sioux City crash led to some of the earliest articles suggesting how mental health teams can provide rapid, effective, and efficient delivery of services to survivors and their families and to the families of those who are killed (Jacobs 1995; Butcher and Hatcher 1988). In the aftermath of a 1994 airline crash near Pittsburgh the National Air Disaster Alliance Foundation (NADA) was formed (www.planesafe.org). Following the terrorist attacks of 11 September 2001, recommendations have also been published which address management of psychological trauma from an airline disaster (O'Flaherty 2011). Merlin, Butcher and Cortacans (2009) described the services provided following the US Airways 1549 Hudson River crash and provided recommendations for responding to an air disaster.

Rizzo (2003) reviewed the provision of mental health services following aviation disasters. Courses are offered for individuals interested in working with air disaster survivors, families, and responders by the International Critical

Incident Stress Foundation (http://www.icisf.org), National Organization for Victim Assistance (http://www.try-nova.org), and by the American Red Cross. The Red Cross air disaster teams are composed of mental health professionals and clergy. In addition the Red Cross offers courses for non-professionals to be involved in the NADA/Family Support Teams.

Summary

Aircraft accidents may make those directly involved, as well as bystanders, out of balance, at least for a certain time period. During this period significant areas in their lives may be negatively impacted. It is crucial that appropriate interventions be tailored to those who can benefit from them. The aviation population can be particularly difficult to assist as aircrew personnel normally tend to suppress any signs of weakness or vulnerability. As a result, aircrew are at risk of investing energy in concealing their reactions at the cost of a decline in performance and motivation. At this time, a data-driven, empirically-derived program for responding to the psychological needs following an air disaster is not available. For the time being, delivery of mental health services following an aviation mishap will continue to be based upon theories of traumatic stress, research on recovery from traumatic events, and expert consensus.

References

Adler, A.B., Bliese, P.D., McGurk, D., Hoge, C.W. and Castro, C.A. 2009. Battlemind debriefing and Battlemind training as early interventions with soldiers returning from Iraq: Randomization by platoon. *Journal of Consulting and Clinical Psychology*, 77(5), 928–940. Doi: 10.1037/a0016877.

Airliners.net. 2000. *Alaska Airlines Swift Crash Response*. Available at: http://www.airliners.net/aviation-forums/general_aviation/read.main/84416/ [accessed 2 September 2012].

Aitken, R.C.B. 1972. A study of anxiety assessment in aircrew. *British Journal of Social and Clinical Psychology*, 11(1), 44–51

Au, T.M., Marino-Carper, T.L., Dickstein, B.D. and Litz, B.T. 2012. Military roles in postdisaster mental health, in *Military Psychology: Clinical and Operational Applications*, 2nd Edition, edited by C.H. Kennedy and E.A. Zillmer. New York: Guilford, 156–184.

Aviation Disaster Family Assistance Act of 1996, H.R. 3923, 104th Cong., 2nd Sess. Available at: http://www.gpo.gov/fdsys/pkg/BILLS-104hr3923rh [accessed 1 September 2012].

Ayalon, O. 1989. Psychotherapy in survivors of terrorist victimization. In *Brief Psychotherapy, Background, Techniques and Application*, edited by H. Dasberg, J.A. Itzigsohn and G. Shefler. Jerusalem: TheMagnes Press, 206–229.

Bandura, A. 1977. Self-efficacy: Toward a unifying theory of behavioral change. *Psychological Review*, 84(2), 211–215.

Bandura, A. 1986. *Social Foundations of Thought and Action: A Social Cognitive Theory*. Englewood Cliffs: NJ: Prentice Hall.

Barnea, I. and Lurie, O. 1989. *The descent from the Olympus: The effect of accidents on aircrew survivors*. Paper presented at the Aerospace Medical Panel Symposium held in The Hague, The Netherlands, from 24–28 October 1988. Available at: http://www.dtic.mil/cgi-bin/GetTRDoc?AD=ADA212884 [accessed 1 September 2012].

Bensimon, M. 2012. Elaboration on the association between trauma, PTSD and posttraumatic growth: The role of trait resilience. *Personality and Individual Differences*, 52(7), 782–787.

Butcher, J.M. and Hatcher C. (1988). The neglected entity in air disaster planning. *American Psychologist*, 43(9), 724–729.

Campise, R.L., Geller, S.K. and Campise, M.E. 2006. Combat stress, in *Military Psychology: Clinical and Operational Applications*, edited by C.H. Kennedy and E.A. Zillmer. New York: Guilford, 215–240.

Devilly, G.J. and Cotton, P. 2003. Psychological debriefing and the workplace: Defining a concept, controversies and guidelines for intervention. *Australian Psychologist*, 38(2), 144–150.

Doran, A.P., Hoyt, G. and Morgan, C.A. 2006. Survival, evasion, resistance, and escape (SERE) training: Preparing military members for the demands of captivity, in *Military Psychology: Clinical and Operational Applications*, edited by C.H. Kennedy and E.A. Zillmer, 241–261.

Everly, G.S., Flannery, R.B. and Eyler, V.A. 2002. Critical incident stress management (CISM): A statistical review of the literature. *Psychiatric Quarterly*, 73(3), 171–182.

Ferren, P.M. 1999. Comparing perceived self efficacy among adolescent Bosnian and Croatian refugees with and without posttraumatic stress disorder. *Journal of Traumatic Stress*, 12(3), 405–420.

Flannery, R.B. and Everly, G.S. 2004. Critical incident stress management (CISM): Updated review of findings, 1998–2002. *Aggression and Violent Behavior*, 9(4), 319–329.

Fowlie, D.G. and Aveline, M.O.1985. The emotional consequences of ejection, rescue and rehabilitation in Royal Air Force aircrew. *British Journal of Psychiatry*, 146, 609–613.

Jacobs G.A. 1995. The development of a national plan for disaster mental health. *Professional Psychology*: *Research and Practice*, 26, 543-549.

Jones, D.R. 1982. Emotional reactions to military aircraft accidents. *Aviation, Space and Environmental Medicine*, 53(6), 595–598.

Jones, D.R. 1986. Flying and danger, joy and fear. *Aviation, Space, and Environmental Medicine*, 57(2), 131–136.

Kennedy, C.H. 2012. Ethical dilemmas in clinical, operational, expeditionary, and combat environments, in *Military Psychology: Clinical and Operational Applications*, 2nd Edition, edited by C.H. Kennedy and E.A. Zillmer, 360–390.

Kushnir, T. and Lurie, O. 2002. Supervisors' attitudes toward return to work following MI or CABG. *Journal of Occupational and Environmental Medicine*, 44(4), 331–337.

Kwok, S. and Wong, D. 2000. Mental health of parents with young children in Hong-Kong: The role of parenting stress and parenting self efficacy. *Children and Family Social Work*, 5(1), 57–65.

Lazarus, R.S. 1984. Puzzles in the study of daily hassles. *Journal of Behavioral Medicine*, 7(4), 375–389.

Lazarus, R.S. 1994. *Emotion and Adaptation*. New York: Oxford University Press.

Lazarus, R.S. and Folkman, S. 1984. *Stress, Appraisal and Coping*. New York: Springer.Lurie, O. 2007. Cognitive inoculation in preparing the human factor for combat in an air force unit. *Israel Journal of Psychotherapy*, xxi, 2.

Meichenbaum, D. 1993. Stress inoculation training: A 20 year update, in *Principles and Practice of Stress Management*, 2nd Edition, edited by P.M. Lehrer and R.L. Woolfolk. New York: Guilford Press, 373–407.

Meichenbaum, D. 2009. Stress inoculation training, in *General Principles and Empirically Supported Techniques of Cognitive Behavior Therapy*, edited by W.T. O'Donohue and J.E. Fisher. Hoboken, NJ: John Wiley & Sons, Inc., 627–630.

Merlin, M.A., Bucher, J. and Cortacans, H.P. 2009. US Airways Flight 1549 Hudson River crash: The New Jersey experience. *American Journal of Disaster Medicine*, 4(4), 189–191.

Mitchell J.T. 1983. When disaster strikes. *Journal of Emergency Medical Services*, 8, 36–39.

Mitchell J.T. 1986. Critical incident stress management. *Response*, 5, 24–25.

Neria,Y. and Solomon, Z. (1999) Prevention of Post-Traumatic Stress Disorder in: Saigh P.J. Bremner ID. (Eds.) Post traumatic Stress Disorder: A comprehensive approach to research and treatment. Boston. Allyn & Bacon.

O'Flaherty J. 2002. Handling an airline disaster: Suggestions from psychological and logistical standpoints. *Occupational Medicine*, 17(2), 315–338.

Omer, H. and Alon, N. 1994. The continuity principle: A unified approach to disaster and trauma. *American Journal of Community Psychology*, 22(2), 273–287.

Popplow, J.R. 1984. After the fire-ball. *Aviation, Space, and Environmental Medicine*, 55(4), 337–338.

Rizzo, W.C. 2003. *Mental Health Services Provided in the Aftermath of Aviation Disasters*. Dissertation May, 2003, Widener University. Available at: http://www.planesafe.org/planesafe_archive/pdfs/Dissertation_AfterAirCrashDisRizzo.pdf [accessed 29 August 2012].

Salmon, T.W. 1919. The war neuroses and their lessons. *New York Journal of Medicine*, 51, 993–994.

Shalev A.Y. 1994. Debriefing following traumatic exposure, in *Individual and Community Responses to Trauma and Disaster: The Structure of Human Chaos*, edited by R.J. Ursano, B.G. McCaughey and C.S Fullerton. Cambridge: Cambridge University Press, 201–219.

Solomon, Z. 1993. *Combat Stress Reaction: The Enduring Toll of War*. New York: Springer.

Taylor, M.K., Mujica-Parodi, L.R., Potterat, E.G., Momen, N., Dial Ward, M.D., Padilla, G.A., Markham, A.E. and Evans, K.E. 2009. Anger expression and stress responses in military men. *Aviation, Space, and Environmental Medicine*, 80(11), 962–967.

Zeller, A.F. 1973. Psychological aspects of aircrews involved in escape and evasion activities, *Aerospace Medicine*, 44(8), 956–960.

Chapter 14
Aviation Mishap Prevention and Investigations: The Expanding Role of Aviation Psychologists

Peter B. Walker, Paul O'Connor and William L. Little

Introduction

As aircraft have become increasingly more reliable, there has been a dramatic reduction in the number of aircraft accidents. In the earliest days of aviation, technical failures were common. However, over the years there have been great advances in material technology, fuels and oils, aerodynamics, meteorology, radio communications, and navigation facilities, which have all helped in reducing the number of accidents (David 1996). Further advances have been made through improvements in procedures and standards. Although there has been a reduction in the overall frequency of accidents over time, the number of accidents and the proportion due to human error has increased. In fact, it has been estimated that as many as 70 to 80 percent of all aviation mishaps can be traced back to some form of human error (Wiegmann and Shappell 2003). Human error causes of crashes in aviation have been commonly linked to a failure in interpersonal communication, crew coordination, decision-making, situational awareness, leadership, and organizational factors (Flin, O'Connor and Mearns 2002) and are usually the culmination of a chain of events as opposed to discrete singular occurences (Merlin, Bendrick and Holland 2012).

Take for example Air Florida Flight 90 on 13 January 1982. The flight, a Boeing 737, was a scheduled flight to Fort Lauderdale, Florida, from Washington National Airport, Washington, DC. There were 77 passengers and five crewmembers on board. The flight's scheduled departure time was delayed about one hour 45 minutes due to a moderate to heavy snowfall which necessitated the temporary closing of the airport (National Transportation Safety Board (NTSB) 1998). Shortly after takeoff, the aircraft crashed into the barrier wall of the northbound span of the 14th Street Bridge and plunged into the ice-covered Potomac River. Four passengers and one crewmember survived the crash. Further, when the aircraft hit the bridge, it struck seven occupied vehicles and then tore away a section of the bridge barrier wall and bridge railing. Four people in one of the vehicles were killed.

The NTSB (1998) determined that the probable causes of this accident were the flight crew's failure to use engine anti-ice during ground operation and takeoff, their decision to take off with snow/ice on the airfoil surfaces of the aircraft, and the captain's failure to reject the takeoff during the early stage when his attention was called to anomalous engine instrument readings. Contributing to the accident were the prolonged ground delay between de-icing and the receipt of Air Traffic Control (ATC) takeoff clearance during which the airplane was exposed to continual precipitation, and the limited experience of the flight crew in jet transport winter operations.

As another example, consider the X-31 mishap in 1995. The X-31 was a highly maneuverable jet aircraft which was a joint US and German venture which began in 1990. On 18 January 1995, during a routine test flight in excess of 20,000 feet, the pilot turned on his pitot heat due to significant water vapor and concerns about icing. He informed the test conductor that he had turned on the pitot heat. However, an engineer who was in the control room mentioned that the pitot heat had not been hooked up. This was not relayed to the pilot until just prior to landing and immediately prior to an "increasing series of pitch oscillations until the nose was about 20 degrees past vertical, following by a sharp roll" (Merlin, Bendrick and Holland 2012, 9). The pilot was able to eject but suffered significant injuries. The cause of the mishap was attributed to the culmination of decisions regarding the pitot heat and the type of probe, lack of understanding by personnel of the possible consequences related to inaccurate data measured by the probe, and failure to communicate crucial information to the pilot.

This chapter will begin with a discussion of human factors, arguably the overarching concept under which the field of psychology falls in aviation. A human factors training program called Crew Resource Management (CRM) will be described, and the research that has been done by psychologists to select out people that are not suitable to the rigors of aviation, as well as identifying at-risk aviators will be delineated. Finally, the discussion will shift toward the role that psychologists have played in accident investigations, and describe the research that has been done to develop a mishap investigation tool for capturing the human causes of aviation accidents.

Human Factors

Human factors is defined by the International Ergonomics Society (2000) as "the scientific discipline concerned with the understanding of interactions among humans and other elements of a system, and the profession that applies theory, principles, data, and other methods to design in order to optimize human well-being and overall system performance." This very broad definition of human factors essentially encompasses any situation in which humans interact with technology. In the earlier years of human factors the research was primarily being carried out by experimental psychologists (see also Chapter 1, this volume for additional history). To illustrate, Fitts and Jones (1947) studied problems with control configurations in aircraft accidents. Many of the aircraft accidents during World War II (WWII) were caused by aviators failing to discriminate between the flap and landing gear handles. The handles were close together, and they were exactly the same shape and size. Therefore, during landings, when the pilot attempted to adjust the flaps, it was not uncommon for the landing gear to be raised instead. This work paved the way for improvements in cockpit design and safety, with increasing input from applied cognitive and perceptual psychologists.

As aircraft have become more complex and automation has been introduced, the need to consider how the human fits into the system has become increasingly more important. Although the goal of automation is to improve performance and reduce the number of pilot errors, evidence suggests that rather than reducing errors the automation just changes the types of errors that are made. "We currently have flight deck automation systems that change the task, re-distribute workload for the crew, and present situations that induce an error" (Wood 2004, 1). Therefore, there is a need for careful deliberation on the principles of user-design of modern aircraft cockpits.

Take for example Air Inter Flight 148 on January 20, 1992 (Aviation Safety Network 2008). While on approach into Strasbourg, the Airbus A320 impacted the side of a mountain. The crew (with 12,000 hours combined flying experience) believed that they set the automated flight control system (FCS) to glide downward to the airport at what they thought was a gentle 3.3 degree slope. However, the pilots actually set the airliner to descend at a rate of 3,300 feet per minute, a rate four times what they expected. The impact into the Alps at 2,620 feet killed 87 of the 96 people on-board, including both pilots. Causal factors of the accident were identified as poor communications with Air Traffic Control (ATC) and between the pilots; lack of situational awareness by the crew exacerbated by the crew's over-reliance on the FCS, and the poor design of the FCS vertical mode controls; and high cognitive workload as the crew attempted to reconfigure the FCS for landing.

More recently, neurocognitive techniques have also become more prevalent in the human factors literature. In an attempt to move toward more objective measures of workload and situational awareness, neurocognitive techniques such as electroencephalogram (EEG; Wilson and Russell 2003), amplitude of certain event-related brain potentials (ERPs; Prinzel et al. 2003), and regional cerebral hemodynamic response (Jansma, Ramsey, Coppola and Kahn 2000) are assisting in the modeling of human factors in a variety of aviation environments. Grubb (2010) postulates that the next generation of military fighter aircraft will be specifically designed to defeat threats through superior human factors.

Cockpit design is not the only area in which psychologists have positively influenced air safety. In the next section, we will describe how psychologists of various disciplines have been influential in the development of CRM training programs.

Crew Resource Management

CRM can be defined as "a set of instructional strategies designed to improve teamwork in the cockpit by applying well-tested tools (e.g., performance measures, exercises, feedback mechanisms) and appropriate training methods (e.g., simulators, lectures, videos) targeted at specific content (i.e., teamwork knowledge, skills, and attitudes)" (Salas et al. 1999, 163). The emergency landing of US Airways Flight 1549 on 15 January 2009 is an example of how effective CRM prevented loss of life (National Transportation Safety Board 2010). Shortly after takeoff from LaGuardia Airport in New York City, the aircraft flew into a flock of geese. The crew quickly determined that the best course of action was an emergency landing in the Hudson River. All 155 people aboard the flight survived.

CRM training programs originated in the late 1970s (Cooper, White and Lauber 1980; Droog 2004). At the time, there was widespread trepidation in a number of countries concerning the large number of commercial aviation accidents attributed to pilot error (Helmreich, Merritt and Wilhelm 1999). Specifically, during the late 1970s and early 1980s, there were a number of commercial aviation accidents in which crew error was identified as a significant causal factor (for example, Tenerife 1977; Portland 1978; Washington, DC 1982). Analysis of flight data recorders and cockpit voice recorders, in addition to sophisticated analysis of wreckage, identified a recurrent pattern emerging from accident investigation reports—crews were not properly fulfilling their assigned roles on the flight deck (David 1996).

From accident analyses, incident reports, and research studies both on the flight deck and in the simulator, training packages have been implemented by psychologists from various disciplines to "close the loop" between accidents and training. CRM training is used by virtually all the large international airlines, is recommended by the major civil aviation regulators (for example, Federal Aviation Administration, FAA; and UK Civil Aviation Authority) and is mandated in many countries (Droog 2004).

CRM training varies between agencies, but in general an introductory CRM course is conducted in a classroom for two or three days. Teaching methods include lectures, practical exercises, role playing, case studies, and video films of accident reenactments (Salas et al. 2006), and some training programs include a simulator component with a focus on crew cooperation (Droog 2004). In addition to formal training, CRM is stressed consistently throughout aircrew, pilot, navigator/flight officer, and maintenance crew training programs. Formal refresher training is shorter, typically a half or one-day course focusing on a specific CRM topic. For flight deck crews, CRM skills are practiced and assessed in flight simulator sessions known as line-oriented flight training (LOFT) and line operational evaluation (LOE).

The topics covered in CRM training, "are designed to target knowledge, skills, and abilities as well as mental attitudes and motives related to cognitive processes and interpersonal relationships" (Gregorich and Wilhelm 1993, 173).

A psychologist would recognize many of the concepts that are taught in CRM training as being part of any undergraduate psychology course (see Table 14.1).

However, care must be taken when introducing these concepts to aviators. A main criticism of the early CRM courses was that there was too much "psychobabble." The pilots were unfamiliar with the psychological terms being used, which was detrimental to the acceptance of the training. Thus, it was necessary that the psychologists adapt to the terms and concepts already used in aviation.

It is important to highlight that although CRM training is based upon psychological theories and research, the industry standard is for the training to be delivered by aviators and other flight personnel, as opposed to psychologists. Early on in the application of CRM training, this seemed to be the most powerful method for getting the point across. Nevertheless, many major airlines employ a psychologist to ensure that the latest research is being integrated into the CRM training program. Therefore, the goal of CRM is to provide aircraft crews with the non-technical skills necessary to do their job safely and effectively.

Table 14.1 Examples of the Types of Psychological Concepts Covered in CRM Courses

Social	Communication, group decision-making, conformity, obedience to authority, assertiveness, group dynamics, biases and attributions
Cognitive	Naturalistic decision-making, situational awareness, information processing, perception, mental models, memory, problem solving
Health	Stress, fatigue, human performance
Personality	Personality traits, individual differences
Industrial/ Organizational	Team work, organizational climate/culture, leadership/ followership, human error

Crew Resource Management Evaluation

Given the widespread adoption of CRM training, it might be assumed that there is a large body of evidence documenting the successes of the training. However, there is no simple answer to the question of whether CRM training can fulfill its purposes of increasing safety and efficiency (Helmreich, Merritt and Wilheim 1999). The FAA (2004) states that, "it is vital that each training program be assessed to determine if CRM training is achieving its goals. Each organization should have a systematic assessment process. Assessment should track the effects of the training program so that critical topics for recurrent training may be identified and continuous improvements may be made in all other respects" (Federal Aviation

Authority 2004, 12). There have been a number of literature reviews of the effectiveness of CRM training (O'Connor, Flin and Fletcher 2002; Salas, Burke, Bowers and Wilson 2001; Salas et al. 2006) and one meta-analysis (O'Connor et al. 2008). These studies adopted Kirkpatrick's (1976) training evaluation hierarchy to categorize the evaluation techniques utilized. Kirkpatrick's hierarchy consists of four different levels of evaluation: reactions, learning, behavior, and organizational impact.

> *Level 1: Reactions.* Evaluating reactions are the equivalent to measuring user satisfaction. Did the participants like the training? The previous reviews deduced that CRM participants generally have favorable reactions to the training.
> *Level 2: Learning.* Learning is the second level in the hierarchy, and refers to "the principles, facts, and skills which were understood and absorbed by the participants" (Kirkpatrick 1976, 11). Learning is made up of two components: attitudinal change and knowledge gains. The previous reviews of CRM training evaluation studies concluded that CRM training generally results in a positive change in CRM attitudes (O'Connor, Flin and Fletcher 2002; Salas, Burke, Bowers and Wilson 2001; Salas et al. 2006). A large effect size of CRM training was found at the attitude level (O'Connor et al. 2008). The effect of CRM training on knowledge has not been widely examined. However, literature reviews conclude there are increases in CRM knowledge of participants as a result of attending CRM training (O'Connor, Flin and Fletcher 2002; Salas, Burke, Bowers and Wilson 2001). A medium effect of CRM training was found for knowledge evaluation (O'Connor et al. 2008).
> *Level 3: Behavior.* Evaluation of behavioral changes is the assessment of whether knowledge learned in training actually transfers to behaviors on the job, or a similar simulated environment. Behavioral markers are the most common method used to assess team behavior. Behavioral markers are "a prescribed set of behaviors indicative of some aspect of performance" (Flin and Martin 2001, 96). O'Connor, Flin and Fletcher (2002) and Salas, Burke, Bowers and Wilson (2001) concluded that there is evidence of changes in the behaviors of CRM trainees. However, Salas et al. (2006) stated that the evidence of the effects of CRM training on behavior is not as consistently positive as evaluations carried out at the reactions or learning levels. Nevertheless, in the meta-analysis carried out by O'Connor et al. (2008), a large effect was found at the behavioral level. The size of the effect was greater than that found for attitudes or knowledge. Goeters (2004) found statistically significant improvements in both situational awareness and decision-making in a CRM validation study.
> *Level 4: Organizational impact.* This is the highest level of evaluation in Kirkpatrick's (1976) hierarchy. The goal of any training program is to produce tangible evidence at an organizational level, such as an

> improvement in safety and productivity. There are few studies in the literature reporting evaluations carried out at this level (O'Connor, Flin and Fletcher 2002; Salas, Burke, Bowers and Wilson 2001). As a result of the low mishap rates in the industries that have adopted CRM training, it is difficult to draw conclusive findings regarding the effect of CRM training on the organization.

Psychologists continue to be integrally engaged in the design, development, and evaluation of CRM training. It continues to be adapted to meet the needs of the users based upon recent mishaps and near-misses, published research, and in response to changes in equipment and technology (for example, increased automation). To accurately capture the human factors causes of mishaps one must use valid, reliable human factors coding systems. Further, as CRM training gains continued popularity in high-risk communities beyond aviation, it is imperative that researchers be rigorous in completing, and reporting, CRM training evaluations (O'Connor et al. 2008).

In the next section we will discuss how psychologists "select-out" individuals who should not be flying aircraft, and help identify aviators who are at risk of having an accident.

Personality and the High-risk Aviator

> Most pilots are safe most of the time.
>
> (Kern 1999, 1)

Many aviation psychologists and other health care providers in the aerospace industry will agree that most aviators are very good at managing risk. Indeed, aviation has become one of the safest forms of travel. In fact, the NTSB (2001) recently cited that an individual involved in an aviation accident has a 96 percent chance of survival. Still, there is a very small percentage of pilots and other aircrew that attempt to break the rules of safe and effective flight. Often, it is the job of aerospace psychologists and other aerospace health care providers to identify these individuals and remove them from their airplanes.

Take the following mishap as an example of a combination of personality and decision factors not conducive to aviation (National Transportation Safety Board 2007). At approximately 1900 hours on 1 January 2006, a Robinson R44 helicopter, N442DH, registered to a private individual, was destroyed after colliding with trees and terrain, about six miles southwest of Grand Ridge, Florida. Weather reports from that evening suggested that visibility may have been limited to no greater than 1.75 miles with a ceiling of 300 feet. In addition to the prevailing Instrument Meteorological Conditions (IMC; conditions in which the pilot must fly solely on the basis of information presented by their instruments) at the time of the mishap, no flight plan was filed for the Title 14, CFR Part 91 ferry flight. The pilot did not have an instrument rating, and the helicopter was not equipped with instrumentation for flight into IMC. Furthermore, toxicological samples from the pilot disclosed extremely high levels of methamphetamine. According to a witness at the Brewton Municipal Airport, the pilot landed to refuel the helicopter at the Brewton Airport at approximately 1500 on the day of the accident. The witness, who operated the fuel concession, said he commented to the pilot about the convective weather south of the airport, and asked the pilot what the weather was like along his proposed route. According to the witness, the pilot said he didn't care, as he could land anywhere, and wait out any bad weather.

Historically, researchers have been interested in identifying various personality characteristics as they relate to hazardous decision-making (for a review on Aeronautical Decision Making see, O'Hare 2003). For example, Berlin and his colleagues sought to identify those thought patterns that result in "irrational pilot judgment" (Berlin et al. 1982, 9). These authors felt that the identification of these types of thought patterns might lead to better training. Specifically, it was thought that if pilots could be trained to recognize unsafe tendencies in their own thought patterns, then these pilots could be taught to apply various corrective techniques to prevent a potential mishap. Berlin and his colleagues went on to consult with psychologists and other health care providers in order to obtain informed opinions regarding those thought patterns.

Based on observational data collected from Berlin and his colleagues, five hazardous attitudes and/or thought patterns were identified (antiauthority, invulnerability, macho, impulsivity, and resignation). A number of studies have since been conducted attempting to quantitatively test the hazardous thought model (for example, Lester and Bombaci 1984, Lester and Connolly 1987). Lester and Bombaci (1984) surveyed 35 civilian aviation pilots and sought to correlate self-assessed propensity toward the five hazardous thought patterns with personality dimensions on Cattel's Sixteen Personality Factor Questionnaire (Cattel, Eber and Tatsuoka 1970). While it was initially proposed for five distinct and separate hazardous thought patterns, the data taken from Lester and Bombaci (1984; see also, Lester and Connolly, 1987) suggested that three of the hazardous thought patterns (invulnerability, impulsivity, and macho) might be sufficient for describing the principle thought patterns during irrational pilot decision-making.

More recently, O'Hare (2003) performed a principal components analysis on the data collected by Lester and Connolly (1987) and found that only one factor was needed to account for over half of the variability in responses in that study. Moreover, four of the five scales loaded very highly on the one factor, with the fifth factor (invulnerability) loading at a moderately high level. These findings suggested that the hazardous thought model really represents one isolated construct.

Similar to the work that has been done in general aviation, military aviation has similarly sought more reliable measures to identify individuals with hazardous tendencies. For example, to assist in the identification and remediation of aviators with psychological and emotional issues, Naval and Marine Corps aviation has relied on periodic meetings of Human Factors Councils (OPNAVINST 3750.6R, MCO 5100.29A). These councils have been organized with the intent to provide early identification of at-risk aviators. In addition, the Human Factors Council helps to identify where to put resources to reduce risk and decrease the chances of a mishap and/or incident. To assist supervisors in identifying individuals that may be at risk for a mishap, the instructions governing a Human Factors Council have classified five different personality profiles (Nugget, Overconfident, Best, Poor performer, and Over-stressed aviator). These "caricatures" may be viewed as a compilation of anecdotes from various interactions with these individuals; however, seasoned aviators are usually able to identify these at-risk individuals without hesitation.

The first category of at-risk aviator has been called the Nugget or Transition Aviator, to signify the increased risk associated with a pilot that is still inexperienced with the aircraft. Typically, these individuals are behind their peers in progression and may illustrate a lack of flying skills, poor headwork, and often lack confidence in his or her ability. Also, these individuals may be weak in aircrew coordination which may only compound the difficulty these individuals are facing inside the aircraft. These deficiencies often create a hazardous, and sometimes fatal, combination that must be identified early by their peers.

The second category of at-risk aviator is known as the Senior or Overconfident Aviator. This individual has been out of the cockpit for some time or may be rusty on his or her flying skills. However, because the overconfident aviator has accumulated significant flight hours, that individual often develops a "been there—done that" attitude. Unfortunately, the attitudes developed by the overconfident aviator often result in that individual relying more on experience instead of proficiency. In addition, these individuals often use their more senior rank to "bend" the rules and may fail to recognize their own limits in the cockpit. Perhaps one of the most dangerous behaviors that are exhibited by the senior aviator is the attempt to intimidate the rest of the cockpit crew. Obviously, intimidation from the overconfident or senior aviator leads to poor communication (that is, poor CRM) inside the cockpit.

Perhaps the prototype for the at-risk aviator is the character of Maverick from the movie *Top Gun* (Badalato, Bruckheimer, Simpson and Skaaren 1986). This individual is more commonly labeled the Best Pilot/Aviator/Aircrewman because of their attitude toward handling their aircraft. Often, these individuals are very talented in the aircraft but they overestimate their ability. In addition, these individuals may be highly regarded by their peers due to their high-quality airmanship. However, this individual is known to consistently push the aircraft envelope and to lack proper judgment. More often than not, the Best Pilot will think that rules only apply to the average aviator. While extraversion is a necessary trait for some subsets of pilots (for example, military; Campbell, Moore, Poythress and Kennedy 2009), the Best Pilot may take this too far, engaging in showy, risky behavior, talking down to other pilots and exhibiting a disregard for aircrew coordination and CRM.

The fourth category of at-risk aviator is known as the Consistent Poor Performer. The Consistent Poor Performer has a history of below average performance and is more likely to be found in military aviation settings, where aviation may be a career choice by necessity rather than interest. Therefore, these individuals may be well liked or even excel at military "ground jobs." However, the Consistent Poor Performer barely meets, or shows slow, qualification progress. In the aircraft, the Consistent Poor Performer is an individual that becomes easily distracted and task overloaded. Also, the Consistent Poor Performer will frequently lose situational awareness during the flight. Taken together, these issues result in a complete lack of confidence by the individual.

The final category of at-risk aviator deals with the Overstressed Aviator. This individual has been exposed to a number of major stressors such as the death of a close family member or friend, a recent divorce, a failed relationship, or serious financial trouble. Whatever the major stressor, this individual is at significant risk of being unable to compartmentalize and may become distracted leading to an uncharacteristic breakdown of flight discipline and violations of the rules.

The at-risk aviator categories identified above have been used as schemas for psychologists and supervisors to identify those individuals that pose a serious risk in the air to themselves and others. As the attitudes and behaviors of these individuals are allowed to continue on, the result is a poor safety climate within a squadron or organization.

Mishap Investigations: Psychologists and Their Expanding Role

It has been well established that as high as 70–80 percent of all mishaps in "error intolerant fields" are due to human error (Helmreich 2000; Reason 1990). Often times, the psychologist as a contributing member of the aviation mishap board will be called upon to evaluate these aspects of flight and the likelihood of their contribution to the cause of the mishap. For example, psychologists might be tasked with determining why particular decisions may have been reached in a high-stress environment. In this case, psychologists would rely on their domain expertise to answer questions such as how stress might affect decision-making in a particular environment.

When a mishap occurs, the investigation process begins with an initial notification to members of an aviation mishap board (for a detailed review see Wood and Sweggenis 2006; see also King 1999). The aviation mishap board will usually be composed of a number of subject matter experts (SMEs) in various disciplines. For example, a typical NTSB aviation mishap board might include aerospace engineers, metallurgists, and medical forensic experts. These individuals are separated into various on-site groups with each group focusing on an area of their own expertise. For example, metallurgists and structural engineers may be grouped together and investigate all aspects related to structures of the mishap. Psychologists participating in the investigation process for civilian aviation are typically assigned to the Human Performance team which examines human factors issues as they relate to the mishap.

Military aviation mishaps differ from civilian aviation in that the mishap investigation team will usually be much smaller and each member of the team will be asked to investigate all aspects of the mishap. For example, a designated pilot or flight officer may be asked to analyze structural issues as well as weather. Therefore, the psychologist that participates in a mishap investigation for military aviation will be required to examine more than just human factor's issues with regards to the causes of a particular mishap. However, at all phases of the investigation the psychologist should be concerned with addressing the "human elements" in the error chain.

As the mishap investigation progresses the mishap board typically begins by gathering available evidence and attempts to reconstruct the mishap. During this phase of the investigation the mishap team will start to compile evidence from numerous different sources. These sources might include interviews, radio transmissions, and/or the infamous "black box" recorder. Each of these sources will be critically analyzed according to the source's veracity and ability to provide detailed information about what happened to the aircraft during the moments leading up to the mishap.

The aviation psychologist will be relied on as a valuable resource due to his or her analytical ability and training in human behavior. With respect to witness interviews, the psychologist often understands appropriate phraseology and other details about the interview process that may bias an eye witnesses' account of what had transpired. In addition, the importance of witness interviews often varies with the accident. In many cases, witness interviews can be a vital piece of information especially if there were no survivors and/or no recorded information. However, in a limited number of cases there is significant factual information available and witness interviews may be used as converging or corroborative evidence (Wood and Sweginnis 2006).

In addition to witness interviews, the psychologist may be asked to assess the psychological factors (that is, affective state, motivation, fatigue) of the crew at the time of the mishap. Unfortunately, there is seldom ever irrefutable evidence of psychological issues and how they may have influenced the crew during flight. Therefore, the psychologist must utilize his/her professional training in evaluating a comprehensive list of potential psychological influences. In the end, the psychologist will need to determine if the psychological dimension in question directly contributed to the mishap.

With such a high percentage of mishaps resulting from human error, it should be apparent that there is a requirement for a more coherent investigation process that incorporates many of the psychological principles addressed in Table 14.1 above. To assist the mishap investigation team in identifying human factors as causal factors in a mishap, the NTSB and the Department of Defense (DoD) have adopted a more formal and structured reporting and investigating tool known as the Human Factors Analysis and Classification System (HFACS). The next section will highlight some of the features of this model and discuss how this model has been used by mishap investigation teams to identify those human factors issues that may have been causal to a particular mishap.

Human Factors Analysis and Classification System (HFACS)

While personality factors can on occasion be a causal factor in a mishap, aviation mishaps are rarely the cause of a single individual at the controls of the aircraft. Rather, mishaps are the result of a combination of active and latent failures that result in a breakdown within the organization (Reason 1990). The final unsafe act that led to the mishap is usually the result of the interaction between failures at various levels within the organization. However, it is the combination of the active and latent failures that produce a set of conditions that are conducive to the final unsafe act.

According to Reason's Organizational Model of Human Error (1990), the final unsafe act, sometimes referred to as aircrew/pilot error, is often the focal point of aviation accident investigation. Not surprisingly, aircrew/pilot errors are the most commonly identified causal factors in accident investigations. The most likely reason for this pattern is that these active failures can be directly attributed to the accident itself. For example, a pilot flying into IMC can result in an immediate threat to the safety of the aircraft and the crew.

However, in addition to the active failures that result in a mishap, a number of latent failures may have contributed as causal factors as well. In contrast to active failures, latent failures may be difficult to identify because they may lie dormant or undetected within the organization. Latent failures are errors committed by individuals within the organization at the supervisory and higher levels that have resulted in a set of preconditions leading to the unsafe act. For example, an organization's decision to reduce training and/or flight qualifications in response to budgetary concerns may result in decreased performance for the aircrew. Viewed from this perspective then, the unsafe acts of aircrew are the end result of a long chain of causes whose roots originate in other parts (often the upper echelons) of the organization. Aviation psychologists have been instrumental in the areas of accident investigation by informing the investigation team about both the unsafe acts that have contributed to the active failure and the preconditions that led to that unsafe act—the latent failures within the organization.

In an attempt to develop a standardized investigative and reporting system for accident investigation, Shappell and Wiegmann (1997, 1998, 1999, 2000, 2001) designed HFACS (for a review see Wiegmann and Shappell 2003). Simply put, HFACS provides structure to Reason's Organizational Model of Human Error (1990). The HFACS system was modeled by analyzing human causal factors from various accident reports in a number of highly technical arenas. For the purposes of this chapter, HFACS will be discussed in reference to aviation as a method for reporting and investigating mishaps (Wiegmann and Shappell 2001). While a comprehensive analysis of HFACS is beyond the scope of this chapter, a broad discussion of the model and its utility in accident investigation will help the reader understand the manner in which psychologists have contributed a substantial amount of work in the area of accident investigation to the aviation industry.

The HFACS model (see Figure 14.1 overleaf) identifies four levels for which failure within an organization can occur and result in a mishap: (1) Unsafe Acts; (2) Preconditions for Unsafe Acts; (3) Unsafe Supervision; and (4) Organizational Influences. The layers within the HFACS model are used to represent the original unsafe act that was committed by the operator and the preceding latent failures that produced a set of preconditions that allowed the unsafe act to occur. The remainder of this section will focus on the various levels identified within the HFACS model and provide a brief description of how human error at each level can lead to the final unsafe act.

Unsafe Acts

Reason (1990) dichotomizes Unsafe Acts into two separate categories: errors and violations. The Errors category is used to identify those mental and physical activities that were conducted by the individual at the time of the unsafe act. These activities can range from stick-and-rudder skills to misperceptions and spatial disorientations occurring in the cockpit. The HFACS model extends the work initially done by Reason by increasing the specificity for which errors are identified.

To allow for an increase in granularity that is needed to identify causal factors in an accident investigation, Wiegmann and Shappell (2003) expanded the error category to include three basic error types: Decision Errors, Skill-Based Errors, and Perception Errors. Judgment and Decision-Making Errors can best be understood as behaviors that have been performed as intended by the aviator. However, the HFACS model identifies Judgment and Decision-Making Errors as those acts that have been performed by the aviator in a manner that they either did not have the appropriate knowledge available or just simply chose poorly.

Skill-Based Errors on the other hand, can best be understood as errors resulting from the lack of flying skill necessary to operating an aircraft safely. In an evaluation of the mishap causal factors of major US Naval F/A-18 (a multi-role jet aircraft) and H-60 (a type of helicopter), skill-based errors were the most commonly cited factors (70.2 percent and 81.3 percent, respectively; O'Connor, Alton and Cowan 2010). These skills have often been referred to as stick-and-rudder skills and have been shown to be extremely vulnerable to lapses in memory and/or attention (Wiegmann and Shappell 2003). Many of the skills required to successfully operate an aircraft are based on a number of automatic behaviors that are learned through countless hours of flying and simulation work. Therefore,

Figure 14.1 The Human Factors Analysis and Classification System (HFACS)

Source: Adapted from Wiegmann and Shappell (2001)

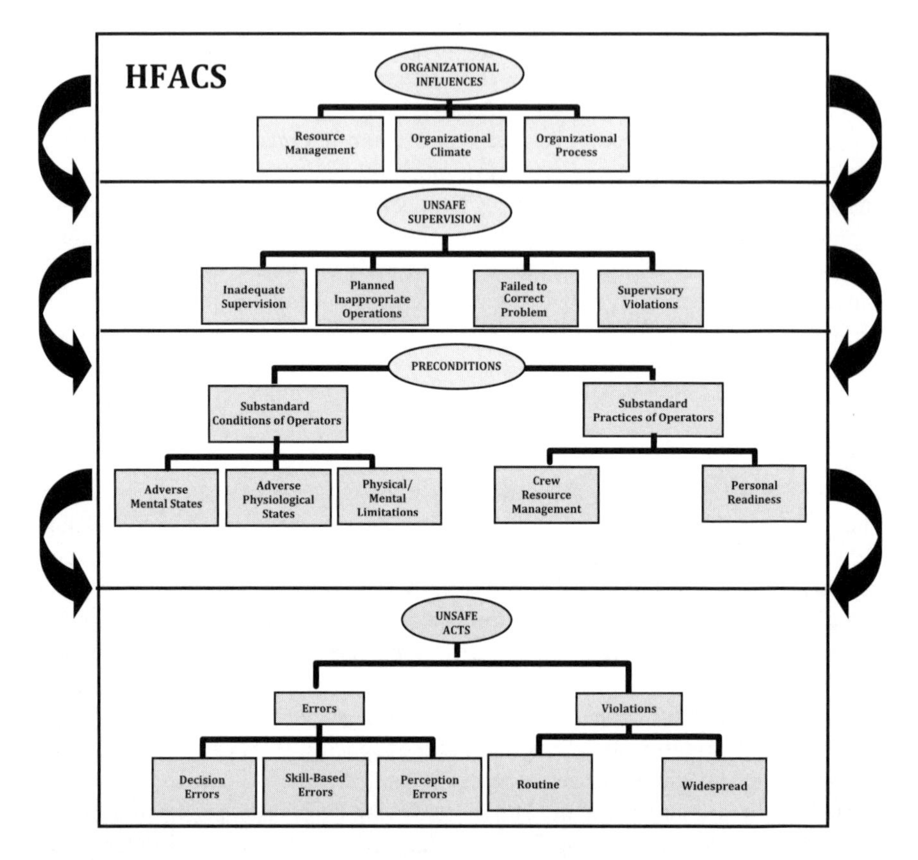

these skills are particularly vulnerable to failures of attention and/or memory (Wiegmann and Shappell 2003).

Finally, Perception Errors occur when sensory input is degraded or "unusual," as is the case when the aviator has suffered from vertigo or other forms of spatial disorientation. In the case of Perception Errors, the individual is often forced to make a decision based on faulty or erroneous information which may result in a mishap. An aviator that has committed a Perception Error may have all of the skills necessary to operate the aircraft efficiently. However, due to environmental or physiological factors the aviator does not possess the necessary information for piloting the aircraft efficiently.

The second category within the Unsafe Acts layer of the HFACS model is used to identify Violations. For the purposes of HFACS, Violations represent a willful disregard for the rules and regulations governing safe flight. This form of Unsafe

Act usually involves punitive results and may involve the loss of flying privileges within that organization.

Preconditions

While the majority of mishaps can be directly traced to the Unsafe Acts committed by the aircrew, accident investigators are sometimes forced to dig a little deeper for some of the underlying reasons that allowed the unsafe act to take place. The Preconditions Category of HFACS attempts to identify some of the underlying reasons the crew participated in the Unsafe Act by looking at a number of environmental and personnel factors of the aircrew.

There are a number of preconditions that may have produced a set of conditions that help to explain why the unsafe act occurred in the first place. HFACS splits the preconditions category into Substandard Conditions and Substandard Practices of the Operator.

Substandard Conditions of the Operator refers to those situations in which the crew is either mentally or physically unprepared to deal with the adverse conditions of flight. Adverse mental states describe those situations in which certain mental conditions might affect overall performance. For example, factors such as a loss of situational awareness, mental fatigue, or complacency may all have a negative impact on performance. In contrast, adverse physiological states describe those instances in which safe flight cannot be achieved due to conditions such as a visual illusion, hypoxia, or some other impairment that might affect the body's physiology. Finally, physical and/or mental limitations may prevent the operator from flying the aircraft safely. For example, there may be an insufficient warning signal in the aircraft or the aircrew may have simply lacked the skill or ability to deal with the emergency. Taken together, these three categories are all used to describe those instances in which a mishap may have occurred due to various conditions that affected the operator's performance.

Substandard Practices of the Operator refers to those behaviors by the aircrew that may have resulted in the Unsafe Acts discussed previously. For example, CRM can impact both communication and crew coordination within the cockpit. A quick review of accident investigation reports will provide ample evidence of the manner in which poor aircrew coordination led to confusion and poor decision-making inside the cockpit. Personal readiness has also been shown to be a major contributing causal factor to a number of aviation accidents. Individuals who fail to prepare themselves for the rigors of high-risk environments have set themselves and their crew up for failure. Certainly these issues need to be identified due to the likelihood an air crewman with these issues may perform an Unsafe Act.

Supervision

As the mishap investigation and analysis continues to unfold, the investigators may begin to uncover safety issues that can be traced back up the supervisory chain of command. Therefore, Reason (1990) has argued that a proper mishap investigation must provide a comprehensive analysis of the supervisors and their influence on the condition of the operators within an organization. Therefore, the HFACS model attempts to identify four areas within the supervisory level that can have an immediate impact on the pilot and the environment in which they operate: Inadequate Supervision, Planned Inappropriate Operations, Failed to Correct a Known Problem, and Supervisory Violations. Each of these categories is briefly addressed below.

Inadequate Supervision within an organization implies that the supervisor and/or management have not provided an atmosphere that is conducive to safe and effective use of that organization's assets. For example, the supervisor may not provide effective guidance, few training opportunities, poor leadership, or provide a proper role model for the individuals within that organization. As these types of issues begin to accumulate within an organization, the aircraft operator may not be as effectively equipped to deal with an emergency procedure should the need arise. For example, a pilot that was not provided annual CRM training may not be as effective at utilizing those resources when the pilot is confronted with an emergency. Therefore, Inadequate Supervision (improper training in this case) may produce an atmosphere that is conducive to unsafe flying.

Unfortunately, the cost to employ and maintain a ready crew is an issue in both the military and civilian world. Therefore, scheduling and planning may place aircrews at risk. Heavy scheduling due to manning requirements should be viewed as unacceptable during normal operations. The category of Planned Inappropriate Operations attempts to account for these supervisory failures by addressing issues such as crew pairing and improper manning. For example, supervisors that continue to pair two individuals with poor communication skills can compromise the safety within that organization. As organizations attempt to become "leaner" and more efficient, crew management can be an extremely perplexing task. However, an organization must be cognizant about scheduling and manning in order to maintain safety.

In addition to the problems addressed above, Supervisory Failures may occur when there is a problem that occurs within that organization and middle management fails to correct the problem. For example, if a supervisor knows about a deficiency in equipment, training, or personnel but fails to act upon these deficiencies, then the supervisor can be viewed as being a contributing causal factor to the mishap. In other words, had the supervisor corrected the equipment problem, provided the proper safety training, or removed the unsafe individual from the flight schedule then the mishap may not have occurred in the first place.

The final category under the supervisory layer involves Supervisory Violations. Although rare, these types of violations refer to an instance where a supervisor may violate the rules in order to manage their assets. For example, a supervisor faced with a limited number of available resources may not provide adequate documentation or may authorize unnecessary hazards within that organization. These types of failures at the supervisory level often contribute to an unsafe culture within that organization that, if left unchecked, may result in a mishap.

Organizational Influences

The final layer of the HFACS model attempts to identify those issues associated with the corporate rules and decisions that govern an organization's everyday activities. Organizational Influences may refer to such rules and regulations as standard operating procedures or the resource and acquisition process within a particular company. These policies and procedures are instrumental in setting the policy for safety within that organization. An organization that approaches its safety policies as a genuine tool for reducing the likelihood of mishaps is more likely to put in place effective procedures at protecting the employees within that organization. However, an organization that views its safety policies with a laissez-faire attitude is likely to be more prone to mistakes leading to the loss of life or damage to equipment.

Summary of Human Factors Analysis and Classification System

Reason (1990) provided a comprehensive model of human error that attempted to identify causal factors of mishaps past the unsafe act. Specifically, Reason's model identified various causes of error that might occur at deeper levels within an organization. However, Reason's model of human error was not able to quantify the types of errors that might occur at the various levels within an organization. HFACS has been implemented by psychologists and human factors professionals as a method to identify or define the sources of error within an organization to facilitate aviation mishap reporting and investigation. It is worth noting that the instantiation of the HFACS model was derived through careful investigation and analysis of hundreds of aviation mishaps. Using this methodology, the HFACS model has been shown to be a reliable tool with regards to mishap reporting (Wiegmann and Shappell 2003).

Conclusions

The field of psychology has influenced the aviation industry both in education for mishap prevention and in mishap investigation and reporting. The purpose of this chapter was to highlight some of the contributions that have been made by psychologists in the aviation industry. Particularly, in the last three decades psychologists have become increasingly involved in many aspects of the aviation industry. If the mishap rate in aviation is to be further reduced, this will not be achieved through engineering solutions. The focus must be on the "human component" of the aircraft. Therefore, it is clear that psychologists will have an increasing role in the aviation industry as long as there are still people flying aircraft.

References

Aviation Safety Network. 2008. *Air Inter, Strasburg, 20 January 1992 Crash Report*. Available at: http://aviation-safety.net/database/record.php?id=19920120-0 June 2009.

Badalato, B. (Executive Producer), Bruckheimer, J. (Producer), Simpson, D. (Producer), Skaaren, W. (Associate Producer) and Scott, T. (Director). 1986. *Top Gun* [Motion Picture]. USA: Paramount Pictures.

Berlin, J.I., Gruber, E.V., Holmes, C.W., Jensen, P.K., Lau, J.R., Mills, J.W. and O'Kane, J.M. 1982. Pilot judgment training and evaluation (DOT/FAA/CT-82/56). Daytona Beach, FL: Embry-Riddle Aeronautical University. (NTIS No. AD-A I 1 7 508).

Campbell, J.S., Moore, J.L., Poythress, N.G. and Kennedy, C.H. 2009. Personality traits in clinically referred aviators: Two clusters related to occupational suitability. *Aviation, Space and Environmental Medicine*, 80, 1049–1054.

Cattel, R.B., Eber, H.W. and Tatsuoka, M.M. 1970. *The Handbook for the Sixteen Personality Factor (16PF) Questionnaire.* Champaign, IL: Institute for Personality and Ability Testing.

Chief of Naval Operations. 2001. *Naval Aviation Safety Program*, OPNAVINST 3750.6R. Washington DC: Author.

Commandant of the Marine Corps. 2004. *Marine Corps Safety Program*. MCO 5100.29A. Quantico, VA: Author.

Cooper, G.E., White, M.D. and Lauber, J.K. 1980. *Resource management on the flightdeck: Proceedings of the NASA workshop on resource management training for airline crews*. Moffett Field, CA: NASA-Ames Research Center.

David, G. (1996). Lessons from offshore aviation: Towards an integrated human performance system, in *Managing the Offshore Workforce*, edited by R. Flin and G. Slaven. Tulsa: PennWell, 219–238.

Droog, A. 2004. The current status of CRM training and its regulation in Europe, in *Aviation Psychology: Practice and Research*, edited by K. Goeters. Aldershot: Ashgate Publishing Limited, 221–230.

Federal Aviation Authority. 2004). *Advisory Circular No 120-51E: Crew Resource Management Training.* Washington, DC: Author.

Fitts, P. and Jones, R. 1947. *Analysis of Factors Contributing to 460 "pilot-error" Experiences in Operating Aircraft Controls.* Memorandum Reported TSEAA-694-12. Dayton OH: Aero Medical Laboratory.

Flin, R. and Martin, L. 2001. Behavioural markers for CRM: A review of current practice. *International Journal of Aviation Psychology*, 11, 95–118.

Flin, R., O'Connor, P. and Mearns, K. 2002. Crew resource management: Improving safety in high reliability industries. *Team Performance Management*, 8, 68–78.

Goeters, K. 2004. Validation of CRM training by NOTECHS: Results of the PHARE ASI project, in *Aviation Psychology: Practice and Research*, edited by K. Goeters. Aldershot: Ashgate, 291–297.

Gregorich, S.E. and Wilhelm, J.A. 1993. Crew resource management training assessment, in *Cockpit Resource Management*, edited by E.L. Wiener, B.G. Kanki and R.L. Helmreich. San Diego, CA: Academic Press, 173–196.

Grubb, J. 2010. Workload, situation awareness, and the right stuff for assessing cockpit design, in *Human Performance Enhancements in High-Risk Environments: Insights Developments, and Future Directions from Military Research*, edited by P. O'Connor and J. Cohn. Westport, CT: Praeger Press.

Helmreich, R.L. 2000. On error management: lessons from aviation. *British Medical Journal,* 320, 781–785.

Helmreich, R.L., Merritt, A.C. and Wilhelm, J.A. 1999. The evolution of crew resource management training in commercial aviation. *International Journal of Aviation Psychology*, 9, 19–32.

International Ergonomics Association. 2000. *What is Ergonomics?* Available at: http://www.iea.cc/01_what/What%20is%20Ergonomics.html [accessed: 11 June 2009].

Jansma, J.M., Ramsey, N.F., Coppola, R. and Kahn, R.S. 2000. Specific versus nonspecific brain activity in a parametric N-back task. *Neuroimage*, 12(6), 688–697.

Kern, A.T. 1999. *Darker Shades of Blue: The Rogue Pilot.* New York: McGraw-Hill.

King, R.E. 1999. *Aerospace Clinical Psychology*. Aldershot: Ashgate.

Kirkpatrick, D.L. 1976. Evaluation of training, in *Training and Development Handbook*, edited by R.L. Craig and L.R. Bittel. New York: McGraw Hill, 18.1–18.27.

Lester, L.F. and Bombaci, D.H. 1984. The relationship between personality and irrational judgment in civil pilots. *Human Factors*, 26, 565–572.

Lester, L.F. and Connolly, T.J. 1987. The measurement of hazardous thought patterns and their relationship to pilot personality, in *Proceedings of the Fourth International Symposium on Aviation Psychology*, edited by R.S. Jensen. Columbus: The Ohio State University, 286–292.

Merlin, P.W., Bendrick, G.A. and Holland, D.A. 2012. *Breaking the Mishap Chain: Human Factors Lessons Learned from Aerospace Accidents and Incidents in Research, Flight Test, and Development.* Washington, DC: National Aeronautics and Space Administration.

National Transportation Safety Board. 1998. *Aircraft Accident Report: Air Florida, Inc., Boeing 737-222, N62AF, Collision with 14th Street Bridge near Washington National Airport.*

National Transportation Safety Board. 2007. *Aircraft Accident Report: MIA06FA039*. Washington DC, May 2010.

National Transportation Safety Board. 2010. *Aircraft Accident Report: Loss of Thrust in Both Engines After Encountering a Flock of Birds and Subsequent Ditching on the Hudson River*. Washington DC, May 2010.

O'Connor, P., Alton, J. and Cowan, S. 2010. A comparison of leading and lagging indicators of safety in naval aviation. *Aviation, Space and Environmental Medicine*, 81, 677–682.

O'Connor, P., Campbell, J., Newon, J., Melton, J., Salas, E. and Wilson, K. 2008. Crew resource management training effectiveness: A meta-analysis and some critical needs. *International Journal of Aviation Psychology*, 18(4), 353–368.

O'Connor, P., Flin, R. and Fletcher, G. 2002. Methods used to evaluate the effectiveness of CRM training: A literature review. *Journal of Human Factors and Aerospace Safety*, 2, 217–234.

O'Hare, D. 2003. Aeronautical decision making: Metaphors, models, and methods, in *Principles and Practice of Aviation Psychology*, edited by P.S. Tsang and M.A. Vidulich. Mawah, NJ: Lawrence Erlbaum Associates.

Prinzel, L.J., Freeman, F.G., Scerbo, M.W., Mikulka, P.J. and Pope, A.T. 2003. Effects of a psychophysiological system for adaptive automation on performance, workload, and the event-related potential P300 component. *Human Factors*, 45(4), 601–613.

Reason, J. 1990. *Human Error.* Cambridge: Cambridge University Press.

Salas, E., Burke, C.S., Bowers, C.A. and Wilson, K.A. 2001. Team training in the skies: Does crew resource management (CRM) training work? *Human Factors*, 41, 161–172.

Salas, E., Prince, C., Bowers, C.A., Stout, R.J., Oser, R.L. and Cannon-Bowers, J.A. 1999. A methodology for enhancing crew resource management training. *Human Factors*, 41(1), 161–172.

Salas, E., Wilson, K.A, Burke, C.S., Wightman, D.C. and Howse 2006. A checklist for crew resource management training. *Ergonomics in Design*, 14, 6–15.

Shappell, S. and Wiegmann, D. 1997. A human error approach to accident investigation: The taxonomy of unsafe operations. *The International Journal of Aviation Psychology*, 7(4), 269–291.

Shappell, S. and Wiegmann, D. April 1998. *Failure analysis classification system: A human factors approach to accident investigation.* SAE: Advances in Aviation Safety Conference and Exposition, Daytona, FL.

Shappell, S. and Wiegmann, D. 1999. Human factors analysis of aviation accident data: Developing a needs-based, data-driven, safety program. *Proceedings of the Fourth Annual Meeting of the Human Error, Safety, and System Development Conference.* Liege, Belgium.

Shappell, S. and Wiegmann, D. 2000. *The Human Factors Analysis and Classification System (HFACS).* (Report Number DOT/FAA/AM-00/7). Washington DC: Federal Aviation Administration.

Shappell, S. and Wiegmann, D. 2001. Unraveling the mystery of general aviation controlled flight into terrain accidents using HFACS. *Proceedings of the Eleventh Symposium for Aviation Psychology*, Ohio State University.

Wiegmann, D.A. and Shappell, S.A. 2001. Human error analysis of commercial aviation accidents: Application of the human factors analysis and classification system (HFACS) within the context of general aviation. *Aviation Space and Environmental Medicine*, 73(2), 266.

Wiegeman, D.A. and Shappell, S.A. 2003. *A Human Error Approach to Aviation Accident Analysis*. Aldershot: Ashgate.

Wilson, G.F. and Russell, C.A. 2003. Real-time assessment of mental workload using psychophysiological measures and artificial neural networks, *Human Factors*, 45(4), 635–644.

Wood, R.H. and Sweggenis, R.W. 2006. *Aircraft Accident Investigation.* Mountain States: Endeavor Books.

Wood, S. 2004. *Flight Crew Reliance on Automation.* Gatwick: Civil Aviation Authority.

Chapter 15
On Becoming an Aeromedical Psychologist

Trevor Reynolds

Introduction

For as long as humans have looked at the sky, wondered "what is up there?" and aspired to fly, the role of psychology in defying the law of gravity has been at least as critical as the science of engineering. Indeed, it could be argued that psychology, encompassing the understanding of human behavior and including such aspects as aspiration, ambition, and risk-taking behavior, laid the table on which engineering and technology made human flight possible.

While the study of human behavior has increasingly been called upon to enhance performance and efficiency in the pursuit of optimal human/machine interface and has been the subject of extensive inquiry, the application of the clinical component of the study of human behavior to flight safety has been less extensively studied or formalized (see Chapter 1, this volume for a review of this evolution). Initially, the industrial psychologists played the predominant psychology role in aviation, creating the first assessment and selection programs in international militaries and commercial aviation agencies (please see Chapters 2 and 3, this volume for more on assessment and selection). However, it soon became clear to flight surgeons that successful (or perhaps more correctly, unsuccessful) flight personnel exhibited other variables which complicated the picture and looked to the field of psychiatry for assistance. Personality variables, significant and sometimes traumatic events associated with flight, disease development affecting mood and cognition, and so on became a major focus for aviation medicine. As with the role of psychologists during World War II (WWII), which moved from a primarily testing and research role to evolving clinical roles, the same happened in aviation. However, unlike other psychological subspecialties (for example, neuropsychology, prescribing psychology, and child psychology), for which there are established training programs and supplemental privileging, there are currently no routine or agreed-upon methods to establish competency as an aeromedical psychologist.

Given the dearth, internationally, of formal training opportunities which focus on the practice of aeromedical psychology, it would be expected that many of the psychological opinions and decisions offered to the aviation community sometimes run the risk of not appreciating, to its fullest consequence, the unique challenges posed to the individual in the aviation environment. Conditions which hold true for general psychological practice do not necessarily hold true in the aviation milieu and uninformed decision-making in the aviation environment truly can be a life or death issue.

This is an interesting, almost paradoxical, problem as professional competence is a well-described concept within the ethics literature (see Nagy 2012 for a comprehensive review) and psychologists engaging in specialty practice are required to "provide services, teach and conduct research with populations and in areas only within the boundaries of their competence, based on their education, training, supervised experience, consultation, study or professional experience" and "psychologists planning to provide services, teach or conduct research involving populations, areas, techniques or technologies new to them undertake relevant education, training, supervised experience, consultation or study" (American Psychological Association 2010). Given that these requirements are also legal requirements of maintaining licensure in most US states and are present in some form within the credentialing actions of international psychology regulatory agencies, the need for formal training and experiential opportunities is magnified. The recent promulgation of the Universal Declaration of Ethical Principles for Psychologists (International Union of Psychological Sciences 2008) has gone some way to further formalizing the values and widening the ambit of relevance of the potential paradox.

Aeromedical psychology is certainly not alone in the pursuit of standards to enable and establish professional competence. Other newer subspecialties (for example, operational psychology) are in the process of achieving these objectives by creating training, holding formal conferences, defining the field, and establishing standards of practice (Kennedy and Williams 2011). This chapter will describe concepts pertaining to the competent practice of aeromedical psychology and provide as an example current efforts in South Africa to implement and provide specialized training and certification.

Traditional Applications of Psychology in Aviation

From early in the history of aviation, and certainly not later than the advent of aviation in military operations during World War I (WWI), there has been a sustained and systematic pursuit of the identification of those skills and attributes which define the successful aviator (see Chapter 1, this volume). This pursuit initially fell largely within the domain of industrial psychology, applying expertise to decisions of recruitment and selection, with mechanical ability, psychomotor speed, processing speed, decision-making, and purposeful behavior being traditionally influential in selection decisions. Later, a variety of other characteristics gained momentum as being decisional factors. Notwithstanding the diversity of expression, it would seem that there is relative consensus, robust over the passage of time, regarding the required attributes of the successful aviator.

Within the commercial sphere it can be argued that a large part of the extensive body of knowledge regarding the requisite characteristics for a successful career as a pilot is a function of the corporate need to insure the best balance between safety and efficiency on the one hand, and profitability and risk amelioration on the other. An argument may also be advanced that there are differences in the personality composition of general/recreational pilots, who possess different motivations for flying as compared to the commercial pilot who does it as a career. Within the militaries, a tolerance for a heightened level of risk and a variety of other skills are required, creating the need for additional selection, monitoring procedures, and flight status requirements.

Modern Roles of Aeromedical Psychologists

Roles of aeromedical psychologists can be conceptualized as broaching two components: initial selection and continued flying. In addition to each agencies assessment and selection procedures (see Chapters 2 and 3, this volume), initial selection includes a variety of physical/medical standards and generally include a disqualification or "select-out" component (that is, does the applicant have a disqualifying condition?) and a "select-in" component (that is, does the applicant possess certain attributes known to result in success?). These require different complements of skills on the part of the evaluating psychologist. For example, within the variety of international militaries and commercial agencies, policies identifying disqualifying mental disorders are set through waiver and certification procedures and usually are fairly straightforward. Many of these evaluations can be done by any credentialed psychologist or psychiatrist as these decisions do not require specialty knowledge of the flight environment (for example, recommendation for disqualification based on the presence of a psychotic disorder). However, aviation jobs (for example, high-level military flying) requiring a complement of personal attributes geared toward success in that environment require evaluations demanding significant specialty training and experience.

The second component of the duties of the aeromedical psychologist are more frequently what is asked of the clinician. Is the individual appropriate for continued flight status? This concept is expressed differently in various agencies and is sometimes referred to as physically qualified, aeromedically adapted, fit to fly, and so on. This question can arise at various stages in the career of the individual with flight status (that is, pilots, navigators, flight officers, air traffic controllers, and aircrew) due to problems experienced in training, transitioning to a new aircraft, following a traumatic mishap, due to a disease process affecting cognition and/or mood, effects of aging, implementation of a mental health treatment regulated by aviation policy, and/or significant life stress (for a guide on conducting the aviation psychological evaluation, see Chapter 4, this volume). How does one become competent to answer these questions, where understanding of the flight environment, the aircraft flown by the pilot being evaluated, pilot personalities, the impact of aging on aviation abilities, the air traffic control environment, a specific military mission, and so on are keys to the evaluation which will be provided and the context in which recommendations and decisions will be made?

Basic Requirements

To address the issue of professional competence in this arena, it is sensible to begin with the basic requirements. As with all psychology subspecialties, aeromedical psychology contains the same core standards. Aeromedical psychologists must meet the minimum academic standards for licensure, and be credentialed to independently practice clinical psychology.

Additional Requirements

However, the basic requirements to work as a psychologist are insufficient for psychologists working with any specialized population. Just as one would not think to provide an evaluation of a high school student without an appreciation of the high school environment, a psychologist must demonstrate an appreciation of the aviation population and the aviation environment. To provide a clinical psychological evaluation of any individual with flight status, other requirements include:

> training and expertise in psychological assessment, possession of the appropriate assessment instruments and normative data to evaluate airmen and other aircrew, familiarity with the literature on aviation psychology and aviation neuropsychology, knowledge of the occupational demands placed on aircrew and … medical certification standards, knowledge of … legal requirements related to fitness-for-duty evaluations and sufficient preparation to serve as an expert witness in any legal proceedings that may arise from the evaluation. (Kay 2002, 233)

The key questions then become, how does one obtain this knowledge and these additional requirements?

There are very few formal clinical aviation psychology training programs and those that exist, exist largely through the militaries, given the involvement of psychologists in military aviation since the world wars. Psychologists providing services for commercial aviation agencies or assessing commercial pilots frequently report receiving their expertise through prior military service, mentorship from an experienced aeromedical psychologist or psychiatrist, on-the-job training and/or by becoming a private pilot themselves. As specialized psychological services have expanded and become more formalized in recent decades, these types of individualized initiatives have become insufficient.

The Training Concept

The need for formal aeromedical psychology training has been recognized and is in preparation in various guises. In some militaries, this training is already highly formalized and structured (for example, US Army and Navy; for a comical but accurate depiction of the Navy training, please see Saitzyk 2012). However, within the civilian/commercial domain, this has generally not been the case, though programs for the specialized training and certification of medical doctors have existed since almost the advent of aviation. Consequently, one proposal in which to provide the formal structure and training needed by aeromedical psychologists is to adapt the already existing, systematized learning pathway used by Aviation Medical Examiners (AMEs) and flight surgeons.

The Aviation Medical Examiner/Flight Surgeon Model

AMEs and flight surgeons enter into the domain of aviation medicine once they have identified their interest in the field, after qualifying as a medical practitioner. Usually, general practitioners (GPs) who are interested in aviation medicine undertake a basic AME course. AME and flight surgery training vary internationally in length between the militaries and civil aviation but in general the focus is the same, that of the application of medical principles to the aviation milieu as well as education on the various aeromedical policies and procedures. As with all other medical specialities, refresher training and continuing education are required for maintenance of the certification.

The Aviation Psychology Examiner Model

A clear starting point to provide necessary initial training for psychologists is the adaptation of the AME designation, an effort which is ongoing in South Africa. The model, pertaining to Designated Aviation Psychology Examiners (DAPEs), is described in concept below. Of significance is the recognition that while AME courses are widespread internationally, to the best of our knowledge there is no similar course directed toward the application of aeromedical psychology competence specifically within the civil aviation milieu.

A certificate course in Aviation Psychology has been designed which has as an initial component the presentation of a theoretical introduction to aviation medicine, addressing those sub-disciplines particularly pertinent to aviation. Attention is therefore paid to neurology, cardiology, pulmonology, otorrhinolaryngology, and psychiatry, with the purpose of identifying those conditions which may impact medical certification, including the impact of those conditions on an individual's personality, mental health, and cognition. Building on the introductory theory of these medical sub-disciplines, specific attention is focused on the aeromedical psychological knowledge required in order to make informed decisions regarding an applicant's psychological fitness to fly or fitness to resume flying duties.

The current structure of this course is that it is a residential course, presented under the auspices of the division of psychiatry of one of the South African medical schools. A parallel development of this certificate course is an investigation into offering the course as an online learning opportunity, which would then make it more accessible and available to an international market. The intention is that this certificate course would enjoy academic and regulatory recognition and allow for some degree of portability to subsequent studies.

An envisaged written further study is the conceptualization of the Master's degree in Public Health (MPH) in Aviation Medicine, which is accessible to any of the health disciplines for whom recognition of specialisation in aviation would contribute to aviation safety, safe flight operations, and even to specialist aspects of aviation, such as medical evacuations. The development of an intermediate step between a certificate course in aviation psychology and entry into the MPH would allow for further upskilling and development of competence in that particular field of aviation medicine as it pertains to aviation psychology, while at the same time formalizing and codifying the requisite knowledge in the clinical application of aviation psychology in order that the regulatory aviation authorities would be in a position to access pertinent specialist expertise when called upon to make decisions regarding an aviator's fitness to fly or to resume flying duties.

Recognition and Certification

Currently, within the South African regulatory framework there is no example of a post-registration training course (that is, postgraduate training/fellowship) leading to recognition by the Professional Board of Psychology of the Health Professions Council of South Africa. The Health Professions Council is the statutory body which registers and licenses medical practitioners, dental practitioners, psychologists, and allied healthcare professionals. Current registration is the minimum requirement for practicing the particular profession in South Africa, and failure to be duly registered may lead to legal action.

The Certificate Course in Aviation Psychology will be submitted to the Educational Committee of the Professional Board of Psychology, with the request that the course be evaluated and successful completion of the course be appropriately acknowledged, for example, by the awarding of the post-nominal qualifier "Cert Av Psych," or something similar. This will be awarded in acknowledgement of the development and demonstration of competence beyond that minimum required for registration.

A component of the presentation to the Professional Board of Psychology will relate to foreign/non-South African psychologists, who may successfully complete the course, especially in its online format, and who could then apply for the post-nominal acknowledgement.

This model is, in our understanding, not in operation anywhere in the world despite recognized problems regarding the lack of formalized training and credentialing for clinical psychologists desiring work in the aviation environment. This model will empower the international regulatory authorities to rely on demonstrated expertise when making decisions regarding psychological and neuropsychological factors influencing fitness to fly.

Conclusion

While psychology has historically played a substantial role in aviation, from early in the history of flight, the major contribution of psychology has related largely to those human factors optimizing flight safety and efficiency. The role of aeromedical psychology in contributing to decisions regarding an aviator's qualifications for obtaining and maintaining flight status, notably after injury or illness, has been the lesser acknowledged sub-discipline of aviation psychology. It is possible that this has been a function of an absence of training and certification in an environment in which the medical model of certification is particularly influential. The recognition of a need for training and certification has led to the development of the concept of training in aviation psychology which, it is foreseen, will lead to designation as an aviation psychology examiner. Designation will allow for regulatory authorities to access particular expertise and competence in order to make informed decisions regarding fitness to fly, or return to flying duties.

References

American Psychological Association (2010). *Ethical principles of psychologists and code of conduct* (2002, Amended June 1, 2010). Available at: http://www.apa.org/ethics/code/index.aspx [accessed 8 October 2011].

International Union of Psychological Sciences (2008). *Universal Declaration of Ethical Principles for Psychologists*. Available at: http://www.cpa.ca/cpasite/userfiles/Documents/Universal_Declaration_asADOPTEDbyIUPsySIAAP_July2008.pdf [accessed 22 July 2012].

Kay, G.G. (2002). Guidelines for the psychological evaluation of air crew personnel. *Occupational Medicine*, 17(2), 227–245.

Kennedy, C.H. and Williams, T.J. 2011. *Ethical Practice in Operational Psychology*. Washington, DC: American Psychological Association.

Nagy, T.F. (2012). Competence, in *APA Handbook of Ethics in Psychology: Volume 1, Moral Foundations and Common Themes*, edited by S.J. Knapp, M.C. Gottlieb, M.M. Handelsman, and L.D. VandeCreek. Washington, DC: American Psychological Association.

Regulations Defining the Scope of the Profession of Psychology: Regulation R.704 in terms of the Health Professions Act, 1974 (Act 56 of 1974) (2011). *Government Gazette*, 2 September 2011. South Africa: Government Printer.

Saitzyk, A. 2012. Earning Navy psychology wings of gold. *The Navy Psychologist*, 4(2), 4, 10. Available at: http://www.wrnmmc.capmed.mil/ResearchEducation/GME/TheNavyPsychologist/TNP-4-2.pdf (Accessed 12 April 2013).

Index